UNDERSTANDING MICROBIOLOGY

UNDERSTANDING MICROBIOLOGY

By

Dr. S.K. Prasad

School of Studies of Zoology & Biotechnology

Vikram University

Ujjain

DISCOVERY PUBLISHING HOUSE PVT. LTD.

NEW DELHI-110 002

First Published-2010
Reprint : 2012
ISBN 978-81-8356-504-2

Published by:
DISCOVERY PUBLISHING HOUSE PVT. LTD.
4831/24, Ansari Road, Prahlad Street,
Darya Ganj, New Delhi-110002 (India)
Phone: 23279245 • Fax: 91-11-23253475
E-mail: parul.wasan@gmail.com
info@discoverypublishinggroup.com
Website: www.discoverypublishinggroup.com

Printed at:
Dynamic Printers, Delhi

Preface

The present title *"Understanding Microbiology"* provides a structured approach to learning by covering all the important topics in a uniform, systematic format. The book has been comprehensively designed incorporating recent advances in this fast moving field. It is written to provide accessible information on microbiology in compact form for undergraduate students in biology and related life sciences. It will be useful for both beginning students and those who are more advanced. In addition, busy lecturers who require a quick reference compendium will find it useful, particularly for tutional planning. Simple, yet hopefully clear figures and tables are provided throughout the book.

The over-riding goal of this book, and indeed of the whole *Understanding series,* is to present the essential information concering microbiology in a compact, readily accessible form which leads itself to student learning and revision. The convergence of various approaches has generated a rich panorama of detail, the significance of which we are still attempting to unraval. The present text has been written as an introduction to this rapidly growing field.

To make the work more comprehensive and informative, the author has consulted many authoritative books, research journals, abstracts, monographs etc., so there can be no claim to originality except in the manner of treatment.

The author expresses his thanks to his friends and colleagues whose continue inspirations have initiated him to bring out this book.

The author expresses his gratitude to Mr. Wasan and staff of M/s Discovery Publishing House Pvt. Ltd. for their whole hearted co-operation in the publication of this book.

In the mean time, the author will remain sincerely responsible for any shortcomings of the book and be grateful to the readers for their suggestions and constructive criticism for the continuous betterment of the book. He takes this opportunity to appeal to the readers to send their suggestions straightaway to his Publisher.

Author

CONTENTS

1

INTRODUCTION

Microbes are part of our lives in more ways then most understand. They have shaped our present environment and their activities will greatly influence our future. Microbes should not be considered apart from humans, but should be considered apart of our world. Microbiology is one of the most applied of all the biological science. Knowledge gained in this field has led to the development of many concepts, which in turn resulted in action. The more we learn, the better we can improve the quality of human life as well as that of all living forms. Our goal should be to understand the environment and better stabilize it to the benefit of all.

SCIENCE AND THE STUDY OF MICROBES

Most people know that biology deals with the study of plants and animals, and probably recognize it as a science. Textbooks define *biology* as the science that deals with the study of life. This appears to be a simple, straightforward definition until the word *science* and *life* are considered.

Science is the study of collection of knowledge of natural events and materials in an orderly fashion for the purpose of learning the basic laws that govern these events. The information (fact) is collected and the scientist uses certain rules in order to place these facts into a sensible framework. Rules, laws, and principles are developed as scientists begin to see patterns or relationships among a number of isolated facts. Some rules are very old while others are being constructed today. The method of collecting information and the way in which it is organized is really what makes somethings a science. The laws and rules are continually tested by the addition of new bits of information. If the new information can fit into the constructed framework, it reinforces the framework.

Biology is a broad science that draws on chemistry and physics for its foundation and applies these basic physical laws to living things. Because there are millions of kinds of living organisms, there is a very large number of special study areas in biology. Practical biology such as medicine, agriculture, plant breeding, and dentistry is balanced by more theoretical biology—evolutionary biology, molecular genetics, and "recreational" biology like insect-collecting and bird-watching. Biology is a science that deals with living organism and how they interact with the environment around them.

Since the prefix "micro-" means small, microbiology must be the biology of small living things. In fact, *microbiology* is the branch of biology that deals mainly with the study of *microscopic* organisms called microbes, which are composed of only one cell. Organisms not usually included in this study are those that are composed of many cells. Trees, grass, humans, and animals have many cells arranged into tissues, organs, and organ systems. These other life forms are studied in the related sciences of botany and zoology. Because of the small size of microbe, microbiologists must study their cells using special equipment and procedures. Microscopes, test tubes, and chemicals are very important tools to a microbiologist. Much of the work is done to understand better the chemical reactions that occur in living cells. This strong connection between chemistry and biology has led to the development of a separate science called *molecular biology*. Molecular biology focuses on the kinds of molecules found in living cells, their behaviour, and how they work together to make a cell alive. Information gained from each of these three sciences (microbiology, chemistry, and molecular biology) is frequently shared. Since the cell is the basic unit of all life as we know it, the study of microbes provides much information in the fields of genetics (inheritance), physiology (function), and biochemistry (chemistry of life). As a result, there has been a great deal of knowledge gained, and we have come to understand better the influence of microbes on humans and the environment.

There are many places in our environment where microbes and the results of their activities can be seen. Both our natural and man-made worlds contain microbes of many kinds. Each type of microbe has special qualities which enable it to survive in such unique places as the soil, oceans, rivers and streams, ice, water pipes, concrete, hot springs, the human intestine, roots of plants, and even oil wells. Because many types of microbes have become specialized to live in uniqιe environments, microbiologists have specialized the study of certain groups of microbes Microbiology is subdivided into a number of diverse fields. *Bacteriology* is the study of the bacteria, *virology* the study of viruses, *protozoology* the study of protozoans, *phycology* the study of algae, and *mycology* is the study of fungi such as yeast and molds. These fields center around the kinds of organisms under study. Other fields of microbiology center around where the microbes grow and have the most significant effects. These include such areas as medical microbiology, food microbiology, agriculture, waste treatment and industrial microbiology.

Microbes Influencing Our Lives

The science of microbiology can be divided into theoretical and applied fields. Farmers are applied scientists. They use microbiological principles to get the best yields from their farms. Doctors are applied microbiologists too, because their primary interest is to keep people healthy through the use of scientific knowledge. The theoretical scientist does not have in mind any specific use for the new knowledge gained. The purpose is to obtain new information to see how it fits the "old laws," and write "new laws" if necessary.

While investigating the cause of wine spoilage in the vineyards of France, Louis Pasteur became interested in the theoretical problem of whether life could be generated from nonliving material. Much of his theoretical work led to very practical applications. His theory that there were very small organisms causing diseases and decay led to the development of vaccinations and the preservation of food by pasteurization.

At this point, however, it is important to note that the study of microbiology is much like the study of a foreign language. The vocabulary includes a great many new words. Once you learn the meaning of these words and how to use them properly, the science of microbiology will be much easier. Many terms refer to the organism being studied; others refer to the activities of a particular organism and using them improperly could lead to a misunderstanding. For example, the words John Smith tell you who a person is, while the word microbiologists tell you what a person does.

A brief look at each of the applied fields of microbiology may give some understanding of how microbes influence our lives.

Medical microbiology is probably a very familiar field since it deals with microbes which cause diseases in humans, animals, and many plants. When asked of their knowledge of microbiology, people will mention something about a sickness or disease that they have had themselves. The "flu' (influenza), strep throat, tetanus, and malaria are all examples of diseases that are caused by the activities of microbes. In medical microbiology, the microbes that cause illness are called *pathogens*. The term *disease* means a process or event that results in illness or harm to a living organism. The *infectious disease* refer to the fact that some microbes are able to go through the process of entering another organism, growing, and causing harm. Some of these microbes may cause only minor damage, while infection by others may result in quick death.

Some microbes can spread through a population and cause disease. Over one hundred years of work was needed to show that a particular illness was caused by the actions of a particular microbes that was able to be transmitted from one individual to another. The bacteria that cause the disease cholera (*Vibrio cholerae*) were not identified with the symptoms they produced until 1883. Today our understanding that microbes can spread through a population and cause a particular disease is known as the *germ theory of disease*. Using this theory as a pattern for research has led to a greater understanding of the disease process. The germ theory of disease has become a unifying concept in that it defines the role of microbes in the disease process. Before the germ theory, many diseases were suspected to have causes ranging from those caused by "evil spirits" to unknown "miasmas" emanating from the soil or earth. Today we know a particular microbe may cause a number of different illnesses. An infection by one microbe may show different symptoms or signs depending on where it becomes located in the body. When a bacterium is located in the lung it may cause pneumonia, while in the joints it may cause arthritis, or meningitis in the spinal cord or brain. Medical microbiology also deals with disease prevention, the body's resistance to disease-causing organisms, and ways in which sick persons may be helped to recover.

Foods must be transported great distance and stored for long period for the world to have a safe, nutritions food supply. To accomplish this, the role of microbes in food poisoning, spoilage, and preservation must be investigated. Contributions to the field of *food and dairy microbiology* have helped resolve many food problems. When microbes enter a food they may either cause it to spoil, make it dangerous to eat, or change it to another form that it still acceptable as a food. Scientists have had to find out how the microbe enters the food, what action it has, and how to control the microbe. The whole idea of preserving foods is based on preventing microbial growth. *Pasteurization* is one of the best-known methods used to prevent spoilage and the transfer to disease. This heating procedure was first used in the wine industry in an attempt to keep wine from being changed into vinegar. This worked so well that other food industries now use the process Cheeses, beer, and milk are usually pasteurized to reduce the number of harmful microbes in the food.

In some cases, microbes are intentionally mixed with foods to change the food to another form. When certain bacteria are added to milk and encouraged to grow under controlled conditions, the milk is converted to cheese. The food value of cheese is very high, and it is much easier to store than milk. The many different kinds of cheeses found in the world are also the result of the actions of different microbes. The favour, smell, and texture of a cheese is determined by the kinds of microbes that are grown in the milk.

Water in our environment commonly contains microbes. The kinds of microbes and their method of entering the water is of great concern since many waterborne microbes can cause human disease. The field of *water and wastewater microbiology* explores all of these areas. The ultimate source of water is from rain or precipitation. The water moves over the surface of the ground as rivers and streams or through the earth as groundwater. Microbes and chemicals may enter the water from the air as the water passes through it or from untreated or poorly treated sewage being dumped into a body of water. Drinking water and industrial water are pumped from lakes, rivers, and wells. This water must be cleared of harmful bacteria, viruses, and other microbes to control diseases and prevent fouling of industrial equipment. Cholera and typhoid fever are diseases that are able to be transferred in water. Purification plants throughout the country treat water before it is used. Wastewater treatment plants treat water before it is again released into the environment. In some cases, water becomes clouded with large amounts of chemicals, mud, and plant material. Unless this water is fast moving and well aerated, microbes will use these materials for food and produce poisons or foul odors.

Soil contains great numbers of many microbe types. Bacteria, algae, and fungi are commonly found in rich, fertile soils. The more fertile the soil, the more likely it is to contain a thriving population of microbes. An effort to maintain good farm production has led microbiologists to develop the field of *soil and agricultural microbiology*. Microbes are responsible for the decomposition and decay of dead plants and animals. If it were not for microbial decomposition, the earth would be covered with all types of decaying organisms. The molecules in these would stay locked-up and unavailable for reuse by other, younger organisms. The recycling of materials through decay activities in the soil is vital to all life. Sewage treatment plants are, in part, giant microbial cultures for speeding decay of waste products. The waste material put into compost piles is broken down as a result of microbial activity and becomes rich, soil-fertilizing material for plants. The increased amount of solid waste throughout the world along with the increase in human population has made it essential to explore new ways of controlling decay. Agricultural microbiology deals with those microbes associated with animals. For examples, cattle and other animals with rumens, or "second stomachs," and laiden with microbes. These are of great value to the health of the animal. The food they eat, silage, is grain that has been preserved by the action of microbes while the grain has been stored in a silo.

Due to the small size of microbes, simple test tubes and covered dishes can provide enough space to grow them for tests and explore their activities. However, many microorganisms are capable of producing products of value to humans, if they can be produced in very large quantities. Microbiologists, engineers, and business have come together for this purpose and developed the field of *industrial microbiology*. This area involves the large-scale growth of particular microbes. Yeast, bacteria, and molds may be grown in 5, 10, or even 50,000 gallon containers. For example, yeast cells are grown for use in ironized yeast tablets for human use and in even larger quantities as a supplement to farm animal feed. In other instances, the by-product of the growth and activities of a microbe is the most useful. These include enzymes, amino acids, antibiotics, alcohol, and organic acids. Handling large quantities of growing microbes is not easy. Expensive equipment and well-trained personnel are essential to the smooth production. Industrial microbiologists must closely monitor the microbes' chemical activities and be prepared to change or stop the activities of the organism to prevent product loss. Careful attention must also be paid to ensure that only one kind of microbe is being grown. Successful control of these factors has resulted in the development of multimillion dollar microbial industries.

Of Microbiologists and Microbes

A vast amount of chemical and microbiological information has been gained from studies in the various microbiology fields. These efforts have resulted in solutions to many problems, and have also revealed new and more challenging areas of concern. A better understanding of the scope of microbiology may be gained by taking a brief look at the history of microbiology and some of the questions explored by famous microbiologists.

For centuries, one of the most intriguing questions related to the origin of life. Even though the answer still continues to be an area of speculation, research and experimental attempts to explain the origin of life have inadvertently been prime movers in expanding our knowledge of microbiology. In simpler times, the origin of life from nonliving things was never doubted. The Greeks, Romans, Chinese, and many other ancients believed that maggots, lice, frogs, and even mice could spontaneously arise from mud. They thought they saw these events happening every day It was thought that mice could be produced from a sweaty shirt if it was kept in a dark, cool room with several grains of wheat. Many prominent scientists believed in this concept. Only through the efforts of scientific investigators like Redi, Spallanzani, and Pasteur was this classical concept of *spontaneous generation* (*abiogenesis*) discarded.

The argument between the supporters of spontaneous generation and those of *biogenesis* has lasted over 300 years. People believing in biogenesis thought that all living things came from pre-existing life. During this period, the cleverness and imagination of those involved were used to the fullest in the attempt to disprove the others' position. Even though microbes could not be seen, a number of people suspected that such small living things existed. In the first century B.C. Varro suggested that diseases were due to invisible organisms. Later in 1546, Fracastorius wrote thee books, *De Contagione*. The contained the first scientific statements as to how infections were transmitted. Fracastorius was medical adviser to the Council of Trent (Italy), which had to be transferred to Bologna because of an outbreak of typhus fever. He wrote, "Contagion is a precisely similar putrefaction which passes from one thing to another: its germs (seminaria) have great activity, they are made up of a strong and viscious combination, and they have not only a material but also a spiritual antipathy to the animal organism." Fracastorius believed even at this early date that consumption (tuberculosis) was able to transferred from person to person.

He commented, "it is extraordinary to see in families up to the fifth and sixth generation all the members die of phthisis (tuberculosis) at the same age."

But it was Francesco Redi (1668) who set up the first controlled experiment to disprove those who supported spontaneous generation. This is the best type of experimental setup from which to draw conclusions, since there is only one unknown factor in question. Redi used two sets of dishes and varied only one part of the experiment These experiments ended the concept of spontaneous generation for only a short period of time. In 1676, the idea of spontaneous generation was accidentally revived, due to the tinkerings of a Dutch clothier, Anton van Leeuwenhoek. His discovery of "animalcules," little animals, while using lenses reopened the debate. Leeuwenhoek's successful method of grinding lenses allowed him to construct over 200 microscopes (all disposable), capable of magnifying a specimen to 300 times its normal size. The images were to clear that Leeuwenhoek made drawings of what we now know to be bacteria. These were submitted to the influential Royal Society of London and were so impressive that many others began exploring the world of microbes and at tempting to resolve the questions of spontaneous generation.

Where did these new life forms come from? Another scientist, Lazaro Spallanzani (1729-99) benefited from Redi's use of controlled experiments and successfully put to rest the spontaneous generation theory once more. In response to a challenge by Joseph Needham, an English priest and naturalist, Spallanzani devised an experiment that

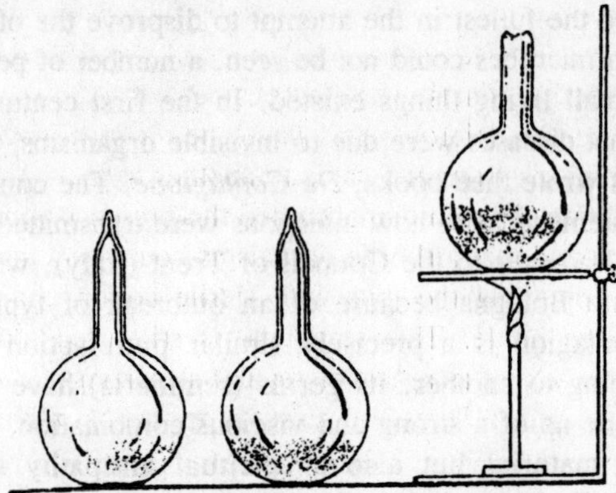

Fig. 1.1. Spellanzani carried the experimental method of Redl one step further. By sealing the flasks after they had been boiled, he demonstrated that spontaneous generation could not occur unless the broth was exposed to the germs in the air.

not only settled the argument, but demonstrated that heating a container and using an airtight seal would prevent spoilage of foods. No "animalcules" were generated in his experiment. This experiment on the origin of life ultimately led to the basic preservative methods used in the commercial canning industry.

Spallanzani's evidence supported the theory of biogenesis until 1775 when Lovosier and Priestly discovered oxygen. This discovery brought on another wave of interest in spontaneous generation. It was suggested that since oxygen was excluded from Spallanzani's experiment, the spontaneous generation of life was prevented. With this new challenge to the theory of biogenesis, Louis Pasteur (1869) began to experiment with the origin of bacteria and their need for oxygen. He successfully defended the biogenic theory. Even though Pasteur's original work was designed to investigate biogenesis, he gathered much other information related to microbes and their activities.

A Founding Father

Louis Pasteur (1822-95) was one of the outstanding scientists of his day. His achievements served as stepping-stones for many others. Pasteur confirmed that certain microbes were directly responsible for the formation of such different kinds of molecules as acetic acid and lactic acid. The wine industry of France relied on him to solve the problem of wine changing into vinegar as the alcohol disappeared how microbes were able to form not only acetic acid but also lactic acid, butyric acid, and alcohol. As a result of these investigations, he developed a process of heating wine to kill harmful microbes and prevent them from ruining the wine (1864). Today, we call the process *pasteurization* and have adapted if for use in controlling the quality of many other food products. Pasteur became interested in the growth or culturing of certain microbes and developed a process of transferring selective organisms from one batch to another to maintain good quality wine. In addition, he also demonstrated the presence of bacteria in the air. Pasteur found that airborne organisms could be kept out of sterile materials by plugging the tops with sterile cotton and still have air circulate in the bottle. This technique is still used today in microbiological laboratories as a part of regular sterilizing techniques. Later, his work expanded to involve disease of animals and humans.

Pasteur was familiar with the work of another famous microbiologist, Robert Koch (1843-1910). Koch had discovered the bacterium that caused anthrax in cattle and demonstrated the progression of the disease. The excitement generated by Koch's work encouraged Pasteur

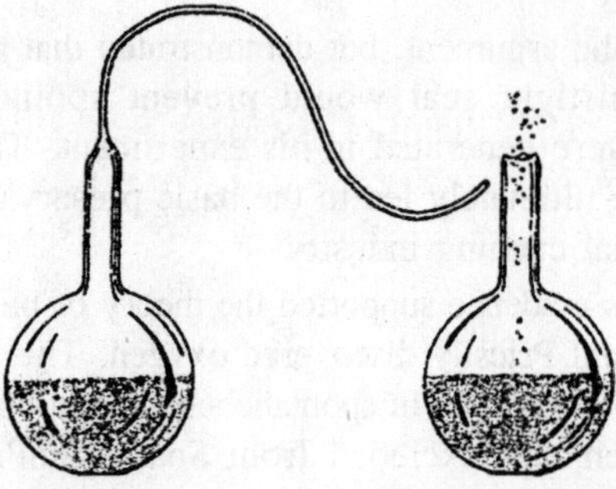

Fig. 1.2. The gooseneck flasks used in Pasteur's experiment allowed for the flow of air into the flasks, but not airborne organisms. Pasteur demonstrated that germ-free air with its oxygen does not support spontaneous generation of life.

to investigate anthrax. As a result, Pasteur demonstrated that he could prevent anthrax in cattle by injecting healthy animals with live anthrax bacteria that had been specially treated to reduce their disease-causing ability. His success with anthrax led him to investigate hydrophobia, or rabies. He had worked for several years to prevent this disease in animals by using much the same techniques that had worked with anthrax. In 1885, his efforts were put to the test when a frightened mother sought out Pasteur for help. Her son, Joseph Meister, had been bitten by a rapid animal and Pasteur began treating the body for rabies. The body survived and Pasteur was again praised for his brilliant work.

Louis Pasteur is known the world over as one of the great men of science. He is credited with starting microbiologists down a path of research in the area of preventive medicine. His concern for the prevention of diseases by inoculation eventually led to the founding of the science of immunology. During this same period in history, Robert Koch began a comparable career investigating microbes. His efforts led to achievements as exciting and important as those of Pasteur. Koch began research into the isolation and identification of individual microbes, especially bacteria. Because of this work, he is regarded by many as the first true bacteriologist. His work eventually led to the founding of the sciences of medical microbiology, bacteriology, and virology. Koch was educated as a physician and chose to practice in a small German town. While serving as a country doctor, he became intensely interested in the cause of the disease anthrax. The disease begins with symptoms similar to a cold and causes itching skin. Blisters or vesicles form which later black and swell. The bacteria may move into the blood and cause fever, shock, and eventually death. The research he did was praised as brilliant even though he used the most basic of techniques and equipment. Koch worked long hours and successfully isolated the bacterium that caused anthrax. He used house mice to show how the disease moved through animals and speculated on the formation of a resistant form of the bacterium, an *endospore*. Later he did special work on the endospores and demonstrated how they were able to survive in dead animals and in soil where they could serve as a source of infection for other animals. He studied the anthrax bacterium in frogs and horses, and saw how the disease organisms became concentrated in the lymph sacs and spleen. This information led to a better understanding of how the lymph system functions to control disease-causing microbes in the body. When Koch presented this information to the Royal Society of London, the praise and support he received was so great that he shifted his work from medicine to research.

Methods and Media

While involved in his research, Robert Koch developed a great variety of tools and techniques to examine bacteria. Most of these we still use today and are the basis of modern laboratory procedures. It was Koch who first adapted staining methods to better see bacteria under the microscope. Basic dyes like methylene blue will chemically bond to bacteria and make them more visible. The hairlike projections, *flagella*, that make some bacteria move were first seen by Koch using special staining process. He used a special oil on his slides to improve lighting at higher magnification and added another lens developed by Abbe to the microscope to better channel the light through the lenses. We use both oil and the Abbe condenser lens on microscopes in microbiological work today. The technique of smearing bacteria on a clean glass slide followed by a slight heating to kill and stick the cells to the glass was developed by Koch. In 1881, he presented the Royal Society with another breakthrough so grand that Pasteur met with him after the presentation with the greeting, "C'est un grand progre's, Monsieur!" ("This is great progress, sir!").

Until this point, bacteria had been grown in the lab in broth or on the surface of potato slices sterilized by hot air (also invented by Koch!). Koch found that by adding the protein gelatin to beef broth it could be solidified and the bacteria more clearly seen and separated. Because this medium could contain a greater variety of nutrients than potato medium, it was possible to grow a greater variety of bacteria. The major problem with his method was that a number of bacteria could digest gelatin. When grown on the gelatin media, these bacteria would form little puddles which made it difficult to identify individual characteristics. This problem was solved when a coworker's wife told Koch about a solidifying agent she used in cooking called *agar-agar*. The chemical comes from seaweed, algae, and was the perfect answer to the problem because most microbes cannot digest agar-agar. Today this material is used worldwide to solidify bacteriological media. It is so common that many use only the term "agar" to refer to the solid nutrient agar-agar growth medium.

However, nobody is perfect, not even Robert Koch! In 1882, Pasteur developed a theory and a method for vaccinating animals against anthrax. He presented his ideas at a scientific meeting which included Koch. Koch vigorously opposed Pasteur's idea in person and though correspondence. The argument between them continued even while Pasteur began field testing his vaccine and method. Koch's letters of opposition soon faded as Pasteur's successes continued to rise. However, this minor setback did not discourage Koch. He continued his work and discovered the bacterium that causes cholera (*Vibrio cholerae*) and isolated the bacterium that causes tuberculosis (*Mycobacterium tuberculosis*). The following appeared in the May 1882 edition of *Scientific American* and recounts Robert Koch's efforts, which confirmed that the bacterium *Mycobacterium tuberculosis* is the cause of the disease once known as the "wasting disease."

"Professor Tyndall has communicated to the London *Times* as account of results obtained by Dr. Koch of Berlin in the investigation of the etiology of tubercular disease, as set forth by him in an address delivered on March 24 before the Physiological Society of Berlin. In pursuing these investigations Dr. Koch subjected the diseased organs of a great number of men and animals to microscopic examination, and he found in all cases the tubercles were infested with a minute, rod-shaped parasite, which, by means of a special dye, he differentiated from the surrounding tissue. Transferring directly, by inoculation, the tuberculous matter from diseased animals to healthy ones, he in every instance reproduced the disease. Dr. Koch has examined the matter expectorated from the lungs affected with phthisis and found in it swarms of bacilli, whereas in matter expectorated from the lungs of persons not thus affected he has never found the organism. Guinea pig infected with expectorated matter that had been kept dry for two, four or eight weeks were smitten with tubercular disease quite as virulent as that produced by fresh expectoration."

Koch also worked with the bacterium that causes boils (*Staphylococcus aureus*), and with the cause of cattle plague (a virus infection). His single most outstanding contribution to all science has become known as *Koch's Postulates*. The Postulates state: (1) a particular organism can always be found in association with a particular disease, but not in a healthy individual; (2) the organism can be grown in the laboratory by itself; (3) this pure growing culture will produce the same disease when placed back into a new, susceptible animal; and (4) it is possible to recover the organisms from this sick animal and grow them in pure culture. By using these postulates as guidelines, many microbiologists have proven that a particular organism is in fact the cause of a particular disease.

Jenner and the Milkmaid

Among other researchers involved in the study of disease prevention was the well-known British physician Edward Jenner (1749-1823). Jenner first developed the technique of vaccination in 1795, well before the time of Pasteur or Koch. This was the result of a twenty-six year study of two diseases, cowpox and smallpox, Cowpox was known as *vaccinae*, and from this word Pasteur developed the present terms "vaccination" and "vaccine." Jenner observed that milkmaids rarely became sick with smallpox, but they did develop pocklike sores after milking cows infected with cowpox. This led him to perform an experiment in which he transferred pus material from the cowpock to human skin. Since the two disease organisms are so closely related, the person vaccinated with cowpox developed an immunity to the smallpox virus as well. The reaction to cowpox was minor compared to the more serious smallpox. Public reaction was mixed; some people thought that the process of vaccination was a work of the devil. However, many European rulers supported Jenner by encouraging their subjects to be vaccinated. Napoleon and the Empress of Russia were very influential, and in the United States, Thomas Jefferson had a number of his family vaccinated. Today, almost two hundred years after Jenner developed a vaccination against smallpox, the disease has been eliminated. The World Health Organization announced that smallpox is the first disease to become extinct through human efforts!

The names and personalities of the people described are but a few in the history of microbiology. Many others have made equally outstanding contributions. During the early 1900s, advances were made and formed the foundation for the field of molecular biology. Microbes were very important in this research. Our increased understanding of

Table 1.1. The history of microbiology—an overview.

Date	*Event*	*Historical Perspective*
1st century B.C.	Varro suggests diseases due to invisible organisms.	Caesar conquers Gaul.
1546	Fracastor's *De Contagione* makes first scientific statement of how infections are transmitted.	Henry VIII succeeds Edward VI.
1555	First use of word "physiology" in in modern sense (Jean Fernel)	
1590	First compound microscopes.	Shakespeare writes *Henry VI.*
1658	Kircher sees "innumberable worms" under microscope.	
1660	Royal Society founded.	Rembrandt: 'The Syndicus of the Cloth Hall."
1665	*Philosophical Transactions* of Royal Society first published. Robert Hooke's *Micrographia*. First drawing of cell (Hooke).	Delaware becomes a separate colony.
1676	Van Leeuwenhoek discovers "little animalcules"; perfects lenses to magnify 300 times.	
1688	Redi publishes book on spontaneous generation of maggots from putrid flesh.	
1720	Bradley's germ theory.	London's Royal Academy of Music names George Frederick Handel as its director and presents Handel's oratorio *Ester*.
1740	Buffon's "organic molecules" as infective agents floating in the air.	
1765	First drawing of cell division (Trembley).	
1765	Spallanzani shows that Buffon's "organic molecules" are distinct organisms.	
1770	Hill introduces new methods of staining and preserving specimens for microscopic study.	A spinning jenny that automates part of the textile industry is patented by English weaver-mechanic James Hargreaves. The Black Death strikes Russia and the Balkans in epidemic form.

1796	Jenner inoculates James Phipps with cowpox.	
		Beethoven writes the "First Symphony."
1802	First used for word "biology" (Treveranus).	
1807	First achromatic microscope.	
1821	First international congress of biology (organized by Oken).	
1835	Bassl's theory of "living contagion" in silkworm disease.	
1838	Liebig establishes biochemistry.	The word "protein" is coined by Dutch chemist Gerard Johann Mulder, 36, who adapts a Greek word meaning "of the first importance." Gas ovens are installed at London's Reform Club. Coal or wood is the common cooking fuel in most of the world, but Arab nomads use camel chips, American Indians use buffalo chips, and Eskimos use blubber oil.
1838-39	Schwann and Schleiden found modern cell theory: plants and animals composed of basically identical units.	
1844	Bassi asserts smallpox, bubonic piague, syphilis, spotted fever due to living parasites.	Potato crops fail throughout Europe, Britain, and Ireland as the fungus disease caused by *Phytophthora infestans* rots potatoes in the ground and also those in storage. Irish potatoes are even less resistant than potatoes elsewhere—up to half the crop is lost. Portland is founded in Oregon Territory near the junction of the Columbia and Willamette Rivers. The town is named after the 213-year-old city in Maine as two New Englanders let a flip of a coin decide in favour of Portland rather than Boston.
1845	Siebold recognizes protozoa as single-celled organisms.	
1848	Semmelweiss demonstrates childbed fever is a form of septicemia and	

	becomes a pioneer in aseptic technique.	
1850	Davaine asserts the anthrax is due to "bacterides" which he sees in the blood of dead sheep.	Dockens write *David Copperfield*.
1854	Davaine sees "monads" in stools of cholera patients.	
1857	Pasteur demonstrates that lactic acid fermentation is due to living organisms.	
1858	Virchow's Doctrine "*Omnis cellula e cellula*" declares that "all cells comes from cells."	Iowa State College is founded at Ames. Oregon State University if founded at Corvallis. *Anatomy of the Human Body, Descriptive and Surgical* by London physician Henry Gray, is publihed.
1860	First selective biological staining.	U.S. Civil War begins.
1864	Pasteur invents pasteurization (for wine).	
1867	Lister publishes work on antiseptic surgery.	
1869	Miescher discovers nucleic acid.	Washington D.C.'s Pennsylvania Avenue is paved with wooden blocks for a mile between 1st street and the Treasury Department building at 15th Street.
1869		"Ecology" is coined to means environmental balance by German zoology professor Ernst Heinrich Haeckel, 35, who is the first German advocate of Charles Darwin's organic evolution theory.
1876	Koch gives three-day demonstration of his work on anthrax, in which he had discovered the sequence of development.	
1877	Koch describes techniques of fixing, staining, and photographing bacteria. Also discovers *Bacillus anthracis* as cause of anthrax.	
1879	Albert Neissen discovers *Neisseria gonorrhea* as cause of gonorrhea.	

1880	Laveran sees malarial parasite but is disbelieved.	
1880	Typhoid bacillus and leprosy agent discovered.	
1881	Koch works out method of culturing bacteria on gelatin. Pasteur, spurred by Koch's work, turns to study anthrax; publicly inoculates sheep at Melun with his "attenuated culture."	Distoevsky writes *Brothers Karamazov*.
1882	Koch discovers tubercle bacillus, and enuciates "Koch's Postulates."	
1882	Mechnikov launches phagocytic theory" "cellular theory of immunity."	
1883	First apochromatic microscopes.	
1885	Pasteur inoculates Joseph Meister for rabies. Theodor Escherich discover *E. coli*.	The first ready-to-use surgical dressings are introduced by Johnson and Johnson. Phagocytosis is discovered by Russian zoologist bacteriologist Ilya Ilich Mechnikov, 40.
1887	Buist, Edinburgh infirmary superintendent, sees pox virus and believes it is a form of bacteria.	
1888	Richet confers immunity on rabbits accidentally with serum from an infect dog.	
1897	Buchner discovers that cell-free yeast converts sugar to carbon dioxide and alcohol.	
1897	Buchner demonstrates that cell-free yeast extract will catalyze glucose breakdown. Van Ermengerm discovers *Clostridium botulinum*.	
1898	Beijerinck discovers and names tobacco mosaic virus; viral cause of foot-and-mouth disease demonstrated.	
1898	Benda discovers and names the mitochondria, previously seen by Altmann.	
		First wireless communication between Europe and America.
1902	Richet discovers anaphylaxis.	
1902	Landsteiner investigates agglutination when blood from different human donors is mixed.	

1903	Sir Almroth Wright and others discover "opsonins" (i.e., antibodies) in blood of immunized animals.	
1905	Harden and Young show inorganic phosphate responsible for fermentative ability of yeast juice.	Helen Keller is graduated magna cum laude from Radcliffe and begins to write about blindness.
1906	Schaudinn and Hoffman discover *Treponema pallidum* as cause of syphilis.	
1906	Von Wassemann develops test for syphilis.	
1912	Ehrlich demonstrates first chemotherapeutic agent for a bacterial disease (sphilis).	
		World War I begins.
1915	D'Herelle and Twort independently show existence of bacteriophage—viruses which destroy bacteria.	
1923	Landsteiner shows M and N factors in blood.	
1925	Keilin discover cytochrome.	Al Capone takes over as boss of Chicago bootlegging from racketeer Johnny Torrio, who retires after sustaining gunshot wounds.
1926	Ultracentrifuge (Svedberg).	
1928	Elford demonstrates size of viruses (from 10 to 3000 micrometers).	
		The Great Depression begins.
1929	Lohmann identifies adenosine triphosphate (ATP) as necessary for the phosphorylation of sugar.	
	Fleming describes penicillin.	
1932	Sir Hans Krebs describes and names the citric acid cycle.	
1933	First electron microscope (Ruska).	Popular songs "Basin Street Blues" by Spencer Williams whose work was published in small orchestra part 4 years ago; "Only a Paper Moon" by Harold Arlen, lyrics by E.Y. Harburg, Billy Rose; "Lazybones" by Hoagy Carmichael, lyrics by vocalist Honny Mercer, 23; "Love is the Sweetest Thing" by Ray Noble; "Stormy Weather-Keeps Rainin' All

		the Time" by Harold Arlen, lyrics by Ten Koehler.
1935	Stanley crystallizes virus.	
		World War II begins.
1941	Beadle and Tatum establish "one gene-one enzyme" theory.	
1944	Avery discovers "blueprint" function of DNA.	
1945	Role of mitochondria revealed.	Aerosol spray insecticides begin a revolution in packaging. The commercial "bug bombs" employ a Freon-12 propellant gas developed by two U.S. Department of Agriculture researchers in 1942. They have been used during the war to protect troops from malaria-carrying mosquitoes.
1948	Electrophoretic methods (Tiselius).	
		Korean conflict begins.
1952	Partition chromatography (Synge and Martin).	
1952	Hershey and Chase prove that DNA injuected by bacteriophage is what disorganizes bacterium. Phage becomes the fruit fly of the molecular biologists.	The American Bandstand debuts in January on ABC network station. Host Dick Clark, 22, will continue to emcee the show for more than 30 years.
1952		Films: Fred Zinneman's *High Noon* with Gary Cooper, Grace Kelly; John Huston's *The Red Badge of Courage* with Audie Murphy, Bill Mauldin.
1952	Sexual recombination discovered in bacteria.	
1952	Waksman discovers streptomycin, the first antibiotic effective against tuberculosis.	
1953	Lipmann discovers coenzyme A and its importance for intermediary metabolism.	
1953	Phase-contrast microscope.	
1953	Medawar shows toleranceto grafts can be conferred by inoculating newborn animal or embryo with antibodies from future donor.	
1953	Lederberg and Zinder discover transduction.	

1953	Crick and Watson propose a structure for deoxyribonucleic acid—a double spiral.	
1954	Enders, Weller, and Robbins discover poliomyelitis viruses in cultures of various types of tissue.	
1957	Virus structure determined. Interferon discovered.	*Sputnik*, the first earth satellite, launched by Russia.
1958	Lederberg makes discoveries concerning genetic recombination and the organization of the genetic material of bacteria.	Cocoa Puffs breakfast food, introduced by General Mills, is 43 percent sugar.
		Transatlantic jet service is inaugurated by Pan American World Airways and British Overseas Airways (BOAC).
1959	Kornberg and Ochoa awarded Nobel Prize for discovery of enzymes that produce artificial DNA and RNA.	
1961	Nirenberg, using artificial DNA, synthesizes a protein molecule.	
1962	Role of thymus in immunity established.	
1962	Crick, Watson, and Wilkins make discoveries concerning the molecular structure of nucleic acid and its significance for information transfer in living material.	
		President Kennedy killed.
1964	Bloch and Lynen work out the mechanism and regulation of the cholesterol and fatty acid metabolism.	
		U.S. Marines land in Vietnam.
1966	Rous discovers tumor-inducing viruses.	
1966	Huggins identifies hormonal treatment of prostatic cancer.	
1968	Holley, Khorana, and Nirenberg win a Nobel Prize for interpreting the genetic code and its function in protein synthesis.	U.S. natural gas consumption begins to exceed new gas discoveries and reserves for interstate pipelines begin falling.
		U.S. first class postal rates climb to 6 cents; up from 5 cents per ounce in 1963.
1969	Delbruck, Hershey, and Luria win a Nobel Prize for the replication mechanism and genetic structure of viruses.	
		First man of the moon.
1971	T.O. Diemer identifies viroids.	

1972	Edelman and Porter: Nobel Prize for work concerning the chemical structure of antibodies.	U.S. *Apollo* 16 astronauts Charles M. Duke, Thomas K. Mattingly, and John W. Young blast off April 16 from Cape Kennedy. A human skull found in northern Kenya by Richard Leakey and Glynn Isaac allegedly dates the first humans to 2.5 million B.C.
1974	Claude, DeDuve, and Palade receive a Nobel Prize for discoveries about the structural and functional organization of the cell.	
1975	Baltimore, Dulbecco, and Temin awarded a Nobel Prize for researching the interaction between tumor viruses and the genetic material of the cell.	
1976	Gajdusek and Blumberg did research leading to Nobel Prize for a test to show hepatitis viruses in donated blood and to an experimental vaccine against the disease.	U.S.A. Bicentennial. Carter becomes President.
1978	Arber, Smith, and Nathans given Nobel Prize for discovery of restriction enzymes and their application to the problems of molecular genetics. Austrial wins award for first vaccine against pneumococcal pneumonia.	
1979	Henle identified first virus regularly associated with human cancer.	
1980	First U.S. patent issued for the process of producing biologically functional molecular chimeras; inventors: Stanley N. Cohen and Herbert W. Boyer	
1980	Nobel Prize for Chemistry, Paul Berg. Walter Gilbert, and Frederick Sanger for development of a rapid way to determine the chemical makeup of DNA.	
1982	Epstein and Barr receive award for showing relationship between EBV and Burkitt's lymphoma.	World's Fair in Knoxville, Tennessee.

the microbe has ushered in a period of rapid advancement in biology. One of the first major contributions came in 1952 when Alfred Hershey and Martha Chase demonstrated by using bacteria and viruses that

DNA (deoxyribonucleic acid) is the controlling molecule of cells. Their work with the viruses that infect bacterial cells, *bacteriophage*, was so significant that the *phage* become a standard laboratory research organism. Just one year following this work, James Watson and Francis Crick used this information, and that of others, to propose the now famous double helix molecular structure for DNA. Ten years later, Watson, Crick and a coworker, Wilkins, were awarded a Nobel Prize for the work. In 1958, George Beadle and Edward Tatum won a Prize for their discovery that genes act by regulating definite chemical reactions in the cell, the 'one gene-one enzyme" concept. The chemical reactions of the cell are controlled by the action of enzymes and it is the DNA that chemically codes the structure of these special protein molecules.

At first glance, some research by microbiologists may seem irrelevant or unrelated to everyday life. But it is a rare occasion when such research ideas do not make their way into our lives in some practical, beneficial form. The work of Watson, Crick, Beadle, and Tatum has been applied in hospitals and doctors' offices. Their basic research into DNA has provided the information necessary to develop medicines that control disease-causing organisms and others that regulate basic metabolic processes in our bodies. The ease with which such theoretical research information becomes a part of science and moves into our lives makes microbiology one of the most applied of all the biological sciences. Advances in microbiology were responsible for eliminating tuberculosis as the number one cause of death in humans. Between 1937 and 1967, pneumonia and influenza dropped from the second to the fifth position as a cause of death. Further researchers will be engaged constantly to find cures for diseases and improving life in many other ways.

2

ORIGIN OF MICROBIOLOGY

Before the dawn of civilization in the Mesopotamian regions and farther east some 7000 to 8000 years ago there was little exact knowledge of either the causes or nature of natural phenomena. However, scholarly thinkers and their works were not wholly lacking. By the time writing and written history had been "invented" 5000 to 6000 years ago in Sumeria, Egypt, Syria and adjacent regions, many keen and ambitious minds in the ancient priesthoods, secular upper classes and royal families had learned of the medicinal and poisonous properties of certain plants and of the venoms of certain snakes and insects. They knew how to exploit nature for political and other purposes. For thousands of years after the beginnings of civilization magic, incantation, abracadabra and witchcraft passed for science and usually also for religion. Even as recently as the Middle Ages (c. 500-1400 AD.) and later in the European Renaissance (c. 1400-1700 A.D.) astrology (aided by imaginative charlatans, with weird grimaces and impressive passes) passed for astronomy; alchemy (strongly flavoured with wizardry) masqueraded as chemistry; the most outrageous quackery was accepted, even by royalty, as medicine.

As always, however, honest, imaginative and inquisitive men here and there were still capable of analytical and creative thought and the proposing of working hypotheses to be tested experimentally. They were sometimes reviled, persecuted and tortured for their supposed dealings with "The Evil One." Century after century these pioneer scientists (seekers after experimentally demonstrable truth) began to establish a system of knowledge based on accurate, purposeful observation; logical inference; imaginative hypothesis; and ingenious experiments designed to establish indisputable fact or destroy fallacy.

Because of great difficulties in travel and communications, ancient scholars shared little of one another's learning. As the centuries passed, exploration began and travel became more common, populations increased, and vast interminglings of peoples occurred because of wars and trade. Scientific information thus began to spread from country to country and, more recently, from continent to continent. Instead of a few great scholars who were thought (even by themselves!) to know everything, men began to realize that there were boundless deserts and plains and illimitable dark forests of ignorance only awaiting the axe and plow of the devoted researcher to yield rich crops of wonderful, golden knowledge. Men also realized the awesome truth that knowledge is power to create or to destroy utterly. Eventually scientific thought, experimentation and communications became permissible and even respectable. They also became incalculably profitable and frightening.

Scientists interested in the mysteries of life collected, over the centuries, a considerable mass of reasonably accurate information about such living things as could be seen with the naked eye, and even with "magnifying glasses" (magnifications of about 10 diameters). By 350 B.C. Aristotle and his students had drawn up a systematic, though limited and (as we now know) often erroneous classification of hundreds of plants and animals. Accumulating knowledge of living organisms slowly became arranged into a more or less orderly system and the study of life was eventually dignified with a given name: *biology* (Gr. *bios* = life; *logos* = study or description). Most of biology was at first largely descriptive of outward form and macroscopic (Gr. *makros* = large; *skopion* = to see) anatomy. These descriptions became the basis of classifications and taxonomy—major preoccupations of most early botanists and zoologists. Until the seven-teenth and eighteenth centuries chemistry and physics were almost completely separate fields of study and little used in biology. Life and living substance were commonly thought of as mysterious and beyond physical and chemical analysis.

Beginnings of Microscopy

Until about 1660 A.D. all knowledge of the form, structure and life processes of plants and animals was narrowly restricted to what could be seen with the naked (or very feebly assisted) human eye. Microorganisms were merely "fabulous monsters." Visual limitations of the pitiably restricted eye of man had always stood, like an impenetrable curtain, between man and the fantastic and glittering cosmos of the microscopic world.

Unaided human vision fails to see objects less than about 100 μ (0.004 or 1/254 inch) in diameter or to perceive as separate objects (i.e., *resolve*) particles separated by distances less than about 100 μ. Microscopic linear units are shown in Table 2.1. The development of practical, relatively high-power microscopy about the middle of the seventeenth century was like turning on a 500-watt lamp in a pitch-dark curiosity shop. It gave men the power to see a universe of objects, living and inanimate, so minute that their very existence had never before even been suspected.

Table 2.1. Some linear measures commonly used in microbiology

1 inch	=	2.54 cm.
1 cm.	=	10 mm. 1 mm. = 1000 μ
1 μ	=	0.001 mm. = 0.00003937 or 1/25,400 inch
	=	1000 mμ
I mμ	=	0.001 μ = 10.0 Angstrom (Å)
1 A	=	.0001 μ = 0.0000001 mm. = 1/254,000,000 inch

The First Microscopes

By the end of the thirteenth century simple lenses (magnifying glasses) had already been used in various ways for many years. Such lenses, however, did not magnify very highly. About 1590, a Dutch spectacle maker, Zacharias Janssen, used a second lens to magnify the image produced by a primary lens. This is the basic principle of the compound microscope used by every microbiologist today. Galileo invented an improved compound microscope in 1610. Robert Hooke (1635-1703) made and used a compound microscope in the 1660's and described his fascinating explorations of the newly discovered universe of the microscopic in his classic "Micrographia" (1665), published at request of the Royal Society in London. Although Hooke's highest magnifications were possibly enough to reveal bacteria, he apparently made no observations of them, probably because he studied mainly opaque objects in the dry state by reflected light, conditions that, as will be explained, are not optimal for observation of microorganisms. However, his pictures of "white mould" (probably a Mucor species) are very informative and accurate.

A contemporary of Hooke, and the man mainly responsible for revealing the hitherto unknown and unseen world of microorganisms, did not use a compound microscope. He was the Dutch investigator, Antonj van Leeuwenhoek (1632-1723), a linen merchant by trade and a

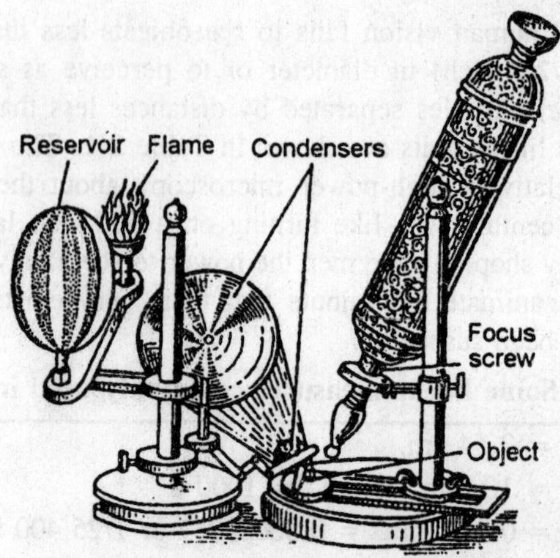

Fig. 2.1. Hooke's compound microscope.

man of public and commercial affairs in the city of Delft. He was not a trained scientist but was self-educated, and amused himself by means of his skill and craftsmanship in glass blowing and fine metal work. He lived in relatively easy circumstances with leisure time for his avocation of making minute, simple but powerful lenses. With these he delighted in examining a great variety of objects: saliva, pepper decoctions, cork, the leaves of plants, circulating blood in the tail of a salamander, seminal fluid, urine, cow dung, scrapings from the teeth and so, on. In many of these he saw living creatures, some of which we now know were protozoa and bacteria but all of which he called "animalcules."

In spite of the fact that his microscopes were not compound he obtained remarkable results with them.

...he showed rare ingenuity and expert craftsmanship in the grinding and mounting of his simple lenses, a skill which he zealously kept to himself; and in spite of the requests of his learned friends, he refused to disclose the secret of his success.

Leeuwenhoek's instruments are not true microscopes at all in the sense in which we think of microscopes, but rather simple magnifying glasses generally consisting of a small, single, biconvex lens. The object, and not the lens, was moved into focus by means of screws.

To adjust the lens to the object was so long and tedious a task that it is not surprising that Leeuwenhoek used an individual lens for each object.... The magnification varied and at best did not exceed two hundred to three hundred diameters.... The size of objects which

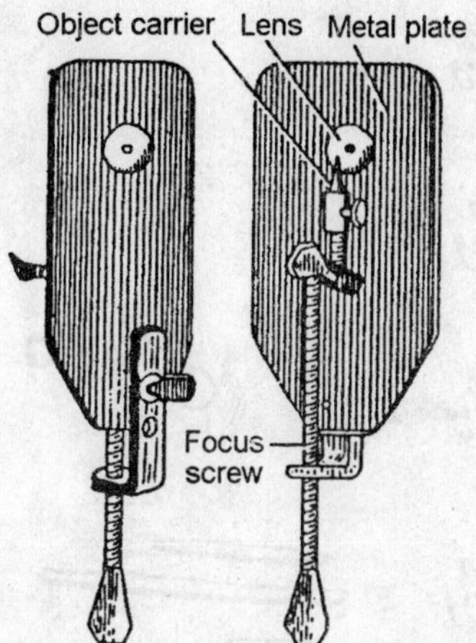

Fig. 2.2. Leeuwenhoek's microscope.

Leeuwenhock examined was determined by comparison. For this purpose he used at various times a grain of sand, the seed of millet or mustard, the eye of a louse, a vinegar eel, and still later hair or blood corpuscles. In this way he secured fairly accurate measurements of a great variety of objects.... he was forced to admit that the sand grain was more than one million times the size of one of the animalcules.

Leeuwenhoek was so interested in the things he observed that, like Hooke, he wrote minutely detailed reports about them to the Royal Society in London, beginning in 1674. He was later elected a fellow of the Royal Society. Some of his observations are at once quaint and epochmaking. For example, after examining material which he scraped from between his teeth, he said:

Though my teeth are kept usually very clean, nevertheless when I view them in a Magnifying Glass, I find growing between them a little white matter as thick as wetted flour; in this substance, though I could not perceive any motion, I judged there might probably be living Creatures.

I therefore took some of this flour and mix it either with pure rain water wherein were no animals; or else with some of my spittle (having no Air bubbles to cause a motion in it) and then to my great surprise perceived that the aforesaid matter contained very small living animals, which moved themselves very extravagantly. The biggest sort had the shape of A. Their motion was strong and nimble, and they darted themselves through the water or spittle, as a Jack or Pike does through the water. These were generally not many in number. The

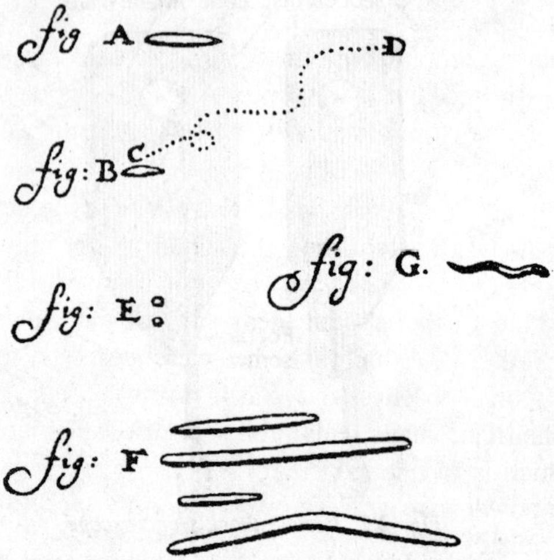

Fig. 2.3. Leeuwenhoek's drawings of bacteria.

second sort had the shape of B. These spun about like a top, or took a course sometimes on one side, as is shown at C and D. They were more in number than the first. In the third sort I could not well distinguish the Figure 2.3, for sometimes it seem'd to be an oval, and other times a circle. These were so small they seem'd no bigger than E and therewithal so swift, that I can compare them to nothing better than a swarm of Flies or Gnats, flying and turning among one another in a small space.

Note that, unlike Hooke, Leeuwenhoek made many of his observations by light transmitted through the object and that the microorganisms were suspended in various fluids, not immobilized or otherwise altered by drying.

Microorganisms and Origin of Life

The ancients knew nothing of microorganisms, of evolution, or of the fact that only living things could beget living things. They believed that creatures like frogs, mice, bees and other animals sprang fully formed from fertile mud, decaying carcasses, warm rain or fog and the like. Van Helmont (1577-1644) devised a method for manufacturing mice. He recommended putting some wheat grains with soiled linen and cheese into an appropriate receptacle and leaving it undisturbed for a time in an attic or stable. Mice would then appear. This observation may still be experimentally confirmed but the conclusions drawn from it differ today.

Belief in spontaneous generation lived on for years, as it had for centuries. For example, an elderly lady of the writer's early acquaintance complained bitterly that she had been cheated by a merchant who sold her a woolen coat which was of such a quality that it turned entirely into moths when left undisturbed in a closed for some months!

In the earlier years, in the absence of exact knowledge of microorganisms or chemistry, there had arisen much skepticism and bitter feeling over the question of the origin of life. One "scientist" who still held to the ancient ideas says of the views of another who doubted, So may we doubt whether, in cheese and timber, worms are generated, or if beetles and wasps in cow dung, or if butterflies, locusts, shellfish, snails, eels, and such life be procreated of putrefied matter which is to receive the forms of that creature to which it is by formative power disposed. To question this is to question reason, sense and experience. If he doubts this let him go to Egypt and there he will find the fields swarming with mice begot of the mud of Nylus, to the great calamity of the inhabitants.

There was a great deal of such acrid discussion by wordy savants of the times, who tried to settle everything by argument. Experimentation was regarded as rather undignified and even smacking of relations with the devil.

Francesco Redi (1626-1679). The experimental method was, however, being invoked here and there. For example, it had always been supposed that the maggots in decaying meat were derived spontaneously from transformations of the putrid meat itself. Francesco Redi, a physician of Arezzo, questioned this hypothesis. He placed meat and fish in jars covered with very fine gauze and saw flies approach the jars and crawl on the gauze. He saw the eggs of the flies caught on the gauze and observed that the meat then putrefied without maggot formation. Maggots developed only when the flies' eggs were deposited on the meat itself. Obviously the meat itself did not turn into maggots. Redi's work was not widely noted, however, and it was not until much later that another series of experiments was made.

Louis Joblot (1645-1723). After Leeuwenhoek's discovery of microorganisms it was thought by many who believed in the Aristotelian doctrine concerning spontaneous generation of life that animal or vegetable matter contained a "vital or vegetative force," capable of converting such matter into new and different forms of life. A popular notion was that geese and lambs could grow from certain kinds of trees. Leeuwenhoek's animalcules were hailed by many as proof of this. In 1710, Louis Joblot observed that hay, when infused in water and allowed to stand for some days, gave rise to countless animalcules, or infusoria (bacteria and protozoa). The hay was thought by some to change into animalcules; anyone today can observe the development of these for himself. Joblot, however, boiled hay infusion and divided it into two portions, placing one in a carefully baked (sterilized) and closed vessel, which he heated thoroughly and kept closed. The other portion was not heated and was kept in an open vessel. The infusion in the open vessel teemed with microorganisms in a few days. In the closed vessel no life appeared as long as it remained closed, thus showing that the infusion alone, once freed of life by heat, was incapable of generating new life. spontaneously.

John Needham (1713-1781). Similar experiments carried out by an English scientist, John Needham, gave conflicting results. Life developed in Needham's heated closed vessels as well as in the open unheated ones. He therefore believed in spontaneous generation. We shall see later that this result was due to insufficient heating which failed to kill heat-resistant forms of bacteria called spores. But nothing was known about spores at that time.

Lazzaro Spallanzani (1729-1799). Spallanzani, an Italian naturalist, published the results of a whole series of the same type of experiments which disagreed with those of Needham. He showed that if heating was prolonged sufficiently and the vessels kept closed to exclude dust and air, no animalcules developed in hay infusions or in any other kind of organic matter, such as urine and beef broth. Needham, in reply, said that the prolonged heating destroyed the "vegetative force" of the organic matter which, he said, was necessary for the spontaneous generation of life. Spallanzani answered Needham's objections by showing that the heated infusions in the closed flasks could still develop animalcules when exposed to air (i.e., when microorganisms were introduced with dust).

In 1775 Lavoisier discovered oxygen and the relation between air and life. This renewed the controversy about spontaneous generation, the objection to Spallanzani's results being that it was the exclusion of air (oxygen) from the flasks which prevented the development of life.

Schulze and *Schwann*. New experiments were performed in which unheated air was admitted freely to the previously heated infusions of meat or hay, but only after passing through sulfuric acid or potassium hydroxide solutions or through very hot glass tubes, the idea being that the air itself introduced the germs of life into the infusions. When the infusions exposed to air so treated failed to develop any life, it was claimed by others that this was not due to a destruction of any germs of life in the air by the sulfuric acid or hot glass, but that the "life-giving" power of the air had been destroyed by these methods, thus preventing spontaneous generation.

Schroder and *von Dusch*. This objection was overcome by Schroder and von Dusch (1854-1861) who performed similar experiments in which the air was not heated or passed through acid or alkali but merely filtered through cotton wool. This method prevented the appearance of animalcules in the heated broth or infusions until the vessels were opened. It was therefore apparent that the method of treatment of the air had nothing to do with the development of animalcules and that these did not develop spontaneously, but that there were particles of living matter floating on dust in the air which not only were killed by heat, acids and alkalis, but which could be caught and withheld by the cotton wool alone. The presence of the microorganisms in the cotton wool was later proved by Pasteur. The experiments of Schroder and von Dusch were the origin of our present-day use of cotton plugs for bacteriological culture tubes and flasks.

In spite of these demonstrations, long and bitter controversies still raged. Schroder and von Dusch were not convinced by their own experiments and admitted the possibility that spontaneous generation might occur under natural conditions.

Louis Pasteur (1822-1895). Pasteur, one of the most famous French scientists, was born in Dole, December 27, 1822. Son of a moderately prosperous tanner who had fought for and been decorated on the battlefield by Napoleon, Pasteur had a great admiration for his father's soldierly accomplishments. He later was moved to many of his best scientific achievements by his patriotic zeal. In his boyhood he was an indifferent student but later became an enthusiastic scholar, devoting his energies to a study of chemistry. He discovered the relationship of the stereoisomeric forms of tartrate crystals and revealed a whole new series of possibilities in physical chemistry.

"Diseases" of Wines

Pasteur, however, was not one to gloat over such successes and rest on his laurels. He sought for other fields of investigation. His choice was guided largely by patriotic motives. An Englishman had written to him:

People are astonished in France that the sale of French wines should not have become more extended here [in England] since the Commercial Treaties. The reason is simple enough. At first we eagerly welcomed those [French] wines, but we soon had the sad experience that there was too much loss occasioned by the diseases [souring] to which they are subject.

Germany was at that time making a much better beer than France, and Pasteur undertook to make France a successful rival in that respect. In order to do so he made a long study of beer manufacture and of the cause of souring and spoilage ("diseases") of beer and wines. As a result of these studies he arrived at far-reaching conclusions. He said at the Academic des Sciences in January, 1864:

Might not the diseases of wines be caused by organized ferments, microscopic vegetations [yeasts, molds, bacteria], of which the germs would develop when certain circumstances of temperature, of atmospheric variations, of exposure to air, would favour their evolution or their introduction into wines? ... I have indeed reached this result that the alterations ["diseases"] of wines are co-existent with the presence and multiplication of microscopic vegetations.

Pasteur had found that acid wines, "ropy" wines, bitter wines, sour beer and so on were caused by the growth in them of undesirable contaminating organisms which produced these so-called diseases.

The solution of the problem, as later proved by Pasteur, lay in preventing the growth of foreign organisms, "wild" yeasts and bacteria, which caused the undesirable conditions. After considerable experimentation along these lines he discovered that the wine did not spoil in transit if it were held for some minutes at a temperature between 50° and 60°C. He said, I have ascertained that wine was never altered by that preliminary operation (heating), and as nothing prevents it afterwards from undergoing... improvement with age—it is evident that this process [heating] offers every advantage.

His experiments were so successful that a practical test of the efficacy of his methods was made. He wrote to a friend:

...experiments on the heating of wines will be made by the Minister of the Navy. Great quantities of heated and of non-heated wine are to be sent to Gabon so as to test the process; at present our colonial crews have to drink mere vinegar.

Pasteur laid down three great principles:

1. Every alteration, either of beer or of wine, depends on the development in it of microorganisms which are ferments or "diseases" of the beer or wine.
2. These "germs or ferments" are brought by the air, by the ingredients or by the apparatus used in breweries.
3. Whenever beer or wine contains no living microorganisms it remains unchanged.

Pasteurization

In the same way that wines could be preserved by heating from various causes of alteration, bottled beer could escape the development of disease ferments by being brought to a temperature of 50° to 55° C. The application of this process soon gave rise to the new word, "*pasteurized*" beer, which became current in technical language. Today, pasteurization of milk (heating at 63°C. for 30 minutes) is routine. The heating kills pathogenic (disease-producing) microorganisms.

Pasteur on Specificity of Disease

Pasteur foresaw the consequences of his studies, and wrote in his book on beer:

When we see beer and wine subjected to deep alterations because they have given refuge to microorganisms invisibly introduced and now swarming within them, it is impossible not to be pursued by the thought that similar facts may, must, take place in animals and in man.

It was obvious from Pasteur's studies that each special kind of fermentation or disease of beer or wine was the result of the growth and activity in it of a special, distinct form of yeast or other microorganism, depending on the type of fermentation or disease under investigation. This furthered an idea, already old, of the specificity of biological action, and supported the view that animal and human diseases also, like different sorts of putrefaction and fermentation, were each caused by a single, specific type of microorganism.

Pasteur of Spontaneous Generation

After Pasteur's views of the nature of fermentation had been made public he became involved in the bitter quarrel over the apparently mysterious appearance of "germs" in fermentable or putrescible liquids like wine, beer, urine and broth. Pasteur carried out many ingenious experiments to answer the various objections and fallacies of previous workers and to show that the animalcules in spoiled beer and wine were merely descendants of microorganisms that had gained access to the fluids from dust in the air and that, by their growth and metabolism, caused fermentation and putrefaction. First, he redemonstrated that living creatures float in the air attached to particles of dust. Then he showed, as Schulze and Schwann had done, that when they could be excluded from various substances such as sterilized broth and urine, these substances did not ferment or putrefy. By using flasks with long open necks having several vertical bends in them, he showed that although unheated, untreated and unfiltered air communicated freely with the interior, the dust was caught by gravity in the bends of the neck, and no life appeared in the infusions. Not until the flask was tilted so that the fluid came into contact with this dust and was allowed to run back into the flask, or until the neck of the flask was broken off close to the body, did growth occur in the fluids. Some of Pasteur's flasks which were sterile in 1864 have been preserved and are still sterile (if they have not been destroyed by wars) after over a century!

Modern Style

In 1864 Pasteur received a prize from the French Academy of Science for his studies that conclusively disproved the Aristotelian doctrine of spontaneous generation. Since that time, studies of organic chemistry have produced a mass of evidence to revive the doctrine of spontaneous generation in a modern form called *chemical evolution*. There is now good reason to believe that life evolved during many millions of years from combinations of a few elements (chiefly C, H, O, N, S and P) by a series of reactions that were compatible with natural laws (second law of thermodynamics or law of entropy) and were actually inevitable under what are called "primitive earth" or *prebiotic* conditions. These are conditions that are thought to have existed during the later stages of the formation of the earth over 4,000,000,000 years ago , eons before the Cambrian period that began as recently as 600,000,000 years ago—practically yesterday, geologically speaking.

Chemical Evolution

Prior to 1828 the formation of organic compounds was believed to be absolutely restricted to living organisms (hence *organic*). In 1828 Friedrich Wohler, a German chemist, produced urea, $O{=}C{=}(NH_2)_2$, an organic substance common in the wastes of many animals, by the simple process of evaporating an aqueous solution of ammonium cyanate, $O{=}C{=}N(NH_4)$, an inorganic substance. The old-time distinction between *organic* and *inorganic* evaporated with the water

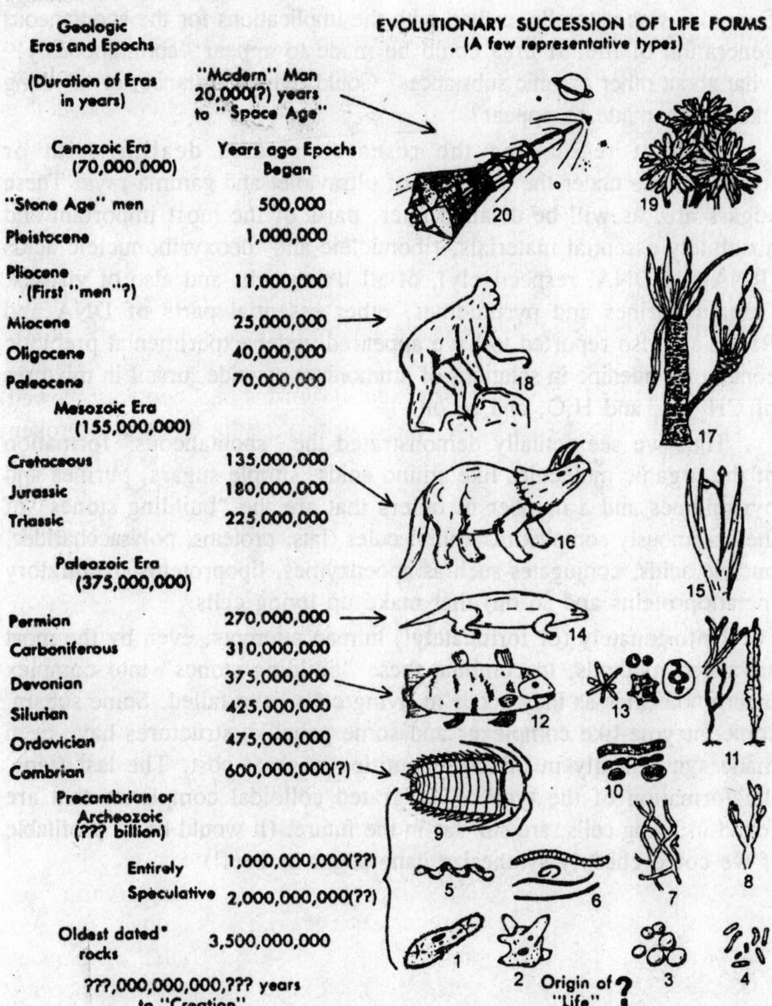

Fig. 2.4. Geologic time scale (left) showing ancient origin of bacteria and other microorganisms in pre-Cambrian times (entirely speculative) at foot of evolutionary scale. Picture 1, 2, protozoa; 3, 4, bacteria; 5, spirochete (protozoa-like bacterium); 6, marine worms; 7, mold-like bacteria; 8, aquatic fungus (Saprolegnia); 9, trilobite (fossil marine arthropod); 10, 13, bacteria-like algae (Cyanophyceae, desmids); 11, higher algae; 12, fossil fish; 14, cotylosaur (fossil reptile); 15, Psilopsida, first vascular land plant (Silurian); 16, Triceratops, a dinosaur; 17, cycad tree (Jurassic); 18, fossil man-like ape; 19, hybrid diasies; 20, man in space.

from Wohler's solution; Wohler had demonstrated that "spontaneous generation" of organic substances from inorganic ones could occur.

Everyone was naturally excited with the implications for the spontaneous generation of life. If urea could be made to appear "spontaneously," what about other organic substances? Could living substance, or anything like it, be made to appear?

Without recounting the research studies dealing with or formaldehyde under the influence of ultraviolet and gamma rays. These sugars are, as will be detailed later, parts of the most important and absolutely essential materials, ribonucleic and deoxyribonucleic acids (RNA and DNA, respectively), of all living cells and also of viruses. Certain purines and pyrimidines, other essential parts of DNA and RNA, are also reported to have appeared under experimental prebiotic conditions: adenine in solutions of ammonium cyanide, uracil in mixtures of CH_4, H_2 and H_2O, and so on.

Thus we see actually demonstrated the "spontaneous" formation of the organic molecules like amino acids, simple sugars, purines and pyrimidines and a number of others that are the "building stones" of the enormously complex macromolecules (fats, proteins, polysaccharides, nucleic acids, conjugates such as apoenzymes, lipoproteins, respiratory metalloproteins and so on) that make up living cells.

Unfortunately (or fortunately!) human attempts, even by the most advanced methods, to combine these "building stones" into complex macromolecules as they occur in living cells have failed. Some sugars, some enzyme-like complexes and some genelike structures have been made synthetically in minute quantities at great cost. The last steps, the formation of the type of integrated colloidal complexes that are found in living cells, are still far in the future. (It would be so profitable if we could cheaply synthesize cane sugar or beef!)

3

ARCHAEOBACTERIA

As a group the *archaeobacteria* are quite diverse, both in morphology and physiology. They can stain either gram positive or gram negative and may be spherical, rod-shaped, spiral, lobed, plate-shaped, irregularly shaped, or pleomorphic. Some are single cells, whereas others are filaments or aggregates. They range in diameter from 0.1 to over 15 μm, and some filaments can grow up to 200 μm in length. Multiplication may be by binary fission, budding, fragmentation, or other mechanisms. Archaeobacteria are just as diverse physiologically. They can be aerobic, facultatively anaerobic, or strictly anaerobic. Nutritionally they range from chemolithoautotrophs to organotrophs. Some are mesophiles; other are hyperthermophiles that can grow above 100°C. Archaeobacteria usually prefer restricted or extreme aquatic and terrestrial habitats. They are often present in anaerobic, hypersaline, or high-temperature environments. Recently archaeobacteria have been discovered in cold environments. It appears that they constitute up to 34% of the prokaryotic biomass in coastal Antarctic surface waters. A few are symbionts in animal digestive systems.

Archaeobacterial Cell Walls

Although archaeobacteria can stain either gram positive or gram negative, their wall structure and chemistry differ form that of the eubacteria. There is considerable variety in archaeobacterial wall structure. Many gram-positive archaeobacteria have a wall with a single thick homogeneous layer like gram-positive eubacteria. Gram-negative archaeobacteria lack the outer membrane and complex peptidoglycan network or sacculus of gram-negative eubacteria. Instead they usually have a surface layer of protein of glycoprotein subunits.

The chemistry of archaeobacterial cell walls is also quite different form that of the eubacteria. None have the muramic acid and D-amino acids characteristic of eubacterial peptidoglycan. Not surprisingly, all archaeobacteria resist attack by lysozyme and β-lactam antibiotics such as penicillin. Gram-positive archaeobacteria can have a variety of complex polymers in their walls. *Methanobacterium* and some other methanogens have walls containing *pseudomurein*, a peptidoglycanlike polymer that has L-amino acids in its cross-links, N-acetyltalosaminuronic acid instead of N-acetylmuramic acid, and β(1→3) glycosidic bonds instead of β(1→4) glycosidic bonds. *Methanosarcina* and *Halococcus* contain complex polysaccharides similar to the chondroitin sulfate of animal connective tissue. Other heteropolysaccharides also found in grama-positive walls.

Gram-negative archaeobacteria have a layer of protein or glycoprotein outside their plasma membrane. The layer may be as thick as 20 to 40 nm. Sometimes there are two layers, a sheath surrounding an electron-dense layer. The chemical content of these walls varies considerably. Some methanogens (*Methanolobus*),

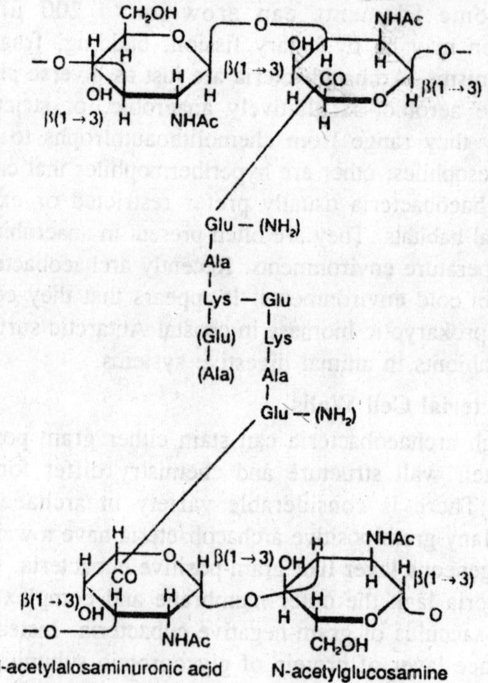

Fig. 3.1. The structure of pseudomurein.

Halobacterium, and several extreme thermophiles (*Sulfolobus*, *Thermoproteus*, and *Pyrodictium*) have glycoproteins in their walls. In contrast, other methanogens (*Methanococcus*, *Methanomicrobium*, and *Methanogenium*) and the extreme thermophile *Desulfurococcus* have protein walls.

Archaeobacterial Lipids and Membranes

Most distinctive feature of the archaeobacteria is the nature of their membrane lipids. They differ from both eubacteria and eukaryotes in having branched chain hydrocarbons attached to glycerol by ether links rather than fatty acids connected by ester links. Sometimes two glycerol groups are linked to form an extremely long tetraether. Usually the diether side chains are 20 carbons in size, and the tetraether chains are 40 carbons. The overall length of the tetraethers can be adjusted by cyclizing the chains to form pentacyclic rings, and biphytanyl chains may contain form one to four cyclopentyl rings. Polar lipids are also present in archaeobacterial membranes: phospholipids, sulfolipids, and glycolipids. From 7 to 30% of the membrane lipids are nonpolar lipids, which usually are derivatives of squalene. These lipids can be combined in various ways to yield membranes of different rigidity and thickness. For example, the C_{20} diethers can be used to make a regular bilayer membrane. A much more rigid monolayer membrane may be constructed of C_{40} tetraether lipids. Of course archaeobacterial membranes may contain a mix of diethers, tetraethers, and other lipids. As might be expected from their need for stability, the membranes of extreme thermophiles such as Thermoplasma and Sulfolobus are almost completely tetraether monolayers.

Genetics and Molecular Biology

Some features of archaeobacterial genetics are similar to those in eubacteria. Their chromosome is a single closed DNA circle. However, the genomes of some archaeobacteria are significantly smaller than the normal eubacterium. *E. coli* DNA has a size of about 2.5×10^9 daltons, whereas *Thermoplasma acidophilum* DNA is about 0.8×10^9 daltons and *Methanobacterium thermoautotrophicum* DNA is 1.1×10^9 daltons. The variation in G+C content is great, from about 21 to 68 mol% and is another sign of archaeobacterial diversity. Archaeobacteria have few plasmids.

Archaeobacterial mRNA appears similar of eubacteria rather than to eukaryotic mRNA. Polygenic mRNA has been discovered, and there is no evidence for mRNA splicing. Archaeobacterial promoters are similar to those in eubacteria.

A normal lipid

Glycerol — Ester bond — Stearic acid

$CH_2—O—C(=O)$

$CH—O—C(=O)$

$CH_2—OH$

Archaeobacterial lipids

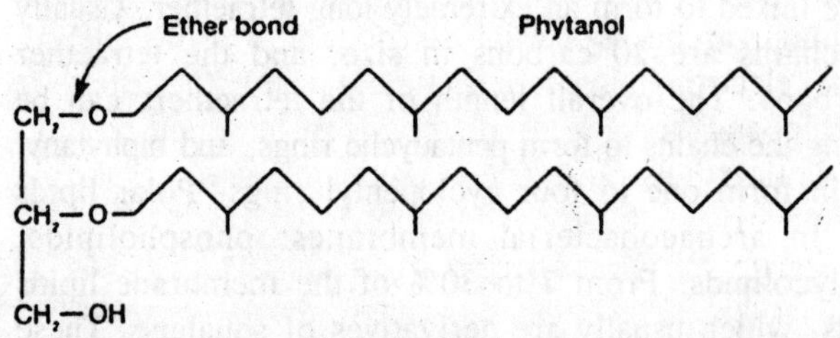

Phytanylglycerol diether

$CH_2—O$ / $CH—O$ / $CH_2—OH$ … $HO—CH_2$ / $O—CH$ / $O—CH_2$

Dibiphytanyldiglycerol tetraether

$CH_2—O$ / $CH—O$ / $CH_2—OH$ … $HO—CH_2$ / $O—CH$ / $O—CH_2$

Tetraether with bipentacyclic C_{40} biphytanyl chains

Fig. 3.2. Archaebacterial membrane lipids.

Despite these and other similarities, there are also many differences between archaeobacteria and other organisms. Unlike both eubacteria and eukaryotes, the TψC arm of archaeobacterial tRNA lacks thymine and contains pseudouridine or 1-methylpseudouridine. The

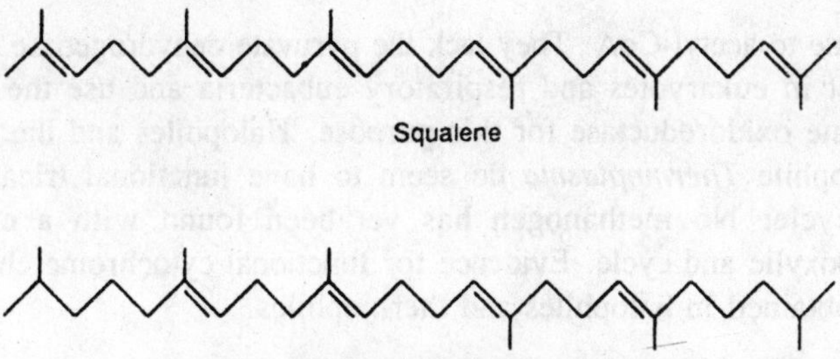

Squalene

Tetrahydrosqualene

Fig. 3.3. Nonpolar lipids of archaeobacteria.

archaeobacterial initiator tRNA carries methionine like the eukaryotic initiator tRNA. Although archaeobacterial ribosomes are 70S like eubacterial ribosomes, electron microscopic studies show that their shape is quite variable and sometimes differs from that of both eubacterial and eukaryotic ribosomes. They do resemble eukaryotic ribosomes in their sensitivity to anisomycin and insensitivity to chloramphenicol and kanamycin. Furthermore, their elongation factor 2 reacts with diphtheria toxin like the eukaryotic EF-2. Finally, archaeobacterial DNA-dependent RNA polymerases resemble the eukaryotic enzymes, not the eubacterial RNA polymerase. They are large, complex enzymes and are insensitive to the drugs rifampin and streptolydigin. These and other differences distinguish the archaeobacteria from both eubacteria and eukaryotes.

Metabolism

Not surprisingly in view of the variety of their life-styles, archaeobacterial metabolism varies greatly between the members of different groups. Some archaeobacteria are organotrophs; others are autotrophic. A few even carry out an unusual form of photosynthesis.

Archaeobacterial carbohydrate metabolism is best understood. The enzyme 6-phosphofructokinase has not been found in archaeobacteria, and they do not appear to degrade glucose by way of the Embden-Meyerhof pathway. Extreme halophiles and thermophiles catabolize glucose using a modified form of the Enter-Doudoroff pathway in which the initial intermediates are not phosphorylated. The halophiles have slightly different modifications of the pathway than do the extreme thermophiles but still produce pyruvate and NADH or NADPH. Methanogens are autotrophs and do not catabolize glucose to any significant extent. In contrast with glucose degradation, gluconeogenesis proceeds by a reversal of the Embden-Meyerhof pathway in halophiles and methanogens. All archaeobacteria that have been studied can oxidize pyruvate to acetyl-CoA. They lack the pyruvate dehydrogenase complex present in eukaryotes and respiratory eubacteria and use the enzyme pyruvate oxidoreductase for this purpose. Halophiles and the extreme thermophile *Thermoplasma* do seem to have functional tricarboxylic acid cycle. No methanogen has yet been found with a complete tricarboxylic and cycle. Evidence for functional cytochrome chains has been obtained in halophiles and thermophiles.

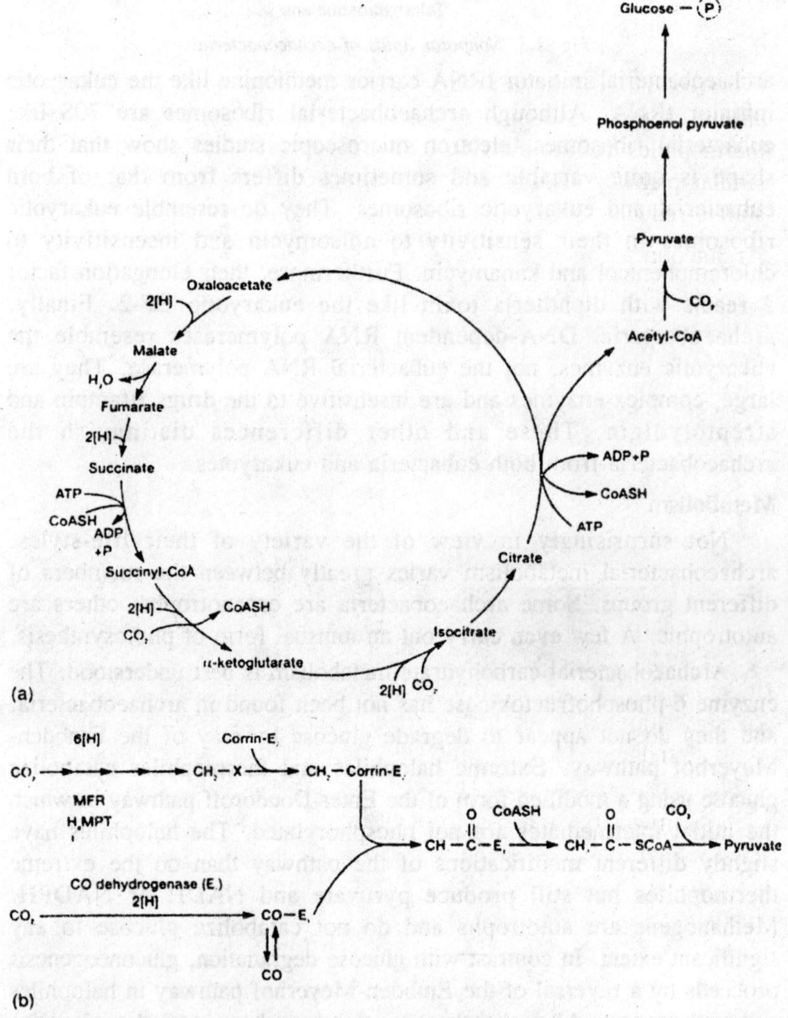

Fig. 3.4. Mechanisms of autotrophic CO_2 fixation.

Very little is known in detail about biosynthetic pathways in the archaeobacteria. Preliminary data suggest that the synthetic pathways for amino acids, purines, and pyrimidines are similar to those in other organisms. Some methanogens can fix atmospheric dinitrogen. Not only do many archaeobacteria use a reversal of the Embden-Meyerhof pathway to synthesize glucose, but at least some methanogens and extreme thermophiles employ glycogen as their major reserve material.

Autotrophy is widespread among the methanogens and extreme thermophiles, and CO_2 fixation occurs in more than one way. Thermoproteus and possibly Sulfolobus incorporate CO_2 by the reductive tricarboxylic acid cycle. This pathway is also present in the green sulfur bacteria. Methanogenic bacteria and probably most extreme thermophiles incorporate CO_2 by the reductive acetyl-CoA pathway. A similar pathway also is present in acetogenic bacteria and autotrophic sulfate-reducing bacteria.

Archaeobacterial Taxonomy

The archaeobacteria are quite distinct from other living organisms. Within the group, however, there is great diversity. Section 25 of *Bergey's Manual* divides the archacobacteria into five major groups: methanogenic archaeobacteria, archaeobacterial, cell wall-less archaeobacteria, and extremely thermophilic S^0-metabolizers.

Methanogenic Archaeobacteria

Methanogens are strict anaerobes that obtain energy by converting CO_2, H_2, formate, methanol, acetate, and other compounds to either methane or methane and CO_2. They are autotrophic when growing on H_2 and CO_2. This is the largest group of archaeobacteria. There are at least three orders and 18 genera, which differ greatly in overall shape, 16S rRNA sequence, cell wall chemistry and structure, membrane lipids, and other features. For example, methanogens construct three different types of cell walls. Several genera have walls with pseudomurein as mentioned earlier. Other walls contain either proteins or heteropolysaccharides.

As might be inferred from the methanogens' ability to produce methane anaerobically, their metabolism is unusual. These bacteria contain several unique cofactors: tetrahydromethanopterin (H_4MPT), methanofuran (MFR), coenzyme M (2-mercaptoethanesulfonic acid), coenzyme F_{420}, and coenzyme F_{430}. The first three cofactors bear the C_1 unit when CO_2 is reduced to CH_4. F_{420} carries electrons and hydrogens, and F_{430} is a nickel tetrapyrrole serving as a cofactor for

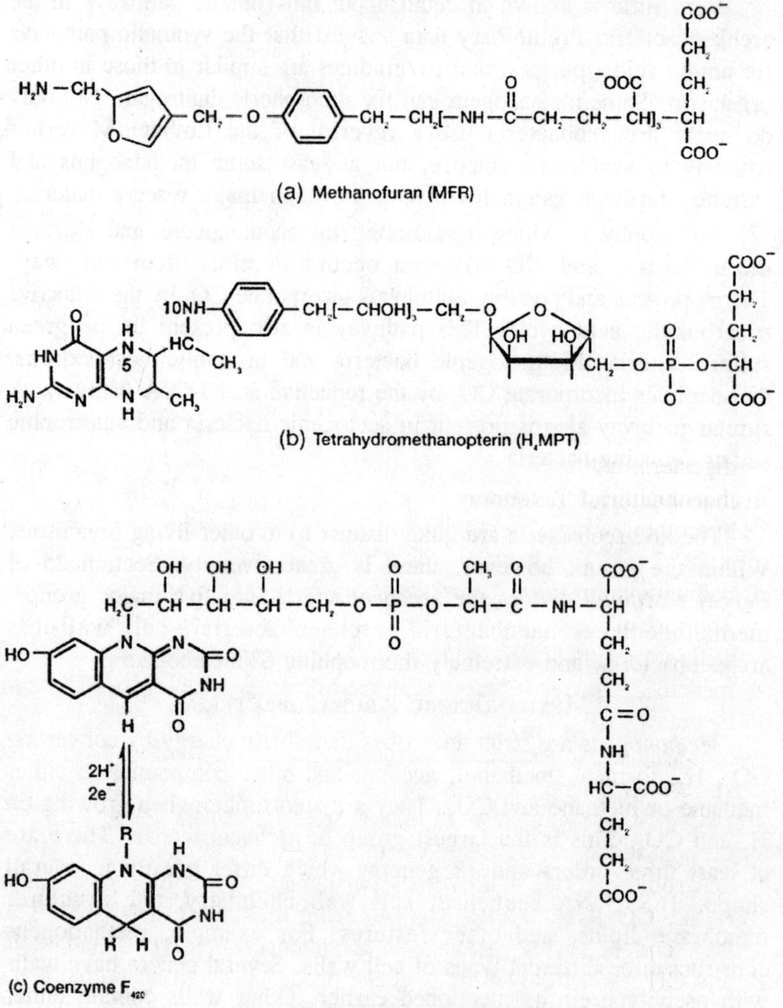

(a) Methanofuran (MFR)

(b) Tetrahydromethanopterin (H_4MPT)

(c) Coenzyme F_{420}

Fig. 3.5. Methanogen coenzymes.

the enzyme methyl-CoM methylreductase. It appears that ATP synthesis is linked with methanogenesis by electron transport, proton pumping and a chemiosmotic mechanism. Some methanogens can live autotrophically by forming acetyl-CoA form two molecules of CO_2 and then converting the acetyl-CoA to pyruvate and other products.

Methanogens thrive in anaerobic environments rich in organic matter: the rumen and intestinal system of animals, freshwater and marine sediments, swamps and marshes, hot springs, anaerobic sludge

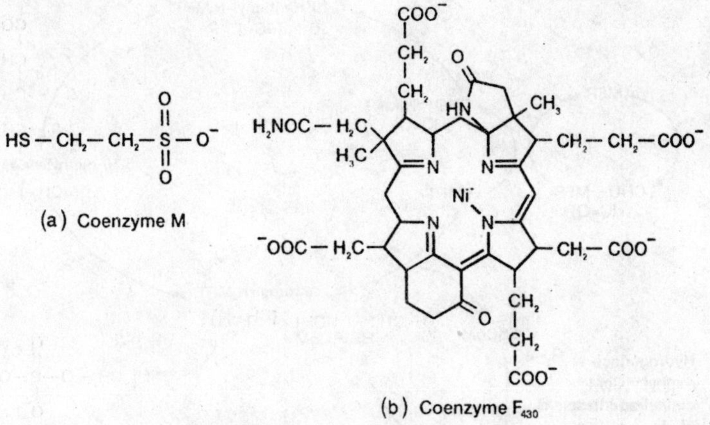

(a) Coenzyme M

(b) Coenzyme F_{430}

Fig. 3.6. Methanogen coenzymes.

digesters, and even within anaerobic protozoa. Recently an extremely thermophilic rod-shaped methanogen has been isolated from a marine hydrothermal vent. *Methanopyrus kandleri* has a temperature minimum at 84°C and an optimum of 98°C; it will grow up to 110°C (above the boiling point of water). Methanogens often are of ecological significance. The rate of methane production can be so great that bubbles of methane will sometimes rise to the surface of a lake or pond. Rumen methanogens are so active that a cow can belch 200 to 400 liters of methane a day.

Methanogenic bacteria are potentially of great practical importance since methane is a clean-burning fuel and an excellent energy source. For many years sewage treatment plants have been using the methane they produce as a source of energy for heat and electricity. An anaerobic digester will degrade particulate wastes like sewage sludge to H_2, CO_2, and acetate. CO_2-reducing methanogens form CH_4 from CO_2 and H_2, while aceticlastic methanogens cleave acetate to CO_2 and CH_4 (about 2/3 of the methane produced by an anaerobic digester comes from acetate). A kilogram of organic matter can yield up to 600 liters of methane. It is quite likely that future research will greatly increase the efficiency of methane production and make methanogenesis an important source of pollution-free energy.

Methanogenesis also can be an ecological problem. Methane absorbs infrared radiation and thus is a greenhouse gas. There is evidence that atmospheric methane concentrations have been rising over the last 200 years. Methane production may significantly promote future global warming. Recently it has been discovered that methanogens can oxidize

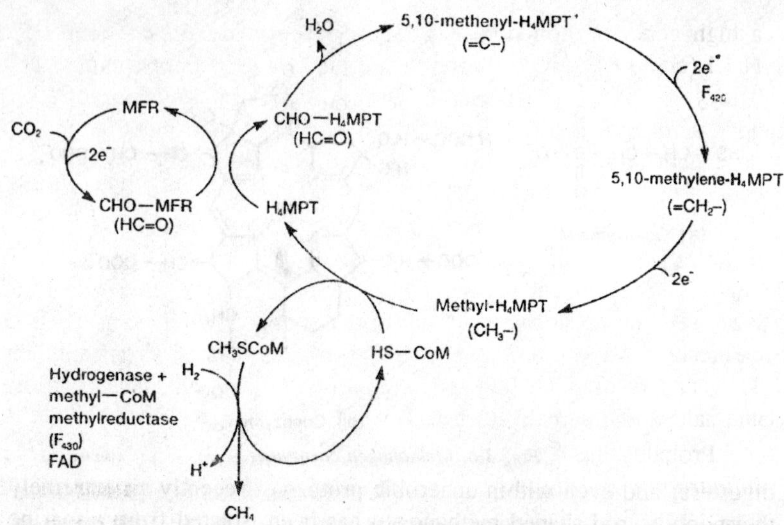

Fig. 3.7. Methane synthesis.

Fe^0 and use it to produce methane and energy. This means that methanogens growing around buried or submerged iron pipes and other objects may contribute significantly to iron corrosion.

Archaeobacterial Sulfate Reducers

This group contains gram-negative, irregular coccoid cells with walls consisting of glycoprotein sub units. At present, only the genus *Archaeoglobus* is known. It can extract electrons from a variety of electron donors (e.g., H_2, lactate, glucose) and reduce sulfate, sulfite, or thiosulfate to sulfide. Elemental sulfur is not used as an acceptor. *Archaeoglobus* is extremely thermophilic (the optimum is about 83°C) and can be isolated from marine hydrothermal vents. The organism is not only unusual in being able to reduce sulfate, unlike other archaeobacteria, but it also possesses the methanogen coenzymes F_{420} and methanopterin. It seems intermediate between the methanogens and the extremely thermophilic S^0-metabolizers.

Extremely Halophilic Archaeobacteria

The *extreme halophiles* are a third major group of archaeobacteria, currently with six genera in one family, the *Halobacteriaceae*. They are aerobic chemoheterotrophs with respiratory metabolism and require complex nutrients, usually proteins and amino acids, for growth. They are either nonmotile or motile by lophotrichous flagella. The most obvious distinguishing trait of this family is its absolute dependence on a high concentration of NaCl. These bacteria require at least 1.5M NaCl (about 8%, wt/vol), and usually have a growth optimum at abut 3 to 4 M NaCl (17 to 23%). They will grow at salt concentrations approaching saturation (about 36%). *Halobacterium's* cell wall is so dependent on the presence of NaCl that it disintegrates when the NaCl concentration drops to about 1.5 M. Thus halobacteria only grow in high-salinity habitats such as marine salterns and salts lakes like the Dead Sea between Israel and Jordan, and the Great Salt Lake in Utah. They also can grow in food products such as salted fish and cause spoilage. Halobacteria often have red-to-yellow pigmentation from carotenoids that are probably used as protection against strong sunlight. They can reach such high population levels that salt lakes, salterns, and salted fish actually turn red.

Probably the best-studied member of the family is *Halobacterium salinarium* (*H. halobium*). This prokaryote is unusual because it can trap light energy photosynthetically without the presence of chlorophyll. When exposed to low oxygen levels, some strains of *Halobacterium* synthesize a modified cell membrane called the *purple membrane*, which contains the protein *bacteriorhodopsin*. ATP is produced by a unique type of photosynthesis without the participation of bacteriochlorophyll or chlorophyll, *Halobacterium* actually has four rhodopsin, each with a different function. As already mentioned, bacteriorhodopsin drives outward proton transport for purposes of ATP synthesis. Halorhodopsin uses light energy to transport chloride ions into the cell and maintain a 4 to 5 M intracellular KCl concentrations. Finally there are two rhodopsins that act as photoreceptors, one for red light and one for blue. They control flagellar activity to position the bacterium optimally in the water column. *Halobacterium* moves to a location of high light intensity, but one in which ultraviolet light is not sufficiently intense to be lethal.

Cell Wall-less Archaeobacteria

Thermoplasma grows in refuse piles of coal mines. These piles contain large amount of iron pyrite (FeS), which is oxidized to sulfuric acid by chemolithotrophic bacteria. As a result of the piles become very hot and acidic. This is an ideal habitat for *Thermoplasma* since it grows best at 55 to 59°C and pH 1 to 2. Although it lacks a cell wall, its plasma membrane is strengthened by large quantities of diglycerol tetraethers, lipopolysaccharides, and glycoproteins. The organism's DNA is stabilized by association with a special histonelike protein that condenses the DNA into particles resembling eukaryotic nucleosomes. At 59°C, *Thermoplasma* takes the form of an irregular filament, whereas at lower temperature it is spherical. The cells may be flagellated and motile.

Extremely Thermophilic S^0-Metabolizers

The last group of archaeobacteria contains extremely thermophilic bacteria, many of which are acidophiles and sulfur dependent. The sulfur may be used either as an electron acceptor in anaerobic respiration or as an electron source by lithotrophs. Almost all are strict anaerobes. They grow in geothermally heated water or soils that contain elemental sulfur. These environments are scattered all over the world. Examples are the sulfur-rich hot springs in Yellowstone National Park, Wyoming, and the waters surrounding areas of submarine volcanic activity. Such habitats are sometimes called solfatara. These archaeobacteria can be very thermophilic and often are classified as hyperthermophiles. The most extreme example is *Pyrodictium*, a bacterium isolated from geothermally heated sea floors. *Pyrodictium* has a temperature minimum of 82°C, a growth optimum at 105°C, and a maximum at 110°C. Both organotrophic and lithotrophic growth occur in this group. Sulfur and H_2 are the most common electron sources for lithotrophs. There are three orders (*Thermococcales*, *Thermoproteales*, and *Sulfolobales*) and at least nine genera. Two of the better-studied genera are *Thermoproteus* and *Sulfolobus*.

Members of the genus *Sulfolobus* are gram-negative, aerobic, irregularly lobed spherical bacteria with a temperature optimum around 70 to 80°C and a pH optimum of 2 to 3. For this reason, they are classed as *thermoacidophiles*, so called because they grow best at acid pH values and high temperatures. There cell wall contains lipoprotein and carbohydrate but lacks peptidoglycan. They grow lithotrophically on sulfur granules in hot acid springs and soils while oxidizing the sulfur to sulfuric acid. Oxygen is the normal electron acceptor, but ferric iron may be used. Sugars and amino acids such as glutamate also serve as carbon and energy sources.

Thermoproteus is a long thin rod that can be bent or branched. The cell wall is composed of glycoprotein. *Thermoproteus* is a strict anaerobe and grows at temperatures from 70 to 97°C and pH values between 2.5 and 6.5. It is found in hot springs and other hot aquatic habitats rich in sulfur. It can grow organotrophically and oxidize glucose, amino acids, alcohols, and organic acids with elemental sulfur as the electron acceptor. That is, *Thermoproteus* can carry out anaerobic respiration. It will also grow chemolithotrophically using H_2 and S^0. Carbon monoxide or CO_2 can serve as the sole carbon source.

4

Morphology of Bacteria

Bacteria belong to the class of organisms known as the *Schizomycetes* (*schizo*, fission, and *mycetes,* fungi). The organisms are single-celled and reproduce normally by transverse or binary fission.

The class Schizomycetes is divided into ten orders. The largest order is the *Eubacteriales*; it includes most of the common bacterial species. Bacteria are typically unicellular plants, the cells being usually small, sometimes ultramicroscopic. They are frequently motile. By means of modern techniques, a true nucleus has been demonstrated in bacterial cells. Individual cells may be spherical or straight, curved or spiral rods. Cells may occur in regular or irregular masses, or even in cysts. Where they remain attached to each other after cell division, they may form chains or even definite trichomes. The latter may show some differentiation into holdfast cells and into motile or nonmotile reproductive cells. Some grow as branching mycelial threads whose diameter is not greater than that of ordinary bacterial cells i.e., about 1 μ. Some species produce pigments. The true purple and green bacteria possess photosynthetic pigments much like or related to the true chlorophylls of higher plants. The phycocyanin found in blue-green algae does not occur in the Schizomycetes. Multiplication is typically by cell division. Endospores are formed by some species of Eubacteriales. Sporocysts are found in Myxobacteriales. Bacteria are free-living, saprophytic, parasitic, or even pathogenic. The latter types cause diseases of either plants or animals.

Filament Formation

Cells that reproduce and divide in a normal manner may be induced to grow in filaments by changing the conditions of the medium.

According to Webb (1953), the division of the bacterial cell follows a complex sequence, which in many respects, resemble that occurring in the cellular reproduction of higher forms. It is now known, for example, that bacterial cell division entails division of the nuclear element, division of the cytoplasm, secretion of new cell wall material, and the separation of the daughter cells.

Some or all of the events of this sequence are readily thrown out of balance, or even completely inhabited. Thus, bacteria, particularly the rod-shaped organisms, may be induced to elongate into filaments by various treatments which apparently inhabit cell division but which do not inhabit growth. Such an effect is produced by various chemical substances, by sub-bacteriostatic concentration of certain antibacterial agents, as, for example, methyl violet, sulfonamides, *m*-cresol, penicillin, irradiation, and higher temperature of incubation.

These changes in morphology induced by chemical substances are usually temporary, since reversion to normal form occurs promptly when the filaments bacteria are subcultured in the absence of the inhibitory agents. Irradiation, on the other hand, may give rise to a temporary or permanent induction of filamentous cells.

From observations such as these the concept has arisen that bacterial growth, in the sense of an irreversible increase in cell substance or volume, and cell division may be considered to come extent as separate and independent processes; at least, in so far as growth may occur either with or without the operation of the cell division mechanism.

Variation in the magnesium (Mg) content of the medium may exert a marked effect on cell division of some bacteria. In a Mg-deficient medium, Gram-positive rods grow in the form of long filaments. Such filaments revert to normal forms when transferred to the same medium supplemented with suitable concentrations of Mg. Filament formation is enhanced by the addition of zinc and cobalt. Inhibition of cell division occurs also in media supplemented with an excess of Mg.

Deibel et al. (1956) produced filamentous *Lactobacillus leichmannii* in the absence of vitamin B_{12}. Reversion to the normal cell form occurred on the addition of either vitaminB_{12} to a medium lacking the growth factor or of an excess of the desoxyriboside thymidine.

Shape of Bacteria

Bacteria exhibit three fundamental shapes: (1) spherical, (2) rod, and (3) spiral or curved rod. All bacteria exhibit pleomorphism in

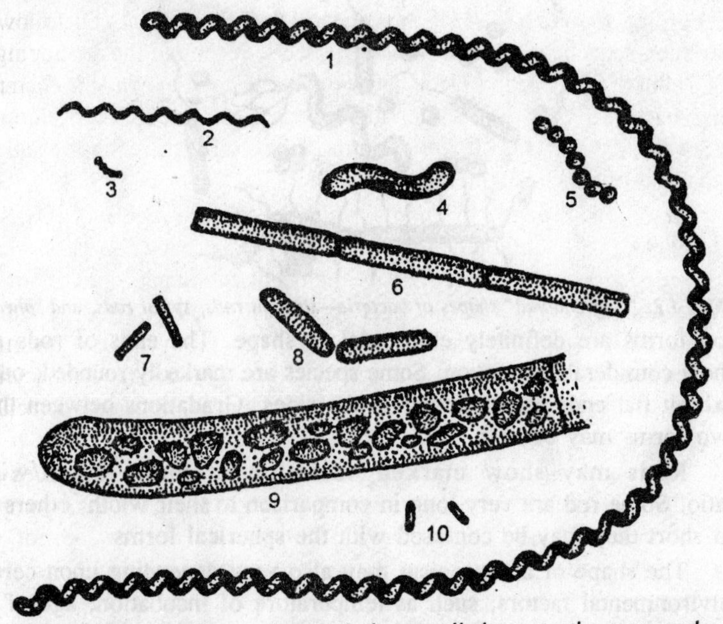

Fig. 4.1. Bacteria of different sizes and shapes, all drawn to the same scale. 1, Spirochaeta plicatilis; 2, Treponema pallidum; 3, Peptostreptococcus parvulus; 4, Spirillum undula; 5, Streptococcus pyogenes; 6, Bacillus anthracis; 7, Mycobacterium tuberculosis; 8, Bacillus megaterium; 9, Beggiatoa alba; 10, Bordetella pertussis.

more or less degree under normal or other conditions, but a bacterial species is still generally associated with a definite cell from when grown on a standard medium under controlled conditions.

The spherical bacteria (singular, coccus; plural cocci) divide in one, two, or three planes, producing pairs or chains, clusters, or packets of cells. Some are apparently perfect spheres; others are slightly elongated or ellipsoidal in shape.

The streptococci divide in only one plane. They grow normally in pairs or chains. Depending upon the species, the distal ends of each pair may be lancet-shaped, or flattened at the adjacent sides to resemble a coffee bean.

The staphylococci divide in two planes, producing pairs, tetrads, or clusters of bacteria, the latter resembling bunches of grapes.

The sarcinae divide in three planes, producing regular packets. These are cubicle masses with one layer of bacteria atop another.

The rod forms also show considerable variation. A rod is usually considered to be a cylinder with the ends more or less rounded. Some

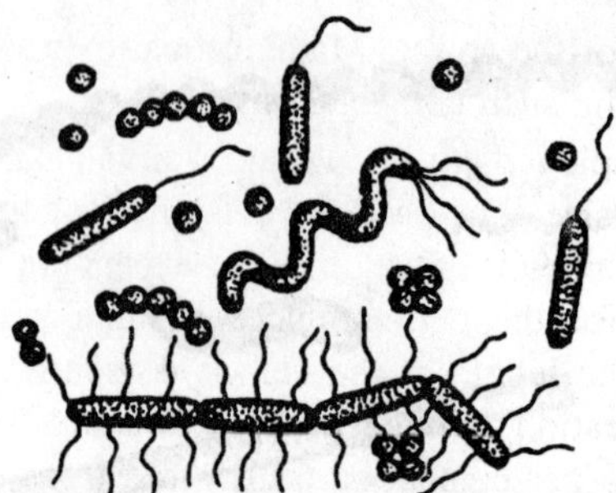

Fig. 4.2. "Conventional" shapes of bacteria—straight rods, spiral rods, and spheres.

rod forms are definitely ellipsoidal in shape. The ends of rods also show considerable variation. Some species are markedly rounded; others exhibit flat ends perpendicular to the sides. Gradations between these two forms may be seen

Rods may show marked variation in their length/width ratio. Some rod are very long in comparison to their width; others are so short they may be confused with the spherical forms.

The shape of an organism may also vary depending upon certain environmental factors, such as temperature of incubation, age of the culture, concentration of the substrate, and composition of the medium. Bacteria usually exhibit their characteristic morphology in young cultures and on media possessing favourable conditions for growth.

Young cells are, in general, larger than old organisms of the same species. As a culture ages, the cells become progressively larger until a maximum is reached, after which the reverse effect occurs. Bacterial variations resulting from changes in age are only temporary; the original forms reappear when the organisms are transferred to fresh medium.

Size of Bacteria

Bacteria vary greatly in size according to the species. Some are so small they approach the limit of visibility when viewed with the light microscope. Others are so large they are almost visible with the normal eye. However, the sizes of the majority of bacteria occupy a range intermediate between these two extremes. Regardless of size, none can be clearly seen without the aid of a microscope.

A spherical form is measured by its diameter; a rod or spiral form by its length and width. Calculation of the length of a spiral organism by this method gives only the apparent length, not the true length. The true length may be computed by actually measuring the length of each turn of the spiral. Mathematical expressions have been formulated for making such computations.

The method employed for fixing and staining bacteria may make a difference in their size. The bacterial cell shrinks considerably during drying and fixing. This will vary somewhat depending upon the type of medium employed for the cultivation. Shrinkage generally averages about one-third of the length of the cell as compared to an unstained hanging-drop preparation. Young cells of *Bacillus megaterium* may shrink from 15 to 25 per cent when transferred from nutrient broth to the same medium containing sodium chloride in 2 *M* concentration.

Measurements show some variation depending upon the staining solution used and the method of application. In dried and fixed smears, the cell wall and slime layer do not stain with weakly staining dyes such as methylene blue but do stain with the intensely staining pararosaniline, new fuchsin, crystal violet, and methyl violet. The great majority of bacteria have been measured in fixed and stained preparations. In some instances dried, negatively stained smears have been used. Therefore, the method employed should be specified when measurements of bacteria are reported; otherwise the results will be of doubtful value.

The unit for measuring bacteria is the micron. It is expressed by the Symbol μ. It is 0.001 mm. or 0.0001 cm. A millimicron is 0.001 μ or 0.000001 mm. It is expressed by the symbol mμ.

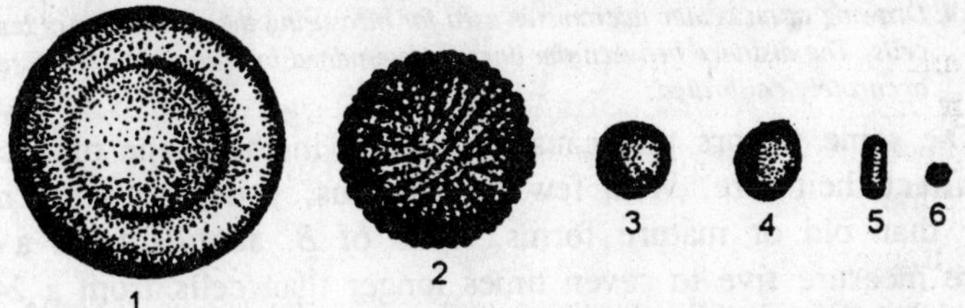

Fig. 4.3. Relative size of microorganisms, etc.: 1, red blood cell; 2, dry spore of Rhizopus; 3, dry spore of Penicillium; 4, yeast cell; 5, Bacillus subtilis; 6, Micrococcus.

Some bacteria measure as large as 80 μ in length; others as small as 0.2 μ. However, the majority of the commonly encountered bacteria, including the disease producers, measure about 0.5 μ in diameter for the spherical cells and 0.5 by 2 to 3 μ for the rod forms. Bacteria producing spores are generally larger than the non-spore-producing species. The size of some common species in dried and stained smears are as follows; *Escherichia coli*, 0.5 by 1 to 3 μ; *Proteus vulgaris* 0.5 to 1 by 1 to 3 μ; *Salmonella typhosa*, 0.6 to 0.7

by 2 to 3 μ; *Streptococcus lactis*, 0.5 to 1 μ in diameter; *S. pyogenes*, 0.6 to 1 μ in diameter; *Staphylococcus aureus*, 0.8 to 1 μ in diameter; *Lactobacillus acidophilus*, 0.6 to 0.9 by 1.5 to 6 μ; *Bacillus subtilis* rods. 0.7 to 0.8 by 2 to 3 μ, spores, 0.6 to 0.9 by 1 to 1,5 μ; *B. megaterium* rods, 0.9 to 2.2 by 1. to 5 μ, spores, 1 to 1.2 by 1.5 to 2 μ; *B anthracis* rods, 1 to 1.3 by 3 to 10 μ, spores, 0.8 to 1.3 to 1.5 μ.

The most commonly employed method for measuring bacteria is by means of an ocular micrometer. Measurements may also be made by using a camera-lucida attached and drawing oculars, or by projecting the real image on a screen and measuring the bacteria.

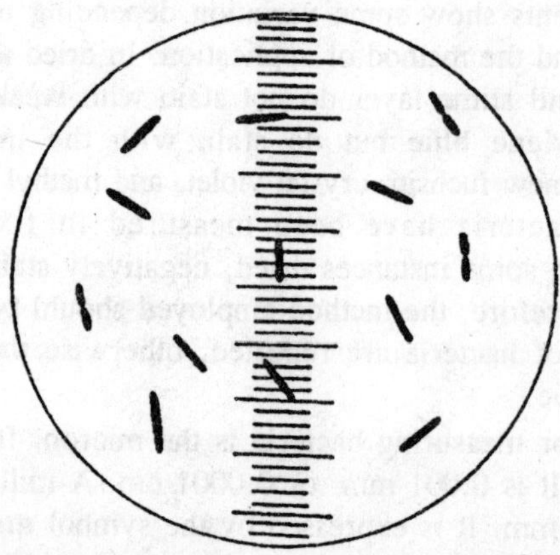

Fig. 4.4. Drawing of an ocular micrometer used for measuring the dimensions of bacterial cells. The distance between the lines is determined by using a stage micrometer accurately calibrated.

The same factors that cause variations in the shape of bacteria also affect their size. With few expectations, young cells are much larger than old or mature forms. Cells of *B. subtilis* from a 4-hr culture measure five to seven times longer than cells from a 24-hr. culture. Variations in width are less pronounced. The organism *Corynebacterium diphtheriae* is a notable exception to the rule of decreasing cell size with age. Variations in cell size with age are due to a variety of factors. The major causes appear to be changes in the environment with the accumulation of waste products. An increase in the osmotic pressure of the medium will also cause a decrease in cell size and may very well be the most important factor.

Bacterial Cell

Bacteria do not show the same morphological picture. Differences in structure exist between species. It is generally agreed that a bacterial cell consists of a compound membrane enclosing cytoplasm and nuclear material and often containing various granules fat globules, and one or more vacuoles. In addition, some species contain resistant bodies known as spores, and some have one or more organs of locomotion called flagella.

The term protoplasm is used to include the thick viscous semifluid or almost jelly-like colourless, transparent material which makes up the essential substance of both the cell body and the nucleus, including the cytoplasm membrane but not the cell wall. It contains a high percentage of water and holds the fine granules in suspension.

Cytoplasmic Membrane

This membrane appears in young cells as an interfacial fluid film, becoming thinker and denser as surface-active material accumulates. It is finally converted into a firm structure. The membrane is believed to be composed mainly of lipide and protein. Polysaccharide has not been demonstrated as a compound. The membrane is acid in reaction because of its content of ribonucleic acid. It stains deeply with basic and neutral dyes over a wide range of pH. The membrane stains Gram-positive in Gram-positive bacteria and acid-fast in acid-fast organisms. It is a semipermeable membrane and is principally responsible for the Gram and acid-fast reactions. When a cell is plasmolyzed by immersion in a hypertonic solution, this membrane is drawn in with the cytoplasm constituents. The thickness of the membrane varies even in a single cell. Measurements on a strain of *Bacillus cereus* at various stages of development ranged from 5 to 10 $m\mu$ in thickness.

Cell Wall

The cell wall is a more rigid structure and is responsible for the form of the bacterial body. It behaves as a selectively permeable membrane and apparently plays a fundamental role in the life activities of the cell. The cell wall has a low affinity for dyes, which means that it is probably not stained in some of the usual staining procedures. It is lightly stained by certain basic dyes such as basic fuchsin and the methyl violets. Where deep staining of the wall is desired, the use of a mordant, such as tannic acid, is necessary. The mordant not only increases the affinity of the cell for dyes, but it may increase the thickness of the wall.

The cell wall accounts for an average of about 20 per cent of the dry weight of bacteria and represents the major structural component. In thickness, it ranges from 10 to 23 $m\mu$, depending upon the species.

According to Salton (1952, 1953), chemical analyses of cell walls have revealed differences in Gram-positive and Gram-negative bacteria. Cell walls of Gram-positive bacteria are lacking in aromatic and certain sulfur-containing amino acids, arginine, and proline. On the other hand, cell walls of Gram-negative bacteria show the presence of aromatic and sulfur-containing amino acids, arginine, and proline.

Gram-negative cell walls are generally richer in lipides than Gram-positive bacteria. Cell walls of a number of Gram-positive and negative bacteria contain the amino acid diaminopimelic acid.

Polysaccharides have been detected in both Gram-positive and Gram-negative bacteria. The polysaccharide is determined as reducing substances after acid hydrolysis. Some polysaccharides yield only one reducing sugar; others yielded two or more sugars.

Gram-positive organisms gave rhamnose, galactose, and glucose; glucose only; rhamnose only; arabinose, galactose, and mannose. Gram-negative bacteria yielded galactose and glucose; galactose, glucose, mannose, and rhamnose.

In addition, all organisms studied contained an amino sugar or hexosamine. Generally, the walls of Gram-positive bacteria are richer in hexosamine than the Gram-negative forms.

Work (1957) pointed to the existence in Gram-positive bacteria of a common basal structure containing the following constituents: a hexosamine component comprising glucosamine and muramic acid and sometimes also galactosamine; a peptide component made up of alanine, glutamic acid, and either diaminopimelic acid or lysine with sometimes also glycine, aspartic acid, or serine; and usually a polysaccharide containing not more than four different sugar residues. Other substances may also be attached to the walls, as for example the protein antigens.

Smithies, Gibbons, and Bayley (1955) reported a relatively high nitrogen content in the walls of several halophilic bacteria which indicated that the cell material was predominantly protein. They contained only small amounts of lipids. The cell walls were lipoprotein.

Barkulis and Jones (1957) found that approximately one-third of streptococcal cell walls was made up of rhamnose and hexosamine. The remaining two-thirds was protein in nature.

Capsules

Extracellular material of a slimy or gelatinous nature is formed by many bacteria, especially those producing mucoid growths. This material may remain firmly adherent as a discrete covering layer on each cell, or it may part freely from the cells. In the former case it is known as a capsule; in the latter, as free slime or gum

Capsules and slime are believed to be distinct from the morphological and biochemical point of view. The capsule is a part of the cell, the slime a secretion. According to Klieneberger-Nobel (1948), capsules are of definite shape, of more or less definite density throughout, and of definite outline, whereas slime envelopes are amorphous and can be drawn out into manifold structures, are concentrated in the vicinity of the bacterial cells, and decrease in density with increasing distance from the cell.

Broth cultures of capsule-producing organisms are usually stringy in texture, and agar colonies exhibit a very moist, glistening surface which is described as mucoid. Capsule formation is dependent upon the composition of the medium but especially the variant phase of the organism. Some disease-producing organisms from large capsules in culture media rich in animal fluids. Others produce prominent capsules when cultures are incubated at low temperatures (4 to 20°C.).

Chemical analysis of capsular material from a number of bacteria show wide difference in composition. For this reason it is impossible to make statements which apply to all bacteria. In some organisms the capsular material appears to be a glycoprotein; in others, a protein-polysaccharide complex; in still others, a polysaccharide framework with the spaces filled in by a larger amount of glutamyl polypeptide.

Capsular material is difficult to distinguish from those gums which flow away from the cells as they are formed. Organisms producing gums do so when grown in sugar solutions. Some organisms produce gums only in the presence of a specific sugar; others produce gums in the presence of any one of several sugars. In the absence of sugar, usually very little, if any, gum is formed. Organisms producing gums this type are the cause of considerable losses in the sugar industry. The increased viscosity produced by the gum interferes with the filtration of the sugar solution.

The species commonly encountered in sugar-cane juice is *Leuconostoc mesenteroides*. The cells are surrounded by a thick, gelatinous, colourless polysaccharide consisting of dextran (glucose polymer).

The formation of gums is of common occurrence by soil bacteria. From 5 to 16 per cent of such forms have been shown to be capable of synthesizing gums from sugars.

Bacterial Protoplasts

When the cell wall is damaged, the protoplasm usually disintegrates. However, methods are available for removing the cell membranes without destroying the vital nature of protoplasm. The term protoplast is used to indicate living protoplasm exclusive of the cell membranes.

Action of lysozyme

Some bacteria are rapidly lysed or dissolved by the action of lysozyme. Weibull (1953) reported that lysozyme possessed a specific depolymerizing action on the cell wall and that this appeared to be only portion of the cell that was affected by such treatment. As a result of the destruction of the wall, the protoplasts were liberated.

Protoplasmic structures are not very stable. Consequently, if protective agents were not employed destruction of the walls was accompanied by a rapid lysis of the protoplasts, followed by the liberation of most of the cell protein and nucleic acid in solution form. This could be prevented by employment of the enzyme in a 0.2 *M* solution of sucrose or cane sugar. After digestion of the cell walls, the living protoplasts rounded up into spheres.

Spiegelman, Aronson, and Fitz-James (1958) found that digestion of protoplasts of *Bacillus megaterium* led to the liberation of nuclear bodies of the protoplasts. Such bodies were collected by centrifugation for 5 min. at 10,000 × g.

Properties of protoplasts

The difficulty in handling and studying protoplasts is their extreme fragility and sensitivity to osmotic shock, shaking, centrifugation, and aeration. Removal of the cell wall does not change the structure and capabilities of the protoplasm. Permeability, respiration, and spore formation appear to be the same for protoplasts and intact cells. Also both can support the development of bacteriophages. Under special conditions the protoplasts grow and probably divide like intact cells. However, there is no evidence that protoplasts form colonies. The metabolism of protoplasts and intact cells appear to be very similar but probably not identical.

Filterability of protoplasts

Sinkovics (1958) reported the spontaneous occurrence of units in aged cultures of *Escherichia coli* which conformed to the description of artificially induced bacterial protoplasts. The disintegration of cells from aged cultures was preceded by swelling of the cell and rupture of the rigid cell wall. Centrifugation of the culture gave a supernate which, after filtration, contained units capable of regeneration when placed in fresh medium. The smallest units capable of regeneration measured about 350 mμ in diameter. The units underwent fusion before cell-wall formation occurred.

The results supported the assumption that aged *E. coli* cultures could survive in the form of units having no cell walls and which, under adequate conditions, regenerated into vegetative forms.

Polysaccharide Structures

Polysaccharides occur (1) in cell walls, (2) extracellularly in capsules and gums, and (3) inside of bacterial cells. Pennington (1949) revealed the presence of polysaccharides by treating bacteria with sodium metaperiodate followed by staining with sulfite-decolorized basic fuchsin. In *Bacillus cereus* the polysaccharide was concentrated in the cytoplasmic membrane as well as in cell wall.

Selective staining of polysaccharide in the cell is said to depend upon the oxidizing action of periodate on such chemical configurations as α, β glycols and α-hydroxyketones. Polyaldehydes generated by this selective oxidation react with sulfite-decolorized fuchsin. Polysaccharide areas in the cell are coloured by the stain.

Nucleus

The question of the presence of a well-defined nucleus in bacteria has been the subject of investigations by bacteriologists almost from the beginning of bacteriology.

Some of the earlier cytologists maintained that bacteria were very primitive organisms devoid of nuclei and consisting simply of cytoplasm, granules and vacuoles. This view was based on their failure to observe a nucleus in a bacterial cell. Others held the view that the nuclear material was present in a diffuse form throughout the cytoplasm. Still others believe that the whole cell should be regulated as a "naked nucleus," corresponding to the nucleus of higher organisms. The naked nucleus is regarded as a primitive form of living matter. Since bacteria have the structural and physiological attributes of true cells, this concept cannot apply to these organisms.

Much of the confusion was caused by the inadequacy of the staining procedures. By means of the HCl-Giemsa staining technique of Piekarski, many observations have been reported demonstrating the presence of chromatinic structures in bacteria.

Robinow (1944) prepared wet smears of *Escherichia coli*. Slides were fixed in osmic acid vapour, dried and immersed in normal HCl for about 9 min. at 53 to 55°C, then washed and stained in 1:20 Giemsa solution for 10 to 60 min., depending on the staining properties of the specimen.

The chromatinic structures in *E. coli* from old cultures were too small to be resolved accurately. After transfer to fresh medium the chromatinic structures increased in size and gave rise to short, often dumbbell-shaped rods or chromosomes, which multiplied by splitting lengthwise in a plane more or less parallel to the short axis of the cell. A single cell of *E. coli* containing one chromatinic body or one or two pairs of these representing primary and secondary division products.

The Smith (1950) techniques consisted of fixing the smear in osmium tetroxide vapour, immersion in HCl, mordanting in dilute formaldehyde, and staining with aqueous basic fuchsin. The method was said to possess certain advantages over the procedure of Robinow.

Another cause of confusion in the recognition of nuclear structures in bacteria was the lack of appreciation for the ages of the cultures. At certain times nuclear structures cannot be seen. In general, most of the early observations of tidy, intelligible "nuclei" were made on preparations from very young cultures, whereas haphazardly scattered, unintelligible granules of chromatin were persistently observed in preparations made from cultures of the same bacteria beyond the logarithmic growth phase.

Recognition of the fact that the configuration of nuclear material in bacteria might change with age removed one of the chief causes of confusion. As Dobell said (1911), My own belief is that the nucleus in bacteria may display not one but many forms during the whole life cycle. Many of the nuclear structures which have been shown to exist in these organisms should, I think, be regarded as temporary states rather than as permanent conditions. The different results which have been reached by different workers when working, apparently, upon the same species, may to some extent find an explanation in this circumstance.

The chromatin bodies, according to Robinow (1956), appear to have the following properties—They are simple structures of relatively low density, not markedly basophilic but reacting positively in the Feulgen test. Normally they lie separately in the cytoplasm, and all those in one bacterium are homologous. Changes in the balance of ions in the cytoplasm may cause the aggregation of several chromatin bodies into a single continuous structure. This effect is reversible.

Growth and division of chromatin bodies are attended by changes of form only, not by visible changes of texture. In the simplest type of body, that which in profile looks like a bar or dumbbell, division begins at one end and causes the successive appearance of V-, U-, and H-shaped phases. The chromatin structures of certain bacteria are netlike or spongelike, and their mode of division is not easily imagined.

Direct division of the chromatin bodies of *E. coli* has been demonstrated by Mason and Powelson (1956) in a series of remarkable phase-contrast photomicrographs. These observations were made on living bacteria. The nuclear areas in the dividing cells appeared to be as clearly defined as the areas in fixed, hydrolyzed, and stained cells.

Vacuoles

Vacuoles have been identified in young bacteria. They are cavities in the protoplasm and contain a fluid known as cell sap. As the cells approach maturity, some of the water-soluble reserve food materials manufactures by the cell dissolve in the vacuoles. Insoluble constituents precipitate out as cytoplasmic inclusion bodies.

Metachromatic Granules

The best-known inclusions bodies in bacterial cells are known as volutin or metachromatic granules. The granules are small in young cells and become larger with the age of the culture. They are believed to originate in the cytoplasm of young cells and to localize in the vacuoles of mature forms. The granules show a strong affinity for basic dyes, indicating that they are acid in character. They are usually considered to be a reserve source of food.

Grula and Hartsell (1954) found the granules to be composed of metaphosphate or another form of inorganic phosphate, some fat, and possibly small amounts of protein. Their presence and size in cells were related to the phosphate concentration of the growth medium in the presence of an energy source and specific divalent ions (Mn and Zn). Older cells possessed larger granules. Their basophilic nature did not depend on either ribonucleic acid (RNA) or deoxyribonucleic acid.

On the other hand, Widra (1959) found metachromatic granules to contain protein-bound lipide, RNA, and polyphosphates.

Fat Globules

Bacteria are capable of storing fat in the form of globules. Fat globules may be demonstrated in 24-hr. cultures and usually reach a maximum in about 48 hr. Some cells may contain only one large globule; others may show the presence of a number of small, scattered globules.

It is generally believed that fat is stored as reserve food material. Globules usually cannot be demonstrated in young, vigorously growing cells. As cells age slow down in activity, fat globules appear in the cytoplasm and may be recognized by appropriate staining.

Motility

Bacteria motion is generally associated with the presence of organs of locomotion known as flagella (singular, flagellum). They were first observed in stained preparations by Cohn (1875). The presence of flagella does not mean necessarily that the organisms are always motile, but it indicates a potential power to move.

Independent bacteria motion is a true movement of translation and must be distinguished from the quivering back-and-forth motion exhibited by small particles suspended in a liquid. This latter type of motion is called Brownian movement and is caused by the bombardment of the bacteria by the molecules of the suspending fluid.

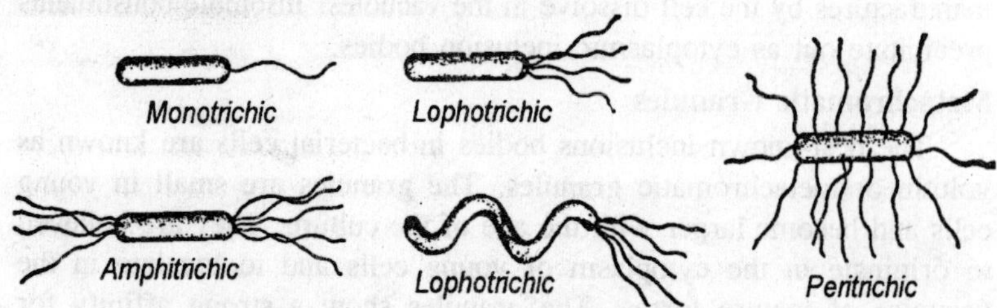

Fig. 4.5. Different distributions of flagella on bacterial cells.

Properties of flagella

Flagella are very delicate organs and easily detached from the cell.. In the stained condition they are long, slender, undulating organs. They are directed backward to the direction of motion at an angle of about 45^0. Reversal of direction occurs by swinging the flagella through an angle of about 90^0. Turning movements take place by swinging the flagella forward on one side only. They propel the organism by a spiral or corkscrew motion.

The thickness of flagella varies from species to species. In *Proteus vulgaris* they measure about 12 mm. This figure is considerably below the shortest wave length of visible light and explains why flagella cannot be seen in hanging-drop preparations or in smears stained by the usual simple procedures. When special staining methods are employed, sufficient dye becomes deposited on the flagella to make their diameters greater than the wave length of visible light. They may then be seen under a light microscope.

Chemistry of flagella

Flagella and bacterial bodies differ in composition. Flagella break up on boiling or when exposed to pH values below 4 or above 11. Their composition is largely protein, having a molecular weight of about 41,000. Weibull (1949) found the flagella of *P. vulgaris* to be composed of 98 per cent protein, traces of carbohydrate and fat, and no phosphorus. The protein contained only 14 known amino acids. It is an incomplete protein, lacking in some of the essential amino acids.

It is well established that the *H* antigens of bacteria are associated with the flagella and the *O* antigens with the bodies . Purified flagella are agglutinated by *H* antiserum but not by o antiserum; *O* antigens are not agglutinated by *H* antiserum. This is another indication that flagella and bacterial bodies differ in composition.

Origin of flagella

Some believe flagella originate from the cell wall; others believe they traverse the cell wall into the protoplasm. Flagella differ chemically both from the cell wall and the protoplasm. Two observations have been made relative to the site of origin of flagella.

Electron micrographs by van Iterson and others show the flagella penetrating the faint outer zones and extending into the cytoplasm. If the outer zone is the cell wall, then the flagella have their origin in the cytoplasm.

Weibull (1953) showed that removal of the cell wall of some bacteria by means of lysozyme produces a spherical protoplast which still retains the flagella of the treated cell. The obvious conclusion is that flagella have their origin in some cell structure deeper than cell wall.

Number and arrangement of flagella

The number and arrangement of flagella vary with different bacteria, but they are generally constant for each species. Some have only one flagellum; others have two or more flagella.

In rod-shaped cells the flagella arise either at one or both poles, or are distributed laterally with the poles being generally bare. In some species flagella are located both laterally and at the poles. A species may show considerable variation in the number and arrangement of flagella. Single *Alcaligenes* cultures may contain forms with a polar flagellum only; some with several lateral flagella; and some with both lateral and polar flagella. Sometimes a species may show cells which are flagellated in one environment and nonflagellated in another.

Leifson, Carhart, and Fulton (1955) reported the presence of four definite types of curvature in the flagella of *Proteus vulgaris*: Individual organisms may have more than one type of flagella, and individual flagella may have one or two types of curves.

Environmental factors, particularly pH, may change the curvature of the flagella on some strains, but not on all strains. In acid media the curly curvature tends to predominate; in alkaline media the normal predominates.

Organisms have been classified on the basis of the number and arrangement of flagella as follows;

Monotrichous—a single flagellum at one end of the cell.

Lophotrichous—two or more flagella at one end or both ends of the cell.

Amphitrichous—one flagellum at each end.

Peritrichous—flagella surrounding the cell.

Staining of flagella

The staining of flagella is a difficult technique, especially in the hands of the beginner. For this reason many methods have been proposed. Regardless of the method employed, the film must first be treated with a mordant to make the flagella take the stain heavily. Mordants consist usually of a mixture of tannic acid and some metallic salt. In some methods the mordant and stain are applied separately; in others they are combined in one solution.

Boltjes (1948) came to the following conclusions on the staining of bacterial flagella—Four factors at least influence the results of staining-the skill of the investigator; the organism studied, the culture medium on which the organism was grown, and the staining method. Of these the first is perhaps the most and the last the least important. The importance of shill is shown by the repeated failure of students to stain flagella although their teacher has no difficulty in demonstrating flagella at the same time and with the same suspension. Further, one usually has success with a new formula only after a number of trials; first attempts to stain an unknown bacterium are often a failure. Another point is that as a rule flagella can be clearly seen only in a rather small part of the preparation. The different colours of the stained bacteria show clearly that during the staining process conditions are not everywhere alike, and since many flagella are torn off during drying, we need not wonder that in most cases only a few bacteria are successfully stained. Notwithstanding all this, it is certain that when one has had some experience with the staining technique, the results are so consistent that there will never be any confusion between bacteria with true polar·flagella, such, for example, as *Pseudomonas fluorescens* or *Vibrio comma,* and peritrichous bacteria like *Proteus mirabilis* and *Salmonella typhosa*. I therefore consider flagella staining to be a reliable procedure.

Pijper theory of motility

In a series of investigations Pijper et al. (1957) questioned the belief that flagella are responsible for motility. He added methyl cellulose to a culture of a motile organism to increase the viscosity of the medium. This treatment decreased the motility of the cells. Under these conditions the cells exhibited a gyratory undulating movement like other aquatic creatures. He concluded that flagella were not organs of locomotion but only artifacts -useless appendages, polysaccharide twirls—the result, not the cause, of bacterial motility. To quote—

That motile bacteria always exhibit a gyratory undulating movement was confirmed by making a slow motion cinemicrographic film of fast-swimming bacteria in broth, and also by examining the same bacteria at lower temperatures, which reduced their speed.

The spirillar motion of bacteria is sufficient to propel them, and their is no need to invoke special motor organs like flagella. There is no evidence to show that the flagella-like appendages of bacteria act as motile organs-in fact all the evidence when critically examined points the other way.

Analysis of the structure of bacteria excludes the possibility that tails, "flagella," or the thin wavy threads are live organs, or that they are in direct communication with the living parts of the cell. There is no evidence from either election picture or stained preparations that it is otherwise.

Not only does the visible gyratory undulating movement of motile bacteria satisfy all requirements for locomotion, but it is possible for bacteria grown under special conditions to swim in this fashion without showing tails or other supported motor organs.

Notwithstanding the findings and conclusions of Pijper, evidence at present appears to be overwhelmingly in favor of flagella as organs of locomotion.

As has already been stated, Weibull (1948, 1949) reported the flagella of *p. vulgaris* to be composed of 98 per cent protein. This contradicts Pijper's statement that flagella are formed from the carbohydrate slime layer that is peeled off into a number of thin, wavy threads.

Several investigators, among them Labaw and Mosley (1955), by means of electron microscopy, demonstrated the presence on *Brucella bronchiseptica* of uniform flagella having an external contour of a counter-clockwise or left-handed triple helix. The average periodicity along the length of the flagella was 19 mμ, with an average diameter of 13.9 mμ.

Motion of Colonies

Several organisms have been described which exhibit colonial motility when grown on a solid medium.

Shinn (1938) prepared lapse-time motion pictures of individual colonies of *Bacillus alvei* grown on agar plates and measured their velocities. The linear motions of colonies measuring 0.2 to 0.5 mm. in diameter averaged about 14 mm. per hr. Comparing this figure with speed of individual cells of other species of motile bacteria gave the following results:

Salmonella typhosa........... 65 mm. per hr.

Bacillus megaterium........... 27 mm. per hr.

B. alvei (colonies)........... 14 mm. per hr.

The colonies exhibited not only linear motion but also a slow rotary movement. The direction of rotation of 200 to 300 colonies observed was counterclockwise, with the exception of two colonies in which it was clockwise.

Turner and Eales (1941*a, b*) reported that the rotation of an aerobe occurred very early during growth. The cells segregated in small groups and aligned themselves concentrically around a common centre to form disk-like plaques one or a few cells thick. The rate of rotation was greater in smaller groups. As multiplication continued, successive layers were gradually built up in terrace fashion and the colony grew in height. The colonies then began to migrate. When a colony migrated, it left a peculiar "track" on the surface of the agar. A small number of cells were left behind, mostly at the edges of the track which formed two parallel lines separated by the width of the moving colony.

Typical migrating colonies pursued curved or spiral paths which were often very elaborate and of relatively great length, even 2 or 3 cm. The direction of rotation was either clockwise or counterclockwise. After wandering for a variable distance, a colony approached the centre of its spiral path with rapidly shortening radius, ceased to migrate, began to rotate around its centre, lost its elongated shape, and increased in size to several times the width of the track at the end of which it was formed.

Endospores

Endospores are bodies produced within the cells of a considerable number of bacterial species. They are more resistant to unfavourable environmental conditions, such as heat, cold, desiccation, osmosis, and chemicals, than the vegetative cells producing them. However, it is debatable if such extreme conditions actually occur in nature. For instance, the resistance of spores to high temperature is a laboratory phenomenon and probably never occurs in a natural environment.

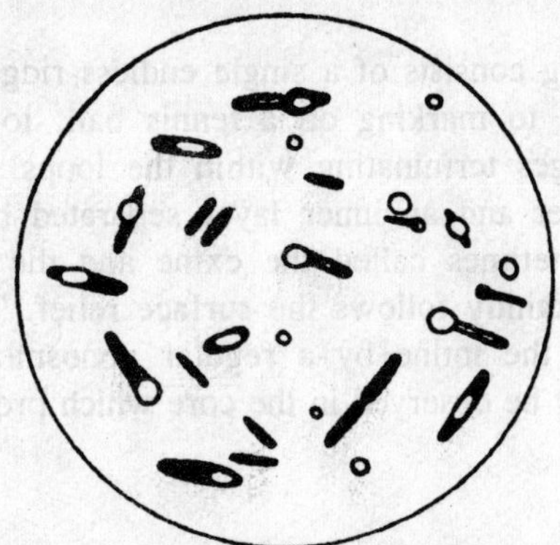

Fig. 4.6. Different types of vegetative cells, endospores, and free spores of bacteria. The spores appear clear, indicating that they do not stain as readily as most vegetative cells.

The bulk of evidence indicates the existence of a close relationship between spore formation and the exhaustion of nutrients essential for continued vegetative growth. Sporulation is a defence mechanism to protect the cell when the occasion arises.

Spore formation is limited almost entirely to two genera of rod-shaped bacteria: *Bacillus* (aerobic or facultatively anaerobic), and *Clostridium* (anaerobic or aerotolerant). With one possible exception, the common spherical bacteria do not sporulate. Some spore-bearing species can be made to lose their ability to produce spores. When the ability to produce spores is once lost, it is seldom regained. Sporulation is not a process to increase bacterial numbers because a cell rarely produces more than one spore.

Morphology of spores

Spores may be spherical, ellipsoidal, or cylindrical in shape. The position of the spore in a cell may be central, subterminal, or terminal. A fully grown spore may have a diameter greater than of the vegetative cell. This causes a bulging of the cell. The resulting forms are known as clostridium if central, and plectridium if terminal. As a rule, each species has its own characteristic size, shape, and position of the spore, but this is subject to variation under different environmental conditions.

Franklin and Bradley (1957), by means of electron microscopy of carbon replicas, reported that the spores of a majority of species of *Bacillus* and *Clostridium* are readily distinguished by surface patterns. The surfaces may be smooth or ribbed, with the ribs usually longitudinal.

The sculpturing consists of a single endless ridge in the form of two loops, similar to marking on a tennis ball, together with two other separate ridges terminating within the loops. The spore coat consists of an outer and an inner layer separated by a space. The outer layer is sometimes called the exine and the inner layer the intine. The intine faintly follows the surface relief. The central core is separated from the intine by a regular nonosmophilic space. A peripheral spot may be observed in the core which probably represents nuclear material.

Parasporal bodies

When sporulation of *Bacillus laterosporus* is complete, the spores are cradled in canoe-shaped bodies. According to Hannay (1957)—On sporulation the slender vegetative rods swell and form larger spindle-shaped cells in which the spores are formed. When the spores mature they lie in a lateral position cradled in canoe-shaped parasporal bodies which are highly basophilic and can be differentiated from the surrounding vegetative cell cytoplasm with dilute basic dyes. On completion of sporulation the vegetative cell protoplasm and the cell wall lyse, leaving the spore cradled in its parasporal body. This attachment continues indefinitely on the usual culture medium and even persists after the spores have germinated. In thin sections of sporing cells the bodies are differentiated from the cell protoplasm by differences in structure. Whereas the protoplasm has a granular appearance, in both longitudinal and cross-sections the parasporal body comprises electron-dense lamellae running parallel with the membranes of the spore coat and less electron-dense material in the interstices of the lamellae. The inner surface of the body is contiguous with that of the spore coat as if it were part of the spore, rather than a separate body attached to the spore. The staining reactions of the parasporal body are not consistent with those of any substance described in bacteria.

Composition of spores

Ross and Billing (1957), by means of refractive index measurements on spores and vegetative cells of *B. cereus*, *B. cereus* var. *mycoides*, and *B. megaterium*, found the values to be very high and comparable with that of dehydrated protein. This suggested that they contained much less water than the vegetative cells.

Strange and Dark (1956) demonstrated the presence of a hexosamine-containing peptide in the spore coats of *B. megaterium* and *B. subtilis*. The breakdown of an insoluble peptide complex might well be one of the first steps of the germination process. It was believed that the release of the hexosamine-amino acid complex was the result of the action of lysozyme present in the spores.

Enzymes of spores

The presence of enzyme systems in Bacillus spores has been reported. Some of these are an inorganic pyrophosphatase that requires manganese for activation, adenine ribosidase that hydrolyzes adenosine, an enzyme that functions possibly to lyse the sporangium and free the spore during germination, highly active alanine racemase that catalyzes the conversion of L-alanine to D-alanine, several glucose dehydrogenases and an aldolase.

Sporulation process

Conditions necessary for sporulation in one species do not necessarily apply to another. The subject appears to be in such a state of confusion that it is impossible to discuss sporulation in terms of generalities.

The conditions which have been reported as favoring sporulation include addition of slats of metals such as manganese, chromium, nickel, etc., to the medium; shaking a culture of vegetative cells of sporing aerobes with distilled water at 37°C; addition of tomato juice to a medium; incubating the cultures at an appropriate temperature; addition of calcium carbonate to a carbohydrate medium to prevent excessive accumulation of acid, and to maintain the pH at 5.5 or above; the necessity of oxygen; addition to the medium of certain amino acids; etc.

Germination of spores

With the exception of some constituents such as high concentration of calcium, dipicolinic acid, and in *Bacillus sphaericus*, α, ∈-diaminopimelic acid, spores are simiiar to the vegetative cells in composition.

When a spore prepares itself for germination, it loses its refractility, which coincides with an imbibition of water. This stage is associated with a loss in heat resistance, stainability, and dry weight. Later the spore coat breaks, followed by the emergence from the spore case of a new germ cell which eventually matures into a vegetative cell.

Spore germination has been defined in various ways. According to Campbell (1957), "Spore germination may be regarded as the change from a heat resistant spore to a heat labile entity which not necessarily be a true vegetative cell." Later development, leading eventually to the formation of a mature vegetative cell, is called outgrowth.

Some conditions which stimulate germination are as follows; (1) Treatment at 90 to 100°C. for 1 to 2 min. stimulates germination. (2) Spores which fail to germinate overcome this dormancy when activated by heat. (3) The use of certain agents such as alanine, glucose, and adenosine stimulates spore germination in most sporing species. In some species other amino acids may be substituted for the alanine. The same applies to glucose. (4) Yeast extract and mixtures of vitamin-free amino acids have also been shown to stimulate germination.

Spores germinate in a variety of ways. There is a considerable degree of constancy in the method of spore germination for each species. Lamanna (1940) classified the modes of germination as follows:

I. Spore germination by shedding of spore coat

Characteristics of this method are

A. Spore does not expand greatly in volume previous to the germ cell breaking through the spore coat. The limit of volume increase of the spore may be considered to be twice its original volume.
B. Spore coat does not lose all its refractive property previous to germination.
C. After the second division of the germ cell, giving a chain of three organisms, the original spore coat, remaining attached to the cell, is visible for a long time after germination.
   1. Equatorial germination
   2. Polar germination
   3. Comma-shaped expansion

II. Spore germination by absorption of the spore coat

Characteristics of this method are

A. The spore expands greatly during germination. A tripling or greater increase of the original volume occurs.

B. The spore loses its characteristic refractiveness during germination, so that it is difficult to say when the spore has disappeared and the germ cell appeared.

C. After the second division of the germ cell, even a thin capsule originally remains, all traces of the spore coat are gone.

Some stains germinating by absorption regularly show a thin capsule remaining about one end of the growing cell. This would appear as a polar germination. In other cases, equatorial capsules are seen. Yet, in all instances, the spore is considered to germinate by absorption inasmuch as the three characteristics of the method are still adhered to.

5

GROWTH OF BACTERIA

In this chapter we shall be concerned with how bacteria are grown, methods by which we can measure their growth, and how a single species of bacteria can be isolated from a mixture of organisms. Various methods that can be used for the physical or chemical sterilization of a given area or object are also discussed. All organisms, whether procaryotic or eucaryotic, require food to live and grow. How an organism assimilates its fold is called its nutrition, and the specific cellular requirements are its nutrients.

The cell also requires a source of energy for the synthesis of cellular constituents and for other life processes, such as motility and active transport. The major metabolic pathways by which bacteria acquire and convert their nutrients into energy and the metabolic products required for growth are discussed.

GROWTH OF BACTERIAL POPULATIONS

Unlike most higher organisms, bacteria are unicellular, which means tat during normal growth the increasing in the population is simply related to the growth of any cell in that population. A bacterial cell in a satisfactory growth medium will grow larger by means of a balanced increase in all of its constituents. After it has doubled in size, cell division can occur, resulting in two identical cells, each of which is capable of growing and dividing at the same rate as the original cell. This means that the number of cells in the population increases in a exponential or logarithmic manner, that is, each doubling of the population occurs in the same time period, but the number of cells that participate in each division doubles for each succeeding generation.

Normal Bacterial Growth Curve

By its very nature, bacterial reproduction by binary fission results in doubling of the number of viable bacterial cells. Therefore, during active bacterial growth, the size of the microbial population is continuously doubling. Once cell division begins it proceeds exponentially as long as growth conditions permit, with one cell dividing to form two, each of these cells dividing so that four cells form, and so forth in a geometric progression. The time required to achieve a doubling of the population size, known as the generation time or doubling time, is the unit of measure of microbial growth rate. The generation time for bacteria can be expressed mathematically as

$$g = t/3.3\ (\log B_t - \log B_0)$$

where g is the generation time, $\log B_t$ is the logarithm to the base 10 of the number of bacteria at time *t*, $\log B_0$ is the logarithm to the base 10 of the number of bacteria at the starting time, and *t* is the time period of growth. By determining cell numbers during the period of active cell division, the generation time can be estimated. A bacterium such as *E. coli* can have a generation time as short as 20 minutes under optimal conditions. Considering a bacterium with a 20-minute generation time, one cell would multiply to 1000 cells in 3.3 hours and to 1,000,000 cells in 6.6 hours.

Table 5.1. Growth rates for some representative bacteria under optimal conditions

Organism	*Temperature (°C)*	*Generation Time (min)*
Bacillus stearothermophilus	60	11
Escherichia coli	37	20
Bacillus subtilis	37	27
Bacillus mycoides	37	28
Staphyolococcus aureus	37	28
Streptococcus lactis	37	30
Pseudomonas putida	30	45
Lactobacillus acidophilus	37	75
Vibrio marinus	15	80
Mycobacterium tuberculosis	37	360
Rhizobium japonicum	25	400
Nostoc japonicum	25	570
Anabaena cylindrica	25	840
Treponema pallidum	37	1980

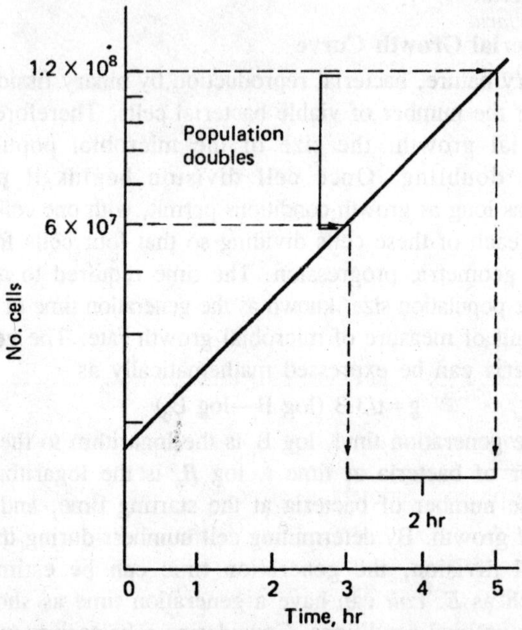

Fig. 5.1. The estimation of the generation time of bacterium based on observed cell numbers during a period of active division.

When a bacterium is inoculated into a new culture medium, it exhibits a characteristic growth curve. The normal growth curve of bacteria has four phases, the lag phase, the log or exponential growth phase, the stationary phase, and the death phase. During the lag phase there is no increase in cell numbers. Rather, during this phase the bacteria are preparing for reproduction, synthesizing DNA and various enzymes needed for cell division. During the log phase of growth, so-named because the logarithm of the bacterial biomass increases linearly with time, bacterial reproduction occurs at a maximal rate for the specific set of growth conditions. It is during this period that the generation time of the bacterium is determined.

If a bacterial culture in the exponential growth phase is inoculated into an identical fresh medium, the lag phase is bypassed, and exponential growth continues. This occurs because bacteria are already actively carrying out the metabolism necessary for continued growth. If, however, the chemical composition of the new medium differs significantly from the original growth medium, the bacteria do go

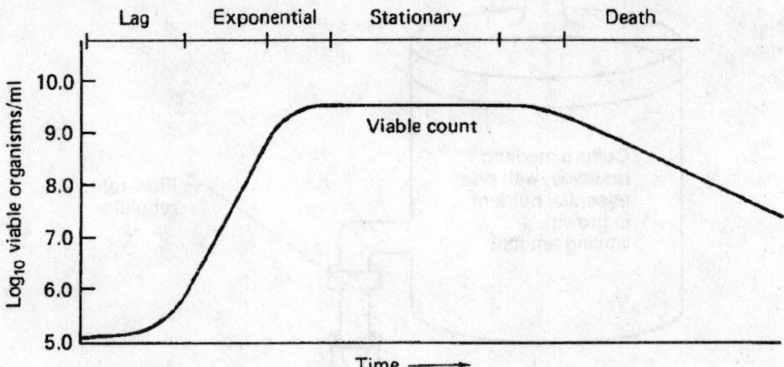

Fig. 5.2. The normal bacterial growth curve has four stages; lag, exponential, stationary, and death.

through a lag phase before entering the logarithmic growth phase when they synthesize the enzymes needed for growth in the new medium.

However, if the bacterium is not transferred to a new medium and no fresh nutrients are added, the stationary growth phase eventually is reached, and there is no further net increase in bacterial cell numbers. During the stationary phase the growth rate is exactly equal to the death rate. A bacterial population may reach stationary growth when a required nutrient is exhausted, when inhibitory end products accumulate, or when physical conditions are restricting. The duration of he stationary phase varies; some bacteria exhibit a very long stationary phase.

Eventually the number of viable bacterial cells begins to decline, signaling the onset of the death phase. The kinetics of bacterial death follows the same exponential kinetics as the logarithmic growth phase. This is true because the death phase really represents the result of the inability of the bacteria to carry out further reproduction. The rate of the death phase need not, however, be equal to the rate of growth during the exponential phase. The rate of death is proportional to the number of the survivors and can be expressed as where *d* is the death rate, *t* is the time period, log B_0 is the logarithm of the number of bacteria at the starting time, and log B_t is the logarithm of the survivors at time *t*.

Batch and Continuous Growth

The normal bacterial growth curve is characteristic of bacteria in batch culture, that is, under conditions when a fresh medium is simply inoculated with a bacterium. A flask containing a liquid nutrient medium inoculated with a bacterium such as *E. coli* is an example of a batch

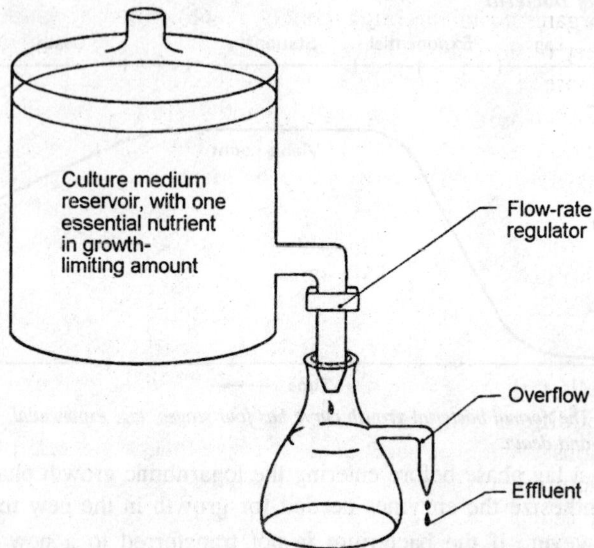

Fig. 5.3. A chemostat is a continuous culture device. In such a device the population density is controlled by the concentration of the limiting nutrient, and the growth rate is controlled by the flow rate, which can be arbitrarily set.

culture. In batch culture growth nutrients are expended and metabolic products accumulate in the closed environment. The batch culture models situations such as occur when a canned food product is contaminated with a bacterium.

Bacteria may also be grown in continuous culture in a chemostat, where nutrients are supplied and end products continuously removed so that the exponential growth phase is maintained. Because end products do not accumulate and nutrients are not completely expended, the bacteria never reach stationary phase. The chemostat is a device in which a liquid medium is continuously fed into the bacterial culture. The liquid medium contains some nutrient in growth limiting concentration, and the concentration of the limiting nutrient in the growth medium determines the rate of bacterial growth. Even though bacteria are continuously reproducing, a number of bacterial cells are continuously being washed out and removed from the culture vessel.

Bacterial Growth on Solid Media

The development of bacterial colonies on solid growth media follows the basic normal growth curve. However, the dividing cells do not disperse; hence the population is densely packed. Under these conditions nutrients rapidly become limiting at the center of the colony, and microorganisms in this area rapidly reach stationary phase. At the periphery of the colony, cells can continue to grow exponentially even while those at the center of the colony are in the death phase. Bacterial colonies generally do not extend indefinitely across the surface of the media but have a well-defined edge. Therefore, individual well-isolated colonies develop from the growth of individual bacterial cells. The fact that the bacteria have reproduced asexually by binary fission means that, barring mutation, at the bacteria in the colony should be genetically identical; that is, each colony contains a clone of cells derived from a single parental cell.

Cytological Approach

Cells that are growing are destined to divide. Cell division takes place at different intervals depending on the species and environmental conditions. Within a short period, often as small as 20 minutes, a bacterium can create a complete duplicate of itself, which is in turn capable of duplicating. This process requires the synthesis of the millions of parts of the parent cell and their organization into the various homo-and heteropolymers which are the functional units of the cell.

In order to understand bacterial growth it is necessary to have some knowledge of the temporal sequence of events during a single cell cycle and an understanding of how these biochemical events are interrelated. Bacteria do not exhibit the characteristic cell-division cycle observed in eucaryotic organisms. Whereas in eucaryotic organisms DNA synthesis is confined to a single phase of the cell cycle, the S phase, in bacteria that are growing exponentially DNA synthesis occurs throughout virtually the entire division cycle. Also, in bacteria the duplication sequences do not necessarily follow one after the other, but instead they overlap, the amount overlapping depending on the growth medium. Bacterial duplication begins with the initiation of DNA replication, followed by the complete duplication and separation of the DNA replicas, and formation of a septum near the equatorial plane so that two cells, each one containing half the DNA content of the mother, are formed.

Techniques that achieve synchrony in a dividing population in a metabolically undisturbed state have proved invaluable in studies of the bacterial division cycle. Autoradiography and marker-frequency analysis have been used for the study of particular aspects of the division cycles in asynchronous cultures; morphologic details related to the division process have been provided by examination of serial ultrathin sections with the electron microscope.

Bacterial Cell Division

Genetic information necessary for duplication of thé bacterial cell is contained within the DNA-containing structures, the bacterial chromosomes. The number of chromosomes in a cell; and the extent of their replication are determined by the physiologic state and age of the cell. Chromosomes of cells in very slowly growing cultures appear to have but a single replication point. Observations of gaps in DNA synthesis in these cells indicate that chromosome replication occupies only part of the cycle. Cells in rapidly growing cultures, on the other hand, appear to contain multiple replication points. Because replication occurs simultaneously at several points along these chromosomes, the entire chromosome can be replicated in a fraction of the time that would be required in there were but a single replication fork, and genetic material can be duplicated at rates that otherwise could not be attained.

Initiation of Chromosome Replication

During division in order to insure the segregation of equal portions of the genetic material among the progeny, chromosome replication and cell division are coordinated in such a manner that the frequency of initiation of replication determines the rate of cell division. As presently conceptualized, a duplication sequence of bacteria begins with the initiation of chromosome synthesis at a unique site, the replicator site of the chromosome. Initiation of chromosome replication requires proteins synthesis and depends on the achievement of a critical cellular mass/DNA ratio. Chromosome replication begins when a fixed amount of protein initiator substances have accumulated regardless of the position of other replication points on the chromosome. Thus, the rate of cell division in a given culture medium is determined by the time required for accumulation of the initiator in that medium. Initiation is not influenced by the presence, absence, position, or rate of movement of replication points on the chromosome.

Following initiation the replication point progresses sequentially along the chromosome, and initiator proteins continue to accumulate. After a time, depending upon the nutritional conditions, another threshold quantity of initiator will have accumulated. This results in the initiation of new rounds of replication, and the movement of new replication points along the chromosome. In cells growing under conditions in which the time for formation of initiator is longer that the time required for a replication point to traverse the chromosome, a new round the replication is not initiated until after the previous replication point has reached the chromosome terminus. In this case, there is a period devoid of DNA synthesis between rounds of chromosome replication. Conversely, in cells growing in media, in which the interinitiation time is shorter than that required for a round of replication, new rounds are initiated before the previous replication point has reached the terminus. In this situation, cells contain chromosomes with three or more replication points, as have been observed in rapidly growing cells. Helmstetter's model of DNA replication is based on the constancy at all growth rates of two periods: (1) the interval between the end of a round of DNA replication and cell division and (2) the time required for a replication point to traverse the genome. The first of these periods could be considered to be analogous to the G_2 period in eucaryotic cells; the second period, analogous to the S period, is the time required for the synthesis of DNA.

Chromosome Duplication

The replication of bacterial DNA begins by the gradual separation of the two strands at the replicator site to form a Y-shaped locus, the replication fork, which represents a growing point. The replicating enzyme system attaches to the DNA at this site and moves along the entire chromosome replicating the parent DNA as the double helix slowly rotates and unwinds. After a replication point reaches the terminus of a chromosome, the two newly replicated sister chromosomes separate form each other. Although the process has been referred to as "nuclear division," it is not strictly analogous to nuclear division in eucaryotic organisms since the chromosomes separate gradually throughout the entire period of replication.

Role of Cell Membrane in Chromosome Separation

Electron microscopy has failed to show any bacterial structure equivalent to the mitotic apparatus of higher cells. It is evident, however, that there must be some mechanism which ensures equitable separation of chromosomes after each replication. The replicon method proposed by Jacob and his associates provides an explanation of how this might be achieved. The model suggests that the cellular DNA is attached to the bacterial membrane where replication occurs by passage of the DNA through an enzyme complex in the membrane. Separation and equipartition of the DNA copies results from the synthesis of new cell membrane in an annular region between attachment points.

This model is supported by both biochemical and morphologic criteria. Biochemical analyses have demonstrated that in gently lysed preparations of bacteria the actively replicating portion of the DNA is

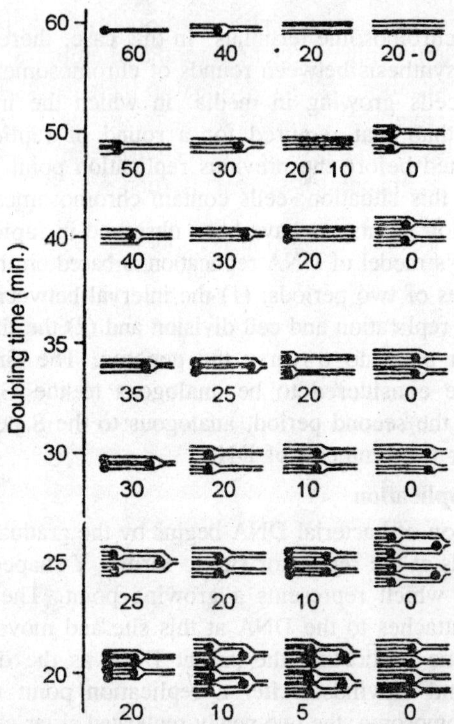

Fig. 5.4. Relationship between chromosome replication and the division cycle in E. coli B/r growing at different rates.

associated with a membrane fraction. Electron microscopic studies have provided a morphologic basis and have shown that in gram-positive organism extensions of the cytoplasmic membrane, the mesosomes, are in contact with the bacterial chromosome. Serial sections through dividing cells have indicated that in the course of replication the mesosomes with thin attached chromosomes split in two and move apart, each carrying with it one of the two sister chromosomes. Evidence that the separation of the mesosomes is achieved by the synthesis of new membrane in the region between them was obtained by following the distribution of a specific marker, potassium tellurite, in the membrane during growth. Potassium tellurite is deposited selectively I the reduced form at the oxidation-reduction sites of the membrane as crystals that are visible in the electron microscope. The observed distribution of crystals supports the hypothesis that new membrane is generated in the region of the point of attachment of the chromosome to the membrane.

In gram-negative bacteria where mesosomal formation is minimal, the chromosome is either attached to the membrane without the mediation of mesosomes, or it is attached to their base. There appears to be no unique membrane attachment site on the *E. coli* genome; either the genome is attached to the membrane at random, or to a number of fixed or changing attachment points.

Terminal Steps in Cell Division

Following the separation of the two daughter chromosomes the cell initiates synthesis of a transverse septum. The temporal relationship between replication of the chromosome and synthesis of the transverse septum is closely regulated. The completion of a round of replication together with a period of protein synthesis which normally occurs concurrent with chromosome replication are both required for cell division. Cell division occurs at a fixed time after the termination of a round of DNA replication. This time period represents the time required for the expression of the many special reactions involved in septum formation, membrane growth, and wall synthesis.

The importance of the coupling between DNA replication and cell division has been emphasized by the finding of a number of mutants in which the regulation of cell division is disrupted. Many of these are conditional lethal mutants that grow normally at 30°C but are unable to form colonies at 41°C. In some of these mutants specific differences in membrane proteins have been observed, some of which correlate well with an interference of septum formation and inhibition of cell division.

Septum formation

Compartmentalization of physiologic division of the cell occurs prior to the actual physical separation of the daughter cells. This is the time at which the cytoplasmic events at one end of the cell become independent of cytoplasmic events occurring at the opposite end. The end of a round of replication occurs well in advance of this compartmentalization.

This stage of growth is characterized by the development of a weak septum followed by the formation of a strong cross-wall. As emphasized previously, the position and timing of septum formation is crucial if each daughter cell is to receive a full compliment of the parental material. The process of septum formation is poorly understood and little is known of the physiologic events which periodically initiate it. Although it is generally agreed that the completion of chromosome replication is necessary for division, other factors such as the presence of polyamines and heat-labile proteins also play an important role.

Cell-wall growth

During balanced growth a greater surface area is normally required to cover a greater volume. In several gram-positive species, cell-wall growth appears to be limited to discrete areas at the equator. These areas have been located in a number of gram-positive organisms by staining the cell surface with fluorescent antibodies and observing the location of the label during subsequent growth. In gram-negative organisms, however, although zonal growth of the murein component also occurs, the lipopolysaccharide of the cell envelope is probably added by diffuse intercalation during cell elongation.

The nascent cross-wall of the growing cell originates within the area enclosed by a mesosome. Subsequent inward growth of the wall is preceded by a corresponding inward proliferation of the mesosome which is formed by a concentric infolding of the plasma membrane. As the mesosome, followed by the growing transverse septum, moves

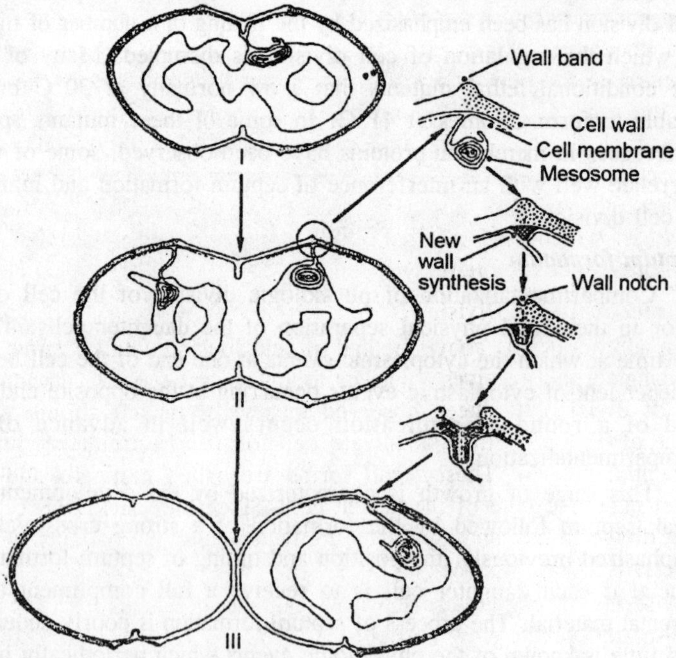

Fig. 5.5. Diagrammatic representation of cell-wall division model for Streptococcus faecalis.

toward the center of the cell, the membranes at its base continue to form around the newly synthesized cell wall.

A notch appears in the external wall directly above the developing cross wall, accompanied by the formation of two new raised bands adjacent to the notch. These bands appear to mark the boundary between new and old wall. Both peripheral wall elongation and centripetal cross-wall extension apparently result from biosynthetic activity in the vicinity of the leading edge of the cross wall. An interplay of murein hydrolyzing and murein-synthesizing enzymes has been invoked to explain the sequence of interrelated events associated with the enlargement of the cell wall. A closed molecule like the structural component of the cell wall, the murein sacculus, can only be modified when covalent bonds are split. The hydrolyases would provide both space and acceptor sites for incoming subunits which in a second step would be condensed into the preexistent murein. The sites at which murein hydrolases are active correspond to the zone where murein is synthesized.

Separation of bacteria

Separation of the two daughter cells takes place some time after the completion of the transverse septum. The separation begins with the appearance of a constriction between the subsequent daughter cells and a thickening of the cell wall in the constriction region. Further thickening of the cell wall accompanies constriction between the cells until the pole of the cell assumes its characteristic rounded shape. The normal thickness of the cell wall, however, is achieved only after constriction is complete and the two daughter cells have completely separated. Separation of the daughter cells is mediated by murein hydrolases which are localized at the site of septum formation, and which are latent in nondividing cells.

Asymmetrical Cell Division

In addition to the above method of cell division which insures that each daughter cell receives a nucleus following cell division, an unusual form of division has been observed in several mutant strains. In a mutant of *E. coli*, small anucleate cell-forms are produced during exponential growth. These small forms are called minicells and are formed near one pole of the parent cell. Since they lack DNA, they do not divide further and are unable to manufacture any polymer whose synthesis is DNA dependent.

A series cf temperature-sensitive mutants of *E. coli K*12 has also been isolated which grow normally at 30°C but cease DNA synthesis at 40°C after a small portion of the chromosome has been replicated.

Most of these mutants are unable to form a septum so that the cells continue to elongate and to produce long filamentous cells without septa. However, a small number of these thermosensitive mutants can still form septa and undergo cell division. The septa in these mutants are not formed near the equatorial plane as in the wild-type parent but to one side of the DNA, so that of the two daughter cells formed, one contains all of the DNA of the mother bacterium and the other none. This type of anucleate cell, however, differs from the minicell in that it is about the same size as a normal wild-type cell.

Measurement of Growth

Isolation of pure cultures

The first step in the measurement of the growth properties of a bacterium is to ensure that one is dealing with a pure culture. This is of special concern in studies of bacteria from clinical samples that contain large numbers of different cells, each of which can respond differently to changes in the growth medium as well as influence the growth of other bacteria in the population. Pure cultures are usually isolated by spreading out a sample of cells on the surface of a suitable medium. A variety of growth media has been developed for the isolation or identification of specific types of bacteria. Some media are selective, meaning that they contain ingredients that inhibit the growth of certain bacteria. An example of this is the inclusion of bile salts and basic dyes in a medium used for the isolation of enteric bacteria; these reagents are able to penetrate the gram-positive cell wall and thus prevent the growth of most gram-positive bacteria. Other media re termed differential, meaning that they permit the differentiation of organisms on the basis of some biochemical property. For example, media used for the isolation of enteric bacteria incorporate lactose and an acid—base indicator into a peptone containing medium. As a result, those organisms that are able to utilize lectose and produce acid from it can be readily distinguished from other enteric bacteria, which grow on the peptone present in the medium and do not produce acid from these peptides.

Almost all media used for the purification of bacteria are solidified by the incorporation of agar, a complex polysaccharide isolated from seaweed. It has the desirable properties of melting at about 95°C, but forming a solid gel at about 45°C. Thus, it can be kept molten at temperatures above 50°C and, if desired, mixed with other heat-labile ingredients prior to being poured into suitable vessels and allowed to harden. Fortunately, few bacteria degrade agar.

Bacteria will grow when placed on an agar surface of a suitable medium either by pouring or by spreading with a wire loop. Wherever a single cell has been deposited, a visible colony will arise containing 10^6 to 10^9 bacteria, all of which are the direct progeny of the original cell at that site. An isolated colony thereby provides the pure culture needed for subsequent studies of its growth and drug-resistance properties. Good practice suggests that such a single colony be restreaked on a similar medium to ensure the absence of any contaminating bacteria.

Determination of bacterial mass

To measure bacterial growth and the effect of antibiotics on a particular bacterial strain, it is necessary to determine both the mass of the cells and the number of living cells in the sample. No single method allows both measurements. Cell mass can be determined from the dry weight, although this requires large samples and considerable time. Somewhat more rapid methods involve measurement of cell volume or the total amount of cell nitrogen of protein, using calorimetric assays. By far, however, the simplest and most frequently employed technique is to measure the optical density or turbidity of a liquid culture. This assay, which can be readily automated, is based on the fact that light scattering by particles is related to the number, size, and mass of the particles. The determination of certain metabolic activities, such as dye reduction by the electron transport components or the release of radioactive CO_2 from a labelled substrate, are also being explored, since such procedures are far more sensitive than those based on turbidity.

Determination of cell number

The total number of cells, both living and dead, can be determined microscopically by using a special glass slide, called a Petroff-Hauser Counter, which has a graduated chamber of known volume. For those who require a frequent determination of cell number, an electronic particle counter, known as a Coulter counter, provides both the number of particles and their size distributions.

The number of living cells in a given sample can be determined by plating a dilution of the original culture onto the surface of an agar medium that has solidified in a Petri dish. The diluted sample may also be mixed with the melted agar medium before pouring it into a Petri dish to harden. The number of colonies that arise after 1 or 2 days' incubation are counted and multiplied by the dilution factor to yield the number of viable cells per milliliter in the original sample.

Figure illustrates such a dilution scheme in which the sample was diluted by a factor of 10^{-6}, and 0.1 ml was plated onto the surface of the plate. In this case, if 90 colonies were found on the plate after incubation, one could calculate that there were 90×10^6 cells per 0.1 ml, or 90×10^7 viable cells per ml of the original sample. Clinical samples expected to have few cells, such as urine or clear water, can be concentrated by centrifugation or filtration rather than diluted prior to plating.

Phases of Growth

Under favourable conditions, most bacteria are able to reproduce rapidly. The generation tine is the time required for the population to double. Examples near both ends of the spectrum would include *Clostridium perfringems*, which has an optimal generation time as short as 8 to 10 minutes, and *Mycobacterium tuberculosis*, in which it is about 15 to 20 hours. As one might expect, the generation time varies according to the growth medium and the species. Because most organisms have generation times in the range of 30 to 120 minutes, bacterial populations can quickly become quite large. For example, one *Escherichia coli* cell with a 30-minute doubling time will grow overnight into more than one billion (10^9) cells. Assuming division at the same rate, it would require only an additional 37 hours for the progeny of that single cell to cover the United States to a depth of 1 meter. These simple examples stress the rapidity of bacterial growth and the necessity for speed in determining the number of cells in a sample. When cells are inoculated into a growth medium, the increase in their number and mass can be plotted on a logarithmic scale as a function of time. Such a growth curve reveals four major phases.

Lag phase

When bacteria are inoculated into fresh medium, reproduction does not usually begin immediately. During this lag phase, the cell mass and size begin to increase long before cell numbers increase. It is during this phase that the synthesis of enzymes and other macromolecules needed for growth in the new medium occurs. The length of the lag phase depends on the kind of bacteria, the age and size of the inoculum, the nature of the medium from which they were taken, and the nutrients present in the new medium.

Log or exponential growth phase

During the exponential growth phase, cell mass and numbers increase in a logarithmic manner with a constant generation time. This is a period of balanced growth during which all cellular constituents are increasing in constant proportion. The rate of cell division, is of course, dependent on the type of organism, the nature of the medium, the temperature, and, for aerobic organisms, the rate of aeration.

The generation time can be determined during the log phase, because during this phase, both cell number and mass increase in a coordinate manner. Thus, measurements of either parameter can be used for this determination, and since each generation results in a doubling of the cell population, the following equation applies:

$$B_t = B_0 \times 2^n$$

where B_t and B_0 are the number of bacteria or their mass at time t and at the initial time 0, respectively, and n is the number of generations that have occurred between time 0 and time t. Taking the logarithm (to the base 10) of both sides of the equation yields.

$$\text{Log } B_t = \log B_0 + n \log 2$$

Since $n = t/g$, where t is the time and g is the generation time in minutes per doubling, the equation above can be rearranged to yield

$$\text{Log } B_t - \log B_0 = \log 2 \times t/g$$

This equation can be rearrange to solve for any unknown when one is given three of the four parameters (log 2=0.301). A frequently employed function is called the "specific growth rate," which is given the symbol, μ, and is in units of doublings per hour.

$$\mu = \ln 2/g = (\ln B_t - \ln B_0)/t$$

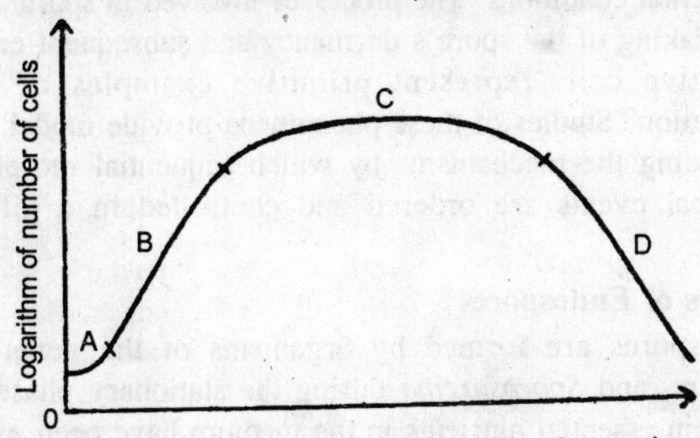

Fig. 5.6. Typical growth curve illustrating the phases of growth occurring when bacteria are inoculated into a culture media. (A) Lag phase, (B) Logarithmic growth phase, (C) Stationary phase, (D) Death phase.

Stationary phase

Logarithmic growth eventually slows because of accumulation of waste products, exhaustion of nutrients, change in pH, or a decrease in oxygen tension. The population then enters the stationary phase in which the number of viable cells remains approximately constant. Usually, there is a division of some cells and the death of others. It is worth noting that entry into the stationary phase or even starvation for a required nutrient does not result in the killing of most bacteria.

Death phase

Eventually the rate of death exceeds the rate of reproduction, and the number of viable cells declines. The length of tie before all cells have died differs markedly for various organisms. Some species have a very short death phase, whereas for others it may take weeks before all of the cells in a culture have died. During this phase, cell often assume unusual shapes, making it difficult to recognize the bacteria in old cultures.

Differentiation in Bacterial Cells

Sporulation and Germination

One of the most unusual properties of certain bacteria is the ability to form endospores. At some point in the vegetative cell cycle of spore-forming organisms growth is arrested and the cell undergoes progressive physiologic and morphologic changes which result in the formation of an endospore. A spore is a dormant structure capable of surviving for prolonged periods, but endowed with the capacity to reestablish the vegetative stage of growth under appropriate environmental conditions. The processes involved in sporulation as well as the breaking of the spore's dormancy and subsequent emergence of a vegetative cell, represent primitive examples of unicellular differentiation. Studies of these phenomena provide model systems for understanding the mechanisms by which sequential morphologic and biochemical events are ordered and controlled in a differentiating system.

Properties of Endospores

Endospores are formed by organisms of the genera *Bacillus*, *Clostridium*, and *Sporosarcina* during the stationary phase of growth after certain essential nutrients in the medium have been exhausted. A single spore is produced within a vegetative cell and differs from it in its morphology and composition, increased resistance to adverse environments, and absence of metabolic activity, Although the thermoresistance of spores has been the major concern of medical microbiologists because of the problems associated with the control of infection caused by spore-producing organisms, spores also have an increased resistance to the lethal effects of desiccation, freezing, radiation, and deleterious chemicals. The primary selective value of the spore, however, lies in its longevity in the soil coupled with its ability to germinate under the proper environmental conditions.

Chemical basis of spore resistance

Resistance to various chemical and physical agents appears at different stages during sporulation, concomitant with changes in the physicochemical composition of the sporulating cell. The resistance to radiation, drying, and chemical agents appears after the cell becomes refractile and depends at least in part upon the properties of the cystine-rich spore coat proteins. Thermal resistance is probably due to the very low content of water, which renders the proteins and nucleic acids more resistant to denaturation. This dehydration occurs late in sporulation at the time of cortex formation and at the time that the spore first appears as a refractile object. The massive synthesis of dipicolinic acid during this stage may be responsible for the dehydration. Dipicolinic acid, a chelating agent, is a spore-specific component present in high concentrations in all bacterial endospores. It is present in the cortex of the mature spore as the calcium salt and accounts for 5 to 15 percent of its dry weight. Both calcium and dipicolinic acid are required for heat resistance.

Metabolic Differences Between Vegetative Cells and Spores

As a bacterial cell passes from the vegetative state to the sporulating state, it undergoes dramatic changes in its morphology and physiology. During sporulation certain compounds such as poly-b-hydroxybutyrate accumulate and are later utilized. Extensive turnover of macromolecules occurs, and some enzyme levels change drastically. New and characteristic spore structures are synthesized, while previously existing structures and degraded.

The pools of small molecules found in spores are distinctly different from those of vegetative cells. There is an accumulation of dipicolinic acid, divalent cations, and high levels of L-glutamic acid. The predominant component of the drastically reduced acid-soluble phosphate pool is 3-phosphoglyceric acid, which constitutes as much as 75 percent of the acid-soluble phosphate in the spore, but no more than 5 percent of the acid-soluble phosphate of the vegetative cell. Spores contain very low levels of ATP, whereas ATP, is a major component of vegetative cells.

During sporulation many differences occur both qualitatively and quantitatively in the pattern of enzyme activities. Some of these differences are associated with mechanisms related to spore formation; some are components of the spore itself. The heat-stable catalase of spores is immunologically distinct from the vegetative cell enzyme; glucose dehydrogenase, a ribonuclease, and spore lytic enzymes are present only in the spore and absent form the vegetative cells. These enzymes as well as the spore coat proteins represent spore-specific gene products.

Initiation of Sporulation

Sporulation is a response to nutritional deprivation. Although limitation of any one of a variety of nutrients can initiate sporulation, the most pronounced effect is exerted by the carbon and nitrogen sources available. The regulation of spore formation in the bacterial cell is negative; the cell makes an inhibitor or repressor from some ingredient in the medium, which prevents the initiation of sporulation. When this ingredient is exhausted, inhibition is released, and the cells sporulate. A metabolizable supply of both carbon and nitrogen are required to inhibit sporulation. If there is a deficiency in the supply of either of these, inhibition will be relieved and sporulation initiated.

Biochemistry of Sporulation

At some period in the development of a cell, the metabolism is irreversibly channeled in the direction of sporulation. There is no single point of commitment for the sporulation process as a whole, but a separate point of commitment for each new specific macromolecule. An examination of thin sections of sporulating cells in the electron microscope has shown that an ordered series of cytologic and structural changes accompanies the physiologic changes associated with sporulation. The morphologic changes observed in *B. subtilis* have been presented as a time scale divided into seven sequently appearing stages. The process is times from the end of exponential growth at t_0 and at hourly periods after this.

Stage 0. This represents the state of the cell at the end of exponential growth at a time when each vegetative cell contains two chromosomes.

Stage I. This stage is characterized by the formation from the more disperse nuclear bodies of a broad axial chromatin filament, which occupies the center of the cell. During this stage the most characteristic events are the production and excretion into the medium of antibiotics and various exoenzymes. The importance of antibiotic and protease production in the formation of spores is emphasized by the finding that mutants unable to produce these substances are almost invariably also asporogenous. Protease's probably play a role in the intracellular turnover of protein by the correlation between sporulation and antibiotic production is more difficult to explain. The importance of antibiotic production per se to the sporulation process may result from the fact that it is merely one of an ordered sequence of biochemical events, some essential and some not to the sporulation process.

Stage II. This stage is characterized by the formation of a forespore septum near one pole of the cell. Septum formation follows the separation of the two discrete chromatin bodies in the axial filament and movement of one of them to the end of the cell, resulting in the segregation of the nuclear material into two compartments, the mother cell and the forespore units. The forespore septum is formed *de novo* and does not result from the rearrangement of pre-existing membrance. During this period there is a marked increase of alanine dehydrogenase, a component important in germination.

Stage III. During this stage the spore protoplast in formed as a consequences of the unidirectional growth of the cytoplasmic membrane of the sporangium. Among the enzymes whose activity is markedly increased during this stage are glucose dehydrogenase and aconitase. Aconitase is an example of an enzyme appearing during the early stages of sporogenesis which is, like sporulation itself, subject to catabolite repression. With the exhaustion of glucose and glutamate at the end of vegetative growth repression is relieved, aconitase is formed, and functional tricarboxylic-acid glyoxylic-acid cycles produced. Heat-resistant catalase and alkaline phosphatase also appear during Stage III.

Stage IV. Cortex formation occurs during this stage at a time when the spores appear for the first time as refractile objects. Metabolic activity of the forespore at the end of State IV is considerably less than that of a vegetative cell, and beyond this stage the forespore probably contributes little to its own development. The activity of both ribosidase and adenosine deaminase increases during Stage IV and the synthesis of dipicolinic acid begins.

Stage V. This is the period of coat formation. Deposition of spore coat protein and the incorporation of cystine lead to a completed coat structure.

Stage VI. Spore maturation occurs during this stage resulting in the fine structure of the cytoplasm of the spore protoplast which makes it more homogeneous and electrondense. There is continued synthesis of or modification of cortical peptidoglycan, and uptake of dipicolinic acid and calcium. During Stage VI alanine racemase, an enzyme important in spore germination, is formed, and a number of ill-defined changes occur which make the spores resistant to organic solvents and to heat.

Stage VII. The mature endospore is liberated from the mother cell. This is initiated by a lytic enzyme synthesized or activated subsequent to maturation or the endospore.

Protein Synthesis during Sporulation

With the appearance of the forespore the cell becomes a compartmentalized structure. In the compartment developing into the forespore, DNA replication is inhibited and the substances synthesized are spore-specific. In the second compartment, the sporangium, synthesis of vegetative DNA continues and products of the vegetative cell continue to be produced. Thus, in the sporulating cell both the vegetative and spore genomes are transcribed. The mRNA contains some of the same mRNA molecules as depressed cells transferred from a rich medium to a poor one, but in addition molecules specific to sporulation. There is no evidence that the mRNA produced late in sporulation is stable and retained in the spore for protein synthesis during outgrowth.

Role of RNA polymerase

The changes that occur during sporulation result from the turn-off of certain vegetative genes and the expression of new classes of genes. Ribosomal RNA genes are not transcribed in sporulating cells. The turn-off, however, of ribosomal RNA genes during sporulation is brought about by a mechanism different from that responsible for stringent control of ribosomal RNA synthesis. The most attractive model which has been proposed suggests that early in passing from the vegetative to the sporulating state, the template specificity of the RNA polymerase is altered. The vegetative σ factor does not function in sporulating cells while the core enzyme, the machinery for polymerizing RNA, is conserved. The reason for failure of the σ factor to function may possibly be explained by differences observed between the β subunits in vegetative and sporulation RNA polymerase cores. The two β polypeptides of vegetative RNA polymerase have a molecular weight of 155,000 daltons. In sporulation RNA polymerase, one of these β subunits has a molecular weight of only 110,000 daltons. Although this modification of vegetative RNA polymerase is critical to sporulation and may be the event responsible for the turn-off of some vegetative genes during sporulation, a more precise explanation is still needed for the mechanism by which the transcription of new genes is directed.

The spore genome is present in vegetative cells but is repressed under conditions of rapid growth. At the end of exponential growth the level of intracellular catabolites is altered, "and the gene governing the synthesis of a "sporulation initiation factor" is derepressed. The nature of this "sporulation initiation factor," however, remains to be clarified.

Germination and Outgrowth

Simultaneous structural and physiologic changes are also manifested during the transformation of a dormant spore into a vegetative cell. The process of spore germination consists of three sequential phase : (1) an activation state which conditions the spore to germinate in a suitable environment; (2) a germination stage, during which the characteristic properties of the dormant spore are lost; and (3) an outgrowth stage during which the spore is converted into a new vegetative cell.

Activation is a reversible process which conditions the spore for germination. Spores do not germinate or germinate very slowly unless activated either by heat or various chemical treatments. Activation probably involves a reversible denaturation of a specific macromolecule. Germination is an irreversible process triggered by the exposure of activated spores to specific stimulants such as amino acids, nucleosides, and glucose. This is the stage during which the dormant stage is ended. During the early stages of germination there is a loss of refractility, swelling of the cortex, and appearance of fine nuclear fibrils. Accompanying these changes is a loss of resistance to deleterious physical and chemical agents, increase in the sulphydryl level of the spore, a release of spore components, and an increase in metabolic activity. The germination of spores is not inhibited by antibiotics which perturb protein and nucleic acid synthesis, indicating that the enzyme responsible for germination are already present in the spore.

During outgrowth there is *de novo* synthesis of proteins and structural components which are characteristic of vegetative cells. During this stage the spore core membrane develops into the cell wall of the vegetative cell. Outgrowth is a period of active biosynthetic activity and is markedly inhibited by interference with the energy supply or by antibiotics which inhibit cell—wall, protein, or nucleic-acid synthesis.

If heat-activated spores are germinated under suitable conditions a high degree of synchrony is obtained. Such a synchronous population provides a model system for the study of the initiation of transcriptional and translational events occurring during differentiation. Since the dormant spore is devoid of functional mRNA the conversion of the spore to a vegetative cell during outgrowth requires transcription and de novo synthesis of gene products. The appearance of these gene products is ordered, determined by the time of transcription of particular genes. Ribosome synthesis starts early; vegetative cell-wall synthesis later, and DNA synthesis begins just before division of the cell. During a stepwise doubling of cell members for several generations, the initiation of certain enzymes occurs at a specific time during each division cycle, and results in a doubling of each enzyme during only a fraction of the total cycle.

Factors Influencing Microbial Growth

The rates of microbial growth and death are greatly influenced by a number of environmental parameters, some environmental conditions favoring rapid microbial reproduction and others not permitting any microbial growth. Conditions permitting the growth of one microorganism may preclude the growth of another. Not all microorganisms can grow under identical conditions. Each micrograms has a specific tolerance range for specific environmental parameters. Outside the range of environmental conditions under which a given microorganism can reproduce, it may either survive in a relatively dormant state or may lose viability; that is, it may lose the ability to reproduce and consequently die.

In both laboratory and natural situations some environmental parameter or interaction of environmental parameters controls the rate of growth or death of a given microbial species. In nature, where condition cannot be controlled and many species coexist, fluctuation environmental conditions favour population shifts because of the varying growth rates of individual microbial populations within the community of a given location. In the laboratory it is possible to adjust conditions to achieve optimal growth rates for a given microorganism. Similarly, in industrial fermentors conditions can be adjusted to optimize microbial growth rates, thereby maximizing the accumulation of desired microbial metabolic products. Many laboratory and industrial applications use pure cultures of microorganisms, facilitating the adjustment of the growth conditions so that they favour optimal growth of the particular microbial species.

Effects of Temperature on Microbial Growth Rates

Temperature is one of the most important of factors affecting both the rates of microbial growth and death. The minimum and maximum temperatures at which a microorganism can grow establish

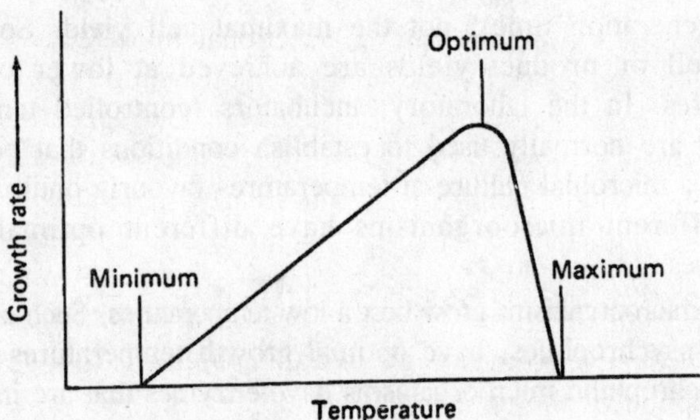

Fig. 5.7. Effect of temperature on growth rate showing growth ranges and optimal temperatures for microorganisms.

the temperature growth range for that microorganism. Most microorganisms are killed when exposed to temperatures above the maximal growth temperature. At very low temperatures, however, microbial metabolism ceases, preserving the microorganisms in a dormant state. Some strains of microorganisms in culture collections are maintained indefinitely by freezing them at very low temperatures, such as in liquid nitrogen; such preservation of reference type cultures is important because it provides necessary standards.

Within the growth range each microorganism has an optimal growth temperature at which the highest rate of reproduction occurs. The optimal growth temperature is defined by the maximal growth rate

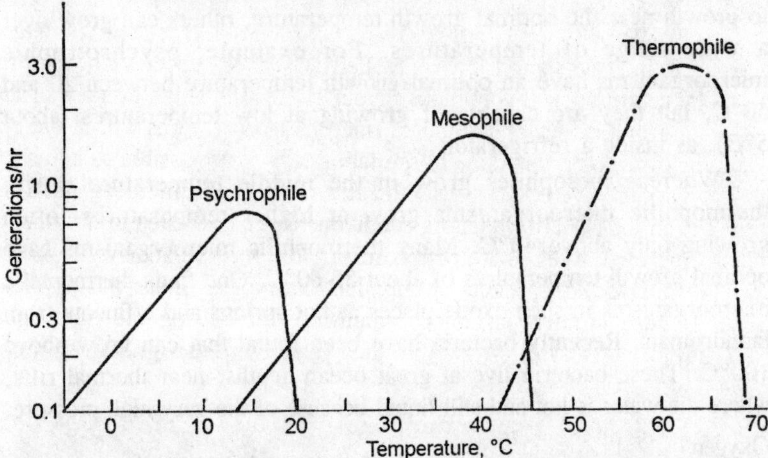

Fig. 5.8. Temperature growth ranges and optima for psychrophilic, mesophilic, and thermophilic organisms.

shortest generation time), not the maximal cell yield. Sometimes, greater cell or product yields are achieved at lower or higher temperatures. In the laboratory, incubators (controlled temperature chambers) are normally used to establish conditions that permit the growth of a microbial culture at temperatures favourig optimal growth rates. Different microorganisms have different optimal growth temperatures.

Some microorganisms grow best a low temperatures. Such organisms, known as psychrophiles, have optimal growth temperatures of under 20°C. Psychrophilic microorganisms have enzymes that are inactivated at even moderate temperatures, about 25°C. Some psychrophilic microorganisms are capable of growing below 0°C as long as liquid water is available. Psychrophilic microorganisms are commonly found in the world's oceans and are also capable of growing in a household refrigerator, where they are important agents of food spoilage.

Other microorganisms, known as mesophiles, have optimal growth temperatures in the middle temperature range between 20 and 50°C. Most of the bacteria grown introductory microbiology laboratory courses are mesophilic. Many mesophiles have an optimal temperature of about 37°C, which corresponds, to human body temperature. Many of the normal resident microorganisms of the human body, such as *E. coli* are mesophiles. Similarly, most human pathogenic microorganisms are mesophiles and thus are able to grow rapidly and establish an infection within the human body. Although some microorganisms are restricted to growth near the optimal growth temperature, others can grow over a wide range of temperatures. For example, psychrotrophic microorganisms have an optimal growth temperature between 20 and 50°C, but they are capable of growing at low temperatures (about 5°C). as inside a refrigerator.

Whereas mesophiles grow in the middle temperature range, thermophilic microorganisms grow at higher temperatures, often growing only above 40°C. Many thermophilic microorganisms have optimal growth temperatures of about 55-60°C. One finds thermophilic microorganisms in such exotic places as hot springs and effluents from laundromats. Recently bacteria have been found that can grow above 100°C. These bacteria live at great ocean depths, near thermal rifts, where the water is hot and still liquid because of the very high pressure.

Oxygen

Another factor that greatly influences microbial growth rates is the concentration of molecular oxygen. Microorganisms can be classified as aerobes, anaerobes, facultative anaerobes, or microaerophiles based on their oxygen requirements and tolerances. Aerobic microorganisms (aerobes) grow only when oxygen is available to support their respiratory metabolism, whereas anaerobic microorganisms (anaerobes), such as *Clostridium*, grow in the absence of molecular oxygen. Anaerobic microorganisms may carry out fermentation or anaerobic respiration to generate ATP. Some anaerobes have very high death rates in the presence of oxygen and such organisms are termed oxylabile anaerobes. Other anaerobic microorganisms, known as oxyduric anaerobes, although unable to grow, have low death rates in the presence of oxygen. In contrast to obligate anaerobes, which grow only in the absence of molecular oxygen, facultative anaerobes, such as *E. coli*, can grow with or without oxygen. As a rule, facultative anaerobes are capable of both fermentative and respiratory metabolism.

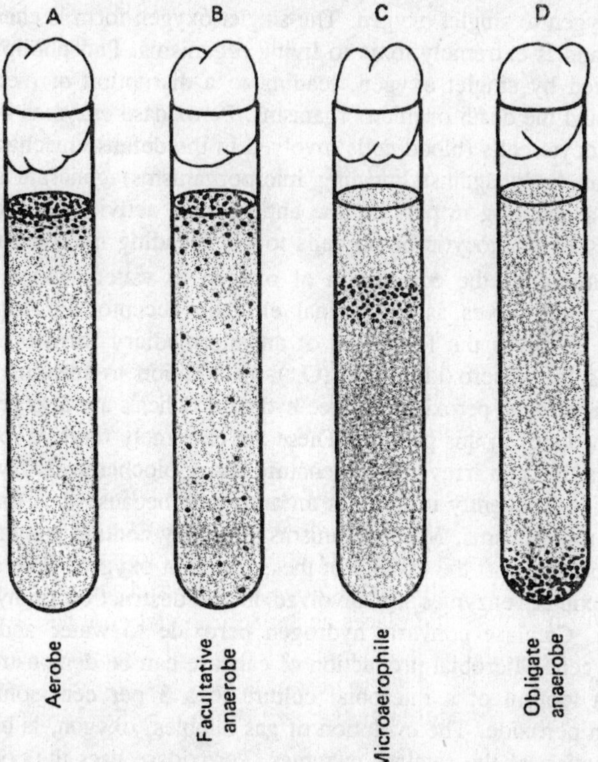

Fig. 5.9. These test tubes show the relationship of the growth of various types of microorganisms to the presence of oxygen.

Although oxygen is required for the growth of many microorganisms, it can also be toxic. Some microorganisms, known as microaerophiles, grow only over a very narrow range of oxygen concentrations. Microaerophiles require oxygen but exhibit maximal growth rates at reduced oxygen concentrations because higher oxygen concentrations are toxic to these organisms. As a consequence of the toxicity of oxygen to microorganisms, even aerobic microorganisms generally posses enzyme systems for detoxifying various forms of oxygen.

Oxygen can exist in a number of energetic states, some for which are more toxic than others. The common state of oxygen is the triplet form in which two of the electrons in the valance shell are unpaired and spin in parallel directions. In singles oxygen, which has a higher energy level than the triplet form, these two electrons have antiparallel spins. Various enzymes are capable of catalyzing the conversion of triplet oxygen to singlet oxygen. The singlet oxygen form is chemically reactive and is extremely toxic to living organisms. Phospholipids can be oxidized by singlet oxygen, leading to a disruption of membrane function and the death of microorganisms. Peroxidase enzymes in saliva and phagocyte cells (blood cells involved in the defense mechanism of the human body against invading microorganisms) generate singlet, oxygen, accounting in part for the antibacterial activity of saliva and the ability of phagocytic blood cells to kill invading microorganisms.

Additionally, the conversion of oxygen to water, which occurs when oxygen serves as a terminal electron acceptor in respiration pathway, involves the formation of an intermediary form of oxygen known as the superoxide anion (O_2^-), in addition to forming singlet oxygen. Hydrogen peroxide and free hydroxyl radicals are also generated as intermediates in this process. These are extremely reactive oxidative chemicals that can irreversibly denature many biochemical. Hydrogen peroxide is frequently utilized as an antiseptic because it is toxic and kills microorganisms. Microorganisms generally contain enzymes that protect them against the toxicity of these forms of oxygen. Both catalase and peroxidase enzymes are involved in the destruction of hydrogen peroxide. Catalase converts hydrogen peroxide to water and triplet state oxygen. Microbial production of catalase can be demonstrated by adding a loopful of a microbial culture to a 3 per cent solution of hydrogen peroxide. The evolution of gas bubbles, oxygen, is evidence for the action of the catalase enzymes. Peroxidase uses the coenzyme NADH to convert hydrogen peroxide to water. In addition to these enzymes, the superoxide radical is converted to hydrogen peroxide and water by the action of the enzyme superoxide dismutase.

Both obligate aerobes and facultative anaerobes usually produce both catalase and superoxide dismutase enzymes. These enzymes permit such microorganisms to use oxygen and continue growing without accumulating toxic forms of oxygen that would kill the organism. In contrast, obligate oxylabile anaerobes generally lack both catalase and superoxide dismutase enzymes. The inability of these organisms enzymatically to remove toxic forms of oxygen probably accounts for the fact that they are obligately anaerobic and sensitive to oxygen. The oxyduric anaerobes generally produce superoxide dismutase but not catalase, evidence that they can survive in the presence of oxygen.

WATER ACTIVITY

All microorganisms require water for growth and reproduction. Water is essential for most biochemical reactions in living systems, and the availability of water has a marked influence on microbial growth rates.

Water activity is an index of the water that is actually available for utilization by microorganisms. In the atmosphere, availability of

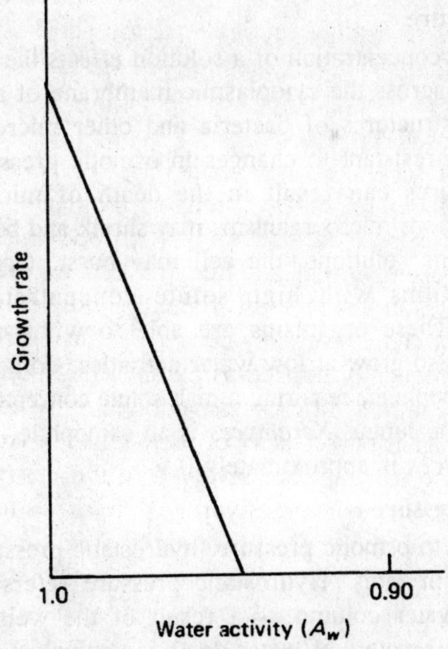

Fig. 5.10. The effect of water activity (A_w) on the growth rate of microorganisms.

water is expressed as relative humidity (RH).RH=100 × A_w; thus a relative humidity of 90 percent corresponds to an *Aw* of 0.90. Pure distilled water has a water activity (A_w) of 1.0. Adsorption and solution factors, however, can reduce the availability of water and thus lower the water activity. Water, for example, may be bound by a solute and hence be unavailable to the microorganisms. Adding high concentrations of sugars, such as sucrose, to a solution lowers the availability of water. For example, maple syrup has a water activity of 0.9. Similarly, adding salt (NaCl) to a solution can lower the availability of water. A saturated solution of sodium chloride has an A_w of 0.8.

Most microorganisms requires a water activity above 0.9 for active metabolism. The relatively low availability of water in the atmosphere amounts for the inability of microorganisms to grow in the air. Microorganisms likewise are unable to grow on dry surfaces, except when there is a relatively high humidity. Microbial growth on surfaces is a problem in tropical zones, where the available water in the atmosphere can support microbial growth, permitting microorganisms to grow on clothing, tents, and numerous other surfaces where microbial growth normally does not occur in temperate regions.

Pressure

Osmotic Pressure

The solute concentration of a solution effects the osmotic pressure that is exerted across the cytoplasmic membrane of a microorganism. The cell-wall structures of bacteria and other microorganisms make them relatively resistant to changes in osmotic pressure, but extreme osmotic pressures can result in the death of microorganisms. In hypertonic solutions microorganisms may shrink and become desiccated; and in hypotonic solutions, the cell may burst. Organisms that can grow in solutions with high solute concentrations are called osmotolerant. These organisms are able to withstand high osmotic pressures and also grow at low water activities. Some microorganisms are actually osmophilic, requiring a high solute concentration for growth. For example, the fungus *Xeromyces* is an osmophile, and the optimum *Aw* for *Xeromyces* is approximately 0.9.

Hydrostatic Pressure

In addition to osmotic pressure, hydrostatic pressure can influence microbial growth rates. Hydrostatic pressure refers to the pressure exerted by a water column as a result of the weight of the water column. Each 10 meters of water depth is equivalent to approximately

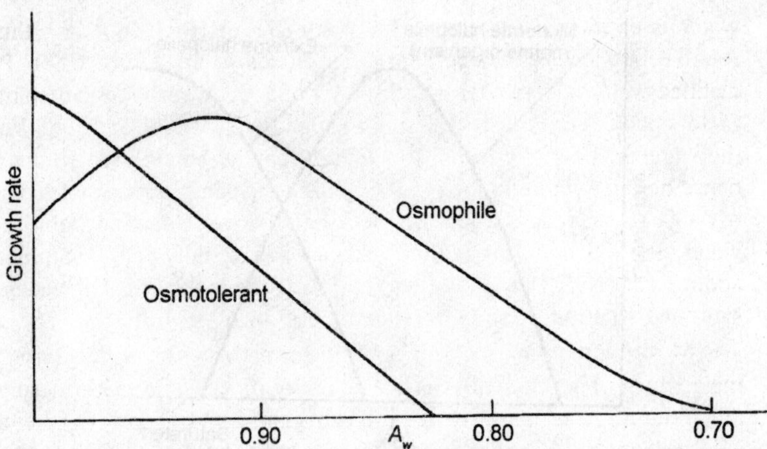

Fig. 5.11. The growth of most microorganisms is restricted by high osmotic pressure, such as those produced by high sugar concentrations.

1 atmosphere pressure. Most microorganisms are relatively tolerant to the hydrostatic pressures in most natural systems but cannot tolerate the extremely high hydrostatic pressures that characterize deep ocean regions. High hydrostatic pressures of greater than 200 atmospheres generally inactivate enzymes and disrupt membrane transport processes. However, some microorganisms—referred to as barotolerant—can grow at high hydro-static pressures and there even appear to be some microorganisms—referred to as barophiles—that grow best at high hydrostatic pressures.

SALINITY

A special terminology is used to describe the salt tolerance and requirements of microorganisms. Some microorganisms, known as halophiles, specifically require sodium chloride for growth. Moderate halophiles, which include many marine bacteria, grow best at salt (NaCl) concentrations of about 3 percent NaCl. Extreme halophiles exhibit maximal growth rates in saturated brine solutions. These organisms grow quite well in salt concentrations of greater than 15 percent NaCl and can grow in places like salt lakes and pickle barrels. High salt concentrations normally disrupt membrane transport systems and denature proteins, and the extreme halophiles must possess physiological mechanisms for tolerating high salt concentrations. Fox example, the extreme halophilic bacterium. *Halobacterium*, possesses an unusual cytoplasmic membrane and many unusual enzymes that require a high salt concentration for activity.

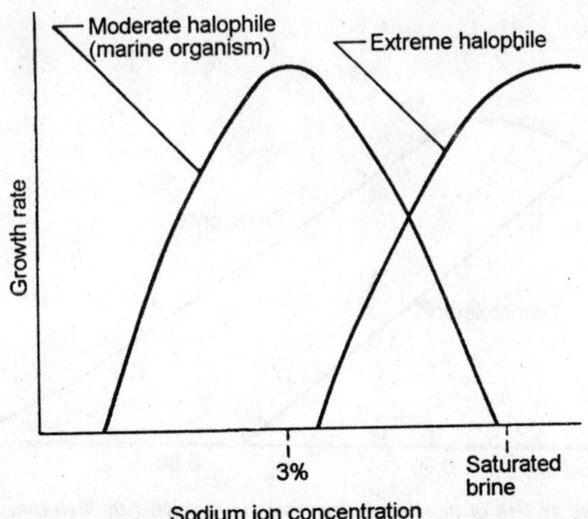

Fig. 5.12. The effect of salt concentration (sodium ion concentration) on the growth of microorganisms of varying salt tolerances.

Most microorganisms, however, do not possess these physiological adaptations and are not toluant of high salt concentrations. The degree of sensitivity to salt varies for different microbial species. Many bacteria will not grow at a salt concentration of 3 percent. Some strains of *Staphylococcus*, however, are salt tolerant and grow at concentrations greater than 10 percent NaCl. This physiological

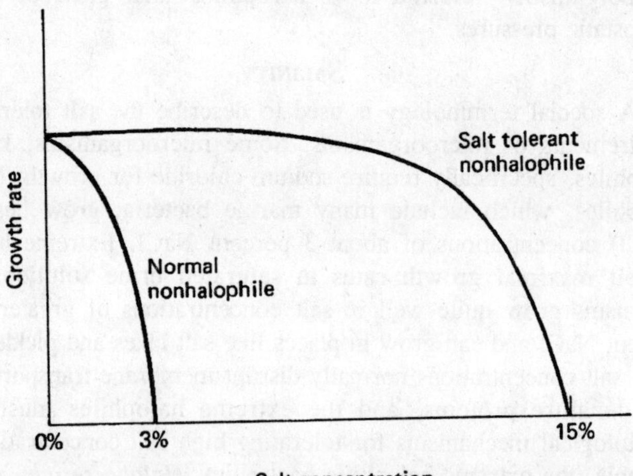

Fig. 5.13. Diagram showing relative salt tolerance. Salt-tolerant microorganisms can grow on the skin and in brines where the growth of most microorganisms is restricted.

adaptation is *Staphylococcus* is important, as some members of this genus on skin surfaces where salt concentrations can be relatively high.

ACIDITY AND pH

The pH of a solution describes the hydrogen in concentration (H^+). The pH is equal to – log (H^+) or 1/log (H^+). A neutral solutions has a pH of 70 acidic solutions have pH values less than 7, and alkaline or basic solutions have pH values greater than 7.

Table 5.2. Table of pH Tolerances of Various Bacteria.

Organism	*pH* *Minimum*	*Optimum*	*Maximum*
Thiobacillus thiooxidans	1.0	2.0—2.8	4.0—6.0
Lactobacillus acidophilus	4.0—4.6	5.8—6.6	6.8
Escherichia coli	4.4	6.0—7.0	9.0
Proteus vulgaris	4.4	6.0—7.0	8.4
Enterobacter aerogenes	4.4	6.0—7.0	9.0
Clostridium sporogenes	5.0—5.8	6.0—7.6	8.5—9.0
Pseudomonas aeruginosa	5.6	6.6—7.0	8.0
Erwinia cartovora	5.6	7.1	9.3
Nitrobacter spp.	6.6	7.6—8.6	10.0
Nitrosomonas spp.	7.0—7.6	8.0—8.8	9.4

Microorganisms vary in their pH tolerance ranges. Fungi generally exhibit a wider pH range growing well over a pH range of 5-9,

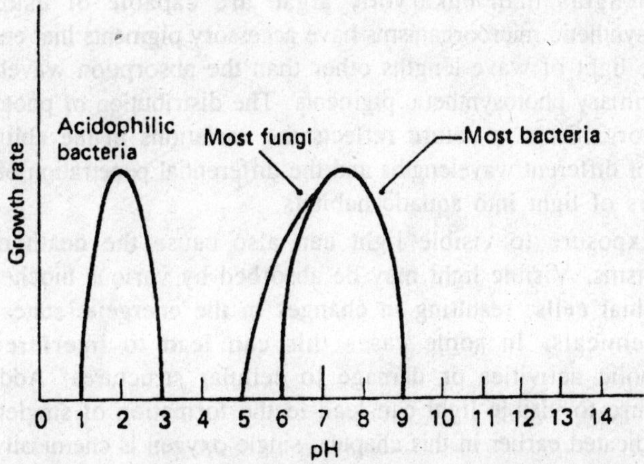

Fig. 5.14. The pH ranges for the growth of fungi and bacteria.

compared in most bacteria, which grow well over a pH range of 6-9. Similarly, some fungi grow well at lower pH values, as low as 0. Some other eukaryotic microorganisms, including protozoa an algae, are able to grow at low pH values; the lower limit for growth of some protozoa is approximately 2 and for some algae approximately 1. Although most bacteria are unable to grow at low pH, there are some exceptional cases. Some bacteria tolerate pH values as low as 0.8. There are even some bacteria, called acidophiles, that are restricted to growth at low pH values. Some members of the genus *Thiobacillus* are acidophilic, and grow only at pH values near 2. Acidophilic bacteria possess physiological adaptations that permit enzymatic and membrane transport activities. The cytoplasmic membrane of an acidophilic bacterium breaks down and cannot function at neutral pH values.

Light Radiation

Although exposure to high intensities of light can be lethal to some microorganisms, photosynthetic microorganisms require light in the visible spectrum to carry out photosynthesis. The rate of photosynthesis is a function of light intensity. At some light intensities, rates of photosynthesis reach a maximum, and although light intensities above this level do not result in further increases in the rates of photosynthesis, light intensities below the optimal level result in lower rates of photosynthesis. The wavelength of light also has a marked effect on the rates of photosynthesis. Different photosynthetic microorganisms are capable of using light of different wavelengths. For example, anaerobic photosynthetic bacteria use light of longer wavelengths than eukaryotic algae are capable of using. Many photosynthetic microorganisms have accessory pigments that enable them to use light of wave-lengths other than the absorption wavelength for the primary photosynthetic pigments. The distribution of photosynthetic microorganisms in nature reflects the variations in the ability to use light of different wavelengths and the differential penetration of different colours of light into aquatic habitats.

Exposure to visible light can also cause the death of micro-organisms. Visible light may be absorbed by various biochemicals of microbial cells, resulting in changes in the energetic states of those biochemicals. In some cases this can lead to interference with metabolic activities or damage to cellular structures. Additionally, exposure to visible light can lead to the formation of singlet oxygen. As indicated earlier in this chapter, single oxygen is chemically reactive and can result in the death of a microorganisms. Some microorganisms

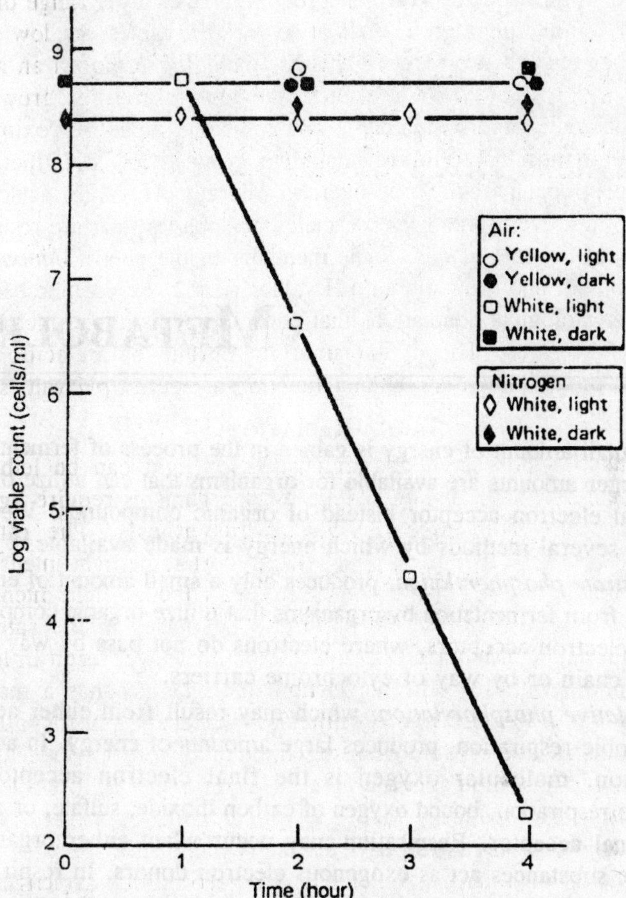

Fig. 5.15. The graph illustrates the effect of sunlight on the viability of colourless and yellow-pigmented Micrococcus lutea.

produce pigments that protect them against the lethal effects of exposure to light. For example, yellow, orange, or red carotenoid pigments interfere with the formation and action of singlet oxygen, preventing its lethal action. Microorganisms possessing carotenoid pigments can tolerate much higher levels of exposure to sunlight than nonpigmented microorganisms. Pigmented microorganisms are often found on surfaces that are exposed to direct sunlight, such as on leaves of trees. Many viable microorganisms found in the air produce coloured pigments. Likewise, some microbial spores that are principally used for aerial dispersal are pigmented.

6

METABOLISM

Only a small amount of energy is gained in the process of fermentation. Much larger amounts are available for organisms that can utilize oxygen as a final electron acceptor instead of organic compounds. We shall describe several methods by which energy is made available.

Substrate phosphorylation, produces only a small amount of energy. It results from fermentation by organisms that utilize organic compounds as final electron acceptors, where electrons do not pass by way of an electron chain or by way of cytochrome carriers.

Oxidative phosphorylation, which may result from either aerobic or anaerobic respiration, produces large amounts of energy. In aerobic respiration, molecular oxygen is the final electron acceptor. In anaerobic respiration, bound oxygen of carbon dioxide, sulfate, or nitrate is the final acceptor. Respiration may occur when either organic or inorganic substances act as exogenous electron donors. In respiration, electrons pass through a series of carriers (cytochromes) in the chain to generate large amounts of energy. If organic compounds furnish reducing electrons, a series of events, grouped into the tricarboxylic cycle, results in electron elimination from these organic compounds.

Electrons are probably availabe without a large number of steps when derived from inorganic exogenous sources.

Photosynthetic phosphorylation results when electrons are removed from chlorophyll by light energy and generate high energy bonds when they return to chlorophyll by way of electron transport chains (cytochromes). Electrons from other substances (H_2O, H_2S, H_2, or organic compounds) may also reduce ionized chlorophyll by way of transport chains and produce phosphorylation. Energy produced by photosynthesis is utilized primarily in CO_2 assimilation in all photosynthetic organisms except one group of bacteria. Members of this group reduce organic compounds into poly-beta-hydroxybutyric acid molecules.

In this chapter we shall review energy relationships of fermentation and outline processes of energy as they become available to microorganisms through aerobic respiration, anaerobic respiration, chemolithotrophic metabolism, and photosynthesis. In fermentations, pyruvate is often an intermediate, and typically, as with the Embden-Meyerhof scheme, no energy is gained beyond pyruvate formation. Where electron acceptors are available (combined or free oxygen), pyruvate may be further degraded with the production of considerable energy. The breakdown of pyruvate products, inorganic exogenous donors, or photosynthesis may furnish electrons, which produce phosphorylation as they are transported by way of cytochrome carriers to reduce oxygen.

Aerobic Respiration

Large Amount of Energy is Gained When Pyruvate is Metabolized to Carbon Dioxide and Water Aerobically

The characteristic pattern of metabolism for many microorganisms is aerobic respiration, a system in which O_2 is the final electron acceptor. Strict aerobic forms can derive sufficient life energy from no other process. In other forms, however, more than one energy-obtaining mechanism is available. When more than one metabolic pathway is available, the one followed by the organism will be determined by substrates, pH, temperature, presence of light, oxygen concentration; and other factors. Certain yeasts and bacteria ferment simple sugars or other organic compounds, provided that sufficient organic electron acceptors are available and O_2 is absent. In the presence of O_2, however, these same cultures oxidize their substrates completely to CO_2 and H_2O (which is formed by the reduction of O_2). This type of organism is *facultative*. In addition, as will be described later, organisms that can use nitrates as electron acceptors in anaerobic respiration can more efficiently use O_2 in the aerobic process.

Alterations of metabolism usually result in changes in both the nature and mode of formation of final metabolic products. The dramatic example of a metabolic change in both intermediary and final products is illustrated by the change from fermentation to aerobic respiration in yeasts, which ferment glucose to ethanol, and bacteria, which ferment it to lactate. In either case if O_2 is available, many additional steps are involved, and the final principal products are CO_2 and H_2O. The presence of O_2 may bring about a cessation or dramatic slowing of the glycolytic process. Although growth is increased, total substrate consumption may be decreased. This phenomenon, known as the *Pasteur effect*, was observed by Pasteur, but it was not fully understood until years later. What actually occurs in the presence of O_2 is *oxidative phosphorylation* of adenosine diphosphate (ADP) and inorganic phosphate to adenosine triphosphate (ATP) to furnish energy for the organisms. As we shall observe shortly, this is a much more efficient source of energy than glycolysis, and much less substrate is required for comparable growth.

As further work produced more knowledge of metabolic pathways, it became apparent that many steps were involved in oxidation, or aerobic respiration. Although all organisms do not follow the same pattern, one very important series of events known as the *tricarboxylic cycle* (Krebs cycle, citric acid cycle) has been established as the pathway of metabolism involved in many aerobic processes. Departure from glycolysis begins with pyruvate. In aerobic respiration, pyruvate is broken down into CO_2 and H_2O by a large number of known steps. Electrons formed in the cycle follow electron transport chains to combine with O_2 and produce considerable energy. The student should remember that this is not the only way aerobic respiration occurs, but it does constitute a main pathway. One molecule of pyruvate will be followed to its conversion to CO_2 and H_2O by this process. Three molecules of H_2O are picked up in the substrate and enzymatically incorporated during the process according to the overall equation

$$CH_3{-}\overset{\overset{\displaystyle O}{\|}}{C}{-}COO^- + 3H_2O \longrightarrow 5(H + H^+) + 3CO_2$$

$$5(H + H^+) + 5(\tfrac{1}{2}O_2) \longrightarrow 5H_2O + 15 \text{ phosphate bonds.}$$

There are many known steps in this process. By the action of carboxylase with its thiamine pyrophosphate (TPP) coenzyme, CO_2 is removed from pyruvate and an acetyl unit,

$$CH_3{-}\overset{\overset{\displaystyle O}{\|}}{C}{-}$$

is formed. This unit becomes attached to coenzyme A (CoA), which loses one H_2 in the process. This H_2 and the one given off by the conversion of pyruvate to the acetyl unit are picked up by nicotinamide adenine dinucleotide (NAD^+) to form NADH and H^+. The formation of NADH and H^+ should be noted carefully because this reaction will be involved in the production of high energy bonds by oxidative phosphorylation, which will presently be described.

The acetyl unit, transferred by CoA, combines with oxalacetate to yield citrate. By the action of aconitase, H_2O is lost, and *cis-aconitate* is formed. Further action converts this compound into isocitrate; H_2O is incorporated in this step. Isocitrate dehydrogenase removes H_2, which is picked up by nicotinamide adenine dinucleotide phosphate ($NADP^+$) to form NADPH and H^+. Oxalosuccinate, which results from this action, is decarboxylated in the next step to form alpha-ketoglutarate.

Removal of H_2 by alpha-ketoglutarate dehydrogenase and CO_2 by lipothiamide pyrophosphate (LTPP) results in the formation of a four-carbon succinyl unit, which combines with CoA to form succinyl-CoA. NAD^+ is reduced to NADH and H^+. Enzymatic action of succinyl-CoA synthetase converts succinyl-CoA to succinate. The CoA is released and a high energy phosphate is generated by the conversion of ADP to ATP. Magnesium is essential to activate these processes. Succinate, by the action of succinate dehydrogenase, is oxidized to fumarate, and the hydrogen pair given off reduces flavin-adenine dinucleotide (FAD) to $FADH_2$. Fumarase converts fumarate to malate by the addition of H_2O, and the malate thus formed is oxidized to oxalacetate by the

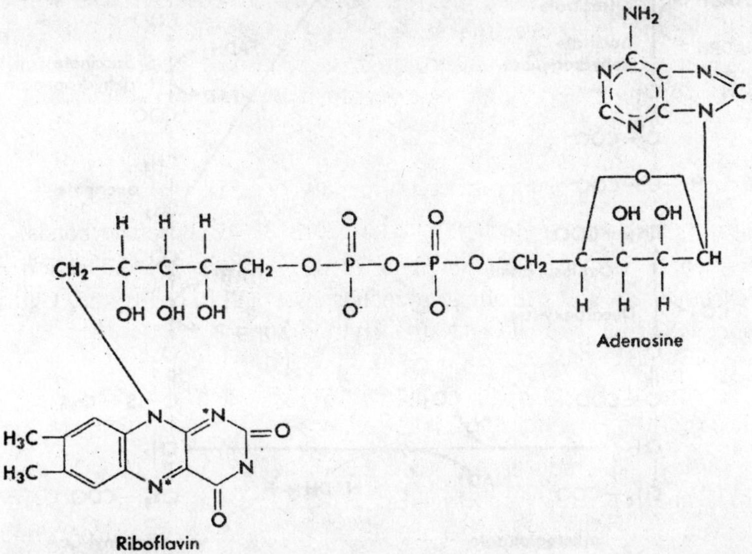

Adenosine

Riboflavin

*Fig. 6.1. Flavin-adenine dinucleotide (FAD). *Place of attachment and removals of hydrogen.*

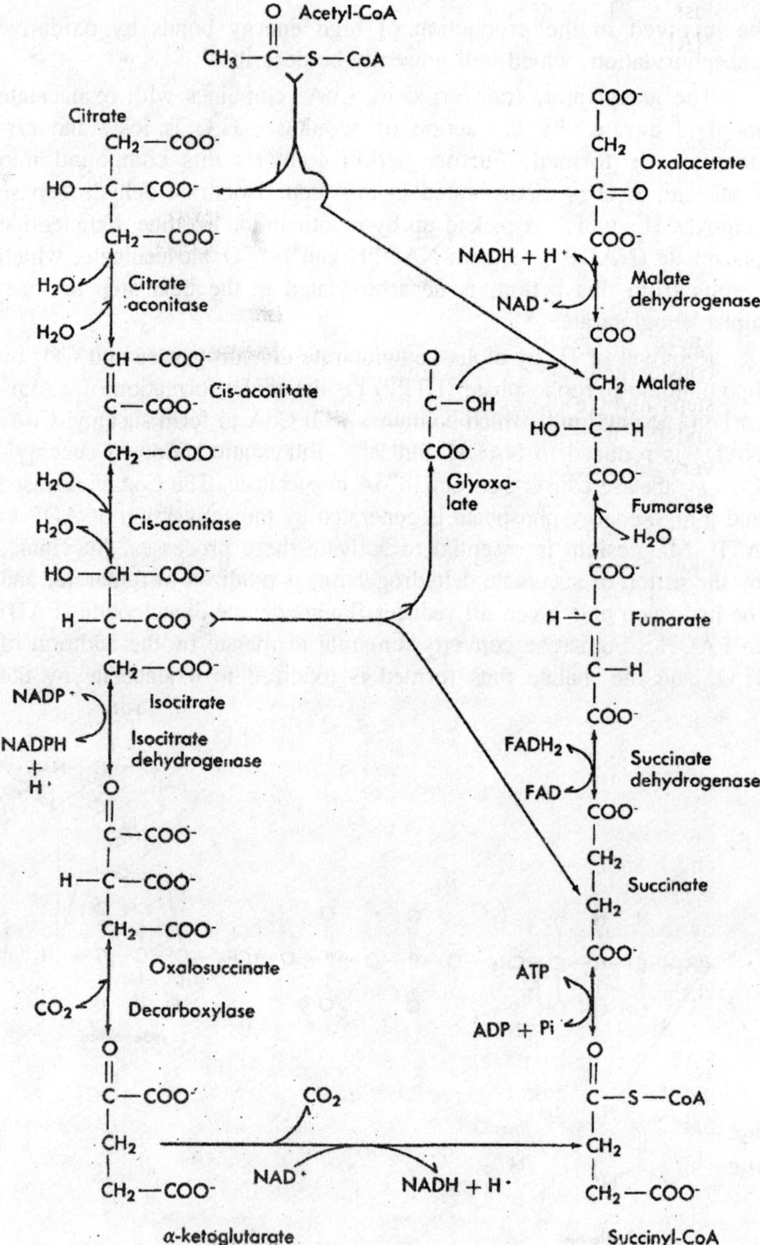

Fig. 6.2. Tricarboxylic (Krebs) cycle.

loss of H_2. NAD malate dehydrogenase catalyzes this reaction, and NAD^+ is reduced to NADH and H^+.

In the metabolism of plants and some microorganisms, the preceding cycle is shortened and isocitrate may go directly to succinate and glyoxalate. Glyoxalate combines with acetyl-CoA to form malate. Steps in the cycle vary in different organisms, but the pattern given is a general representation and is involved in the metabolism of numerous organisms. Macroorganisms, as well as microorganisms, have similar metabolic patterns.

Many steps in the citric acid cycle involve small amounts of energy and are readily reversible. Two steps, however, involve considerable energy and ordinarily proceed in only one direction. These steps involve the formation of citrate from acetyl-CoA and oxalacetate and the formation of succinyl-CoA from alphaketoglutarate. So far, very little energy is apparent from the metabolic steps involved in this cycle, but an important source of energy has been generated provided oxygen is available for a final electron acceptor. Conversion of one molecule of pyruvate, with the addition of three molecules of H_2O, has yielded three CO_2 units and five *hydrogen pairs*. Hydrogen pairs are picked up by coenzyme hydrogen carriers.

FAD Coenzyme is Built Around the Vitamin Riboflavin

The other active form, flavin mononucleotide (FMN) has only a single phosphate esterified onto the riboflavin molecule. The H_2 of reduction and oxidation is attached to or removed from the nitrogens designated by an asterisk. When H_2 is attached in the reduced state, the double bonds disappear. The H_2 picked up by the coenzymes NAD+ and FAD is now passed by a series of carriers, known as the *electron transport chain*, to O_2, and energy to convert ADP to ATP is generated by the reactions involved. Molecular oxygen is the final electron (H_2) acceptor in aerobic respiration, and the O_2 of CO_2, sulfate, or nitrate is the acceptor in the anaerobic process.

Hydrogens of reduced NAD or NADP are passed to oxidized FAD. The mechanism of oxidative phosphorylation is poorly understood, but the generation of considerable energy by the passage of a hydrogen pair from NAD or NADP to FAD and the conversion of one molecule of ADP to ATP in the process have been verified. It has been further established that the passage of a pair of electrons along the transport chain to molecular oxygen generates two additional high energy phosphate bonds. One is generated between FAD and cytochrome c and the other, between cytochrome c and O_2. A number of cytochromes are involved in electron transport, and all contain iron. The H_2 is not transported by the cytochromes, but electrons are carried to O_2 by them. Although many details are omitted, the following general description will present one pattern of ATP generation.

The components of the electron transport chain operate at different potentials, and when electrons move across a large potential, a high energy bond is apparently generated. Thc $NAD^+/NADH + H^+$ system operates at a potential of -0.32 volt and the $FAD/FADH_2$ system, at -0.06 volt. The potential difference is 0.26 volt, and a pair of electrons passing this potential generate energy to convert ADP to ATP. One cytochrome system (b) operates at 0.0 volt and another (c) operates at +0.26 volt. The transport of two electrons from FAD to cytochrome c generates another high energy bond. Cytochrome oxidase next receives the electrons, and they are passed from cytochrome oxidase to molecular oxygen in aerobic respiration. One high energy phosphate is generated by the passage of a pair of electrons from cytochrome to O_2, but it is not clear whether it is generated between cytochrome and cytochrome oxidase or between cytochrome oxidase and O_2. Cytochrome oxidase operates at a potential of +0.53 volt and O_2 at +0.82 volt. In the overall passage of a pair of electrons by way of the carrier chain from NAD+ or NADP+ to molecular oxygen, the potential has changed from -0.32 to +0.82 volt, and three high energy phosphates have been generated; they have changed 3ADP to 3ATP. All energy produced in the electron transport is not transferred into the three high energy phosphate bonds, and the remainder is dissipated as heat or other forms. Again, the student is reminded that this method of oxidative phosphorylation is only one pathway, and organisms vary in the type of cytochromes they contain and in other particulars.

As was indicated earlier, the breakdown of pyruvate in aerobic respiration produces five hydrogen pairs. It will be of interest to compare the amount of energy available to an organism by its ability to utilize O_2 as a final electron acceptor in aerobic respiration and an organic compound like acetaldehyde or pyruvate in fermentation. As an example, let us take an organism that produces lactic acid from glucose fermentatively but can, in the presence of O_2, oxidize the hexose to CO_2. As has been previously stated, each mole of glucose fermented to lactate by the Embden-Meyerhof pathway would produce only two high energy phosphates: 2ADP $\rightarrow$ 2ATP. About one half of the H_2 produced fermentatively would ‘be utilized in the reduction of pyruvate to acetate, and no energy would be involved with the other one half.

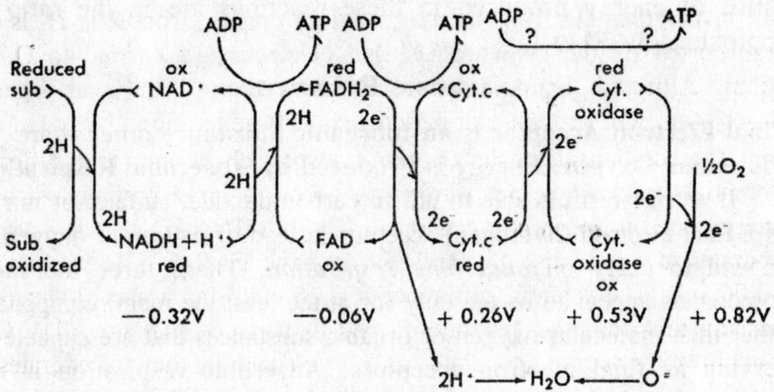

Fig. 6.3. Oxidative phosphorylation by one system, showing production of ATP from ADP and formation of H_2O as an end product.

When pyruvate is oxidized completely, however, the following amounts of energy per mole of glucose are available

1. From pyruvate to acetyl-CoA; NADH + H^+ oxidized (3 ATP)
2. From isocitrate to oxalosuccinate; NADPH + H^+ oxidized (3 ATP)
3. From alpha-ketoglutarate to succinyl-CoA; NADH + H^+ oxidized (3 ATP)
4. From succinyl-CoA to succinate (1 ATP)
5. From succinate to fumarate; $FADH_2$ oxidized (2 ATP)
6. From malate to oxalacetate; NADH + H^+ oxidized (3 ATP)
7. From pyruvate to H_2O + CO_2 (15 ATP)

This amount of energy can be added to the two high energy bonds produced by oxidative phosphorylation of each triose to give a total of 17 ATP for each triose. An additional 3 ATP will come from the oxidation of NADH and H^+, which was reduced in the oxidation of glyceraldehyde phosphate to diphosphoglycerate. This gives a total of 20 ATP for each triose, and because two trioses were formed from, each glucose, there would be 40- for each glucose. Since 2 ATP were utilized in phosphorylating glucose and fructose, there would be a net gain of 38 ATP for each hexose metabolized to H_2O and CO_2.

From these calculations one can see that known energy-yielding steps account for fifteen to twenty times as much energy in aerobic respiration as in fermentation. In overall experiments, however, this factor is too small and complete breakdown of a hexose to CO_2 can be shown to produce about 675 kg calories per mole, and conversion to lactate produces only about 22.5 kg calories per mole. The actual ratios of energy produced in these reactions are in the ratio of approximately 30 to 1.

ANAEROBIC RESPIRATION

Final Electron Acceptor is an Inorganic Substance other than Molecular Oxygen, Energy is Produced by Anaerobic Respiration

If an organism is able to utilize carbon dioxide, sulfate, or nitrate as a final electron (hydrogen) acceptor instead of molecular oxygen, it is said to carry on *anaerobic respiration*. These three inorganic compounds appear to be the only (or at the least the main) compounds other than molecular oxygen or organic substances that are capable of serving as final electron acceptors. Anaerobic respiration is not distinguished from fermentation by some microbiologists because both proceed in the absence of molecular oxygen. Anaerobic respiration, in general, is more limited and individual than fermentation.

It should be emphasized at this point that electrons are transported by means of special cytochrome carriers to their final acceptors in anaerobic respiration. The transport of electrons along special cytochrome chains can produce oxidative phosphorylations by changing ADP to ATP somewhat as in the aerobic system. *The movement of electrons through the transport system is of paramount importance in converting ADP to ATP and thus producing high energy bonds, whether the process occurs in anaerobic metabolism, aerobic metabolism, or photosynthesis.*

Methane bacteria apparently can reduce either CO_2 or certain organic compounds to methane (CH_4). The process is poorly understood, but several compounds may be involved. Strict anaerobic respiration is best seen in the reduction of CO_2 to CH_4, with molecular hydrogen as the reducing agent as follows:

$$4H_2 + CO_2 \longrightarrow CH_4 + H_2O$$

Other hydrocarbons do not appear in this reaction, which indicates that CO_2 is being reduced. The process will not occur unless all traces of molecular oxygen have been eliminated from the system. This reaction is produced by bacteria of four morphological types. Gram-variable *Methanotsarcina* and *Methanococcus* along with two species of *Methanobacterium* were originally identified, and gram-positive rods, *Sarcina* organisms, and organisms from rumens of cattle have been added to the group in later years. These organisms characteristically ferment acetic acid, butyric acid, and alcohols. Although the processes are poorly understood, electron transport from H_2 to CO_2 changes ADP to ATP.

Methyl alcohol also may be metabolized to CH_4. In the overall reaction, three molecules of methanol are apparently reduced to CH_4 for each one oxidized to CO_2. It is not known whether all the CH_4 is first oxidized to CO_2 and then partly reduced to CH_4 or whether some is reduced directly. The overall reaction shows the following results:

$$4CH_3OH \longrightarrow 3CH_4 + CO_2 + 2H_2O$$

Other alcohols may be oxidized partially to their corresponding acids, and CO_2 may be reduced to CH_4, as shown by reactions involving CO_2 with ethanol to produce acetate and CH_4 and with butanol to produce butyrate and CH_4.

$$2CH_3{-}CH_2OH + CO_2 \longrightarrow 2CH_3{-}COOFI + CH_4$$

$$2CH_3{-}CH_2{-}CH_2{-}CH_2OH + CO_2 \longrightarrow 2CH_3{-}CH_2{-}CH_2{-}COOH + CH_4$$

Acetate and benzoate may be fermented to CH_4 in the only known reactions where any compound except CO_2 is utilized by microorganisms in its formation. The CH_4 is derived from the methyl group and CO_2 from the carboxyl group in the early reactions of some experiments utilizing acetate, but later some CO_2 is converted to CH_4. In cases of benzoate fermentation, CH_4 is apparently derived from a carbon beside the carboxyl.

A variation of the conversion of CO_2 to CH_4 yields CH_4 from CO. This process has been shown to proceed in two steps; the first is by the production of CO_2, and the second is by the conversion of CO_2 to CH_4. Certain bacterial strains produce acetate instead of CH_4 as an end product. It appears here that CO_2 and CH_4 combine to give the end products:

$$2CO_2 + 4H_2 \longrightarrow CH_3{-}COOH + 2H_2O$$

The first group of reactions represent true anaerobic respiration because they involve the use of CO_2 as the final electron recipient. Reactions involving acetate and benzoate, however, are not clearly distinguished because both anaerobic respiration and fermentation may be involved.

The production of methane (marsh gas), just described, is a fairly frequent phenomenon. The false firė (ignis fatuus) produced by ignited escaping methane may be seen over swamps at night. There are many stories of travelers who have sought to follow these flickering blue lights, and they account for many ghost stories and eerie reports. An abundance oı methane is produced in sewage decomposition, and it can be utilized commercially when piped from disposal plants.

Electron Acceptor

Organic substrates can be oxidized by O_2 that is combined with sulfur. The O_2 of sulfate serves as the final electron acceptor. This process is carried out by anaerobic sulfate-reducing bacteria. This type of reaction results in the production of hydrogen sulfide (H_2S) and is prevalent around sulfur springs, swamps, and in sewage where organic matter and anaerobic conditions are present; H2S is readily produced. If iron is present, H_2S may combine with it to form the black precipitate ferrous sulfide (FeS), which results in black mud. This type of mud is abundant in many areas of the earth, and the Black Sea has been named for it. Some organisms, capable of sulfate reduction, can utilize hydrogen gas directly in the process. This reduction is carried out by the curved, gram-negative, polar-flagellated *Desulfovibrio:*

$$4H_2 + H_2SO_4 \longrightarrow H_2S + 4H_2O$$

In this process, gas need not be derived from an organic donor, and the final step shown apparently does not need organic substrates because organisms carrying out the process can grow on defined media containing CO_2, H_2, and no organic compounds.

A number of organic compounds may furnish electrons for the reduction of sulfates. Ethyl, propyl, and butyl alcohols, glycols, lactate, glucose, and amino acids have been shown to furnish electrons (H_2) for reduction. Acetate is formed by the oxidation of ethanol as follows:

$$2CH_3\text{—}CH_2OH + H_2SO_4 \longrightarrow 2CH_3\text{--}COON + H_2S + 2H_2O$$

Lactate is decarboxylated and further metabolized to acetate, H_2O, and H_2S. Butyrate is degraded further than ethanol or lactate, with the formation of considerable CO_2 and some acetate.

Oxygen in Nitrate as Electron Acceptor

Some fungi and bacteria can utilize nitrate for final electron acceptors in the absence of molecular oxygen. Because molecular oxygen is utilized when present, however, organisms that are capable of nitrate reduction are facultative. The reduction of nitrate to nitrite, however, utilizes an enormous quantity of substrate for the quantity of energy derived. Nitrite may be reduced further and form ammonia or molecular nitrogen as a final end product, or it may be reoxidized to nitrate by *Nitrobacter*. The reduction of nitrate to molecular nitrogen, by way of nitrite as an intermediate, is known as *denitrification*, and the formation of inorganic nitrogen substances as end products is characteristic of denitrifiers. When the process is carried all the way from nitrate to molecular nitrogen or ammonia, more energy than that derived in the formation of nitrite is available, and the process is more efficient. For the utilization of oxygen of nitrate as an electron (hydrogen) acceptor, a source of combined hydrogen and the absence of molecular oxygen are essential.

Organic substrates usually furnish electrons for this type of reduction, and oxidation of substrates is generally complete. Alcohols, organic acids, and protein have been widely utilized for hydrogen donors in experimental work. Different microbial species may be involved in reactions where different hydrogen sources are supplied. Acetate, for example, is characteristically broken down into carbon dioxide and hydrogen. Carbon dioxide is evolved, and hydrogen combines with nitrates to form nitrogen and water. Glucose is also completely dehydrogenated to carbon dioxide and water by *Pseudomonas aeruginosa*. Pseudomonads are active in denitrification under anaerobic conditions.

If sufficient nitrate is available, anaerobic growth by nitrate-reducing bacteria parallels in some respects that of aerobic growth by other organisms. Actually, in the final analysis, electrons with hydrogen reduce oxygen, but it is not molecular oxygen. The combination of hydrogen with bonded oxygen of nitrate involves nitrate reductase. Enzyme action responsible for nitrate reduction is complicated and requires cytochromes. When the reduction proceeds further than nitrite, a nitrite reductase is involved. The reduction system involves NAD, FAD, cytochrome c, and molybdenum. Electron transport is channeled from cytochrome c through cytochrome oxidase to molecular oxygen, when present. Molybdenum serves to transport electrons between cytochrome c and NO_3^-, whereas cytochrome oxidase would transfer electrons to oxygen. These systems compete in some oxidations. Complete breakdown of organic compounds by aerobic and that by nitrate-reducing systems appear comparable. Electron transport to NO_3^- by way of cytochrome c and molybdenum furnishes considerable energy for nitrate reducers as does transport to oxygen by aerobic forms, although the quantity in nitrate reduction may be smaller than that gained by reduction of molecular oxygen by the same or a comparable organism.

Substrates that are oxidized by molecular oxygen through cytochrome c and cytochrome oxidase should also be oxidizable through cytochrome c and molybdenum to nitrate in the absence of oxygen. As mentioned, many substrates are thus oxidized in nitrate reduction. The presence of nitrate, the absence of oxygen, and proper microbial enzyme systems. are essential for this type of nitrate reduction.

Chemolithotrophic Respiration

Energy is Generated in Chemolithotrophic Organisms by the Transport of Electrons from Exogenous Inorganic Donors

Chemolithothrophic organisms receive electrons from exogenous inorganic donors and derive energy by reducing molecular oxygen with these electrons. As in other cases, electron transport results in the production of high energy phosphate bonds, which are then available for use by the organism. Oxidation-reduction processes in these lithotrophic organisms are poorly understood, but electron sources and end products have been identified in a number of cases. Obtaining electrons appears to be less involved than in cases of organotrophic organisms. Chemolithotrophic organisms, which are in general autotrophic, represent a type of metabolism known nowhere else in nature.

Four groups of chemolithotrophic organisms are usually recognized. Their unusual metabolism stems from the utilization of exogenous inorganic electron donors. Molecular hydrogen, sulfur (or inorganic compounds of sulfur), nitrogen (in ammonia and nitrite forms), and inorganic iron have been recognized as exogenous electron donors. Hydrogen, sulfur, nitrogen, and iron chemolithotrophs have therefore been recognized.

Difficulties in grouping and describing some bacteria that would ordinarily fall in this group are encountered. For example, should organisms be termed *lithotrophic* if molecular hydrogen is utilized but must first be obtained from an organic substance? The answer here depends on whether the organism in question or another organism removes hydrogen from the organic compound. If free molecular hydrogen is utilized, should chemolithotrophy include organisms that can utilize it to reduce carbon dioxide, nitrate, and sulfate? These organisms were described previously under the topic of anaerobic respiration and are usually included in grouping of anaerobic organisms.

A group of bacteria termed *hydrogen bacteria* derive energy from the oxidation of hydrogen by molecular oxygen under certain growth conditions. Hydrogen does not normally appear, however, in appreciable quantities where molecular oxygen is available, but the oxidation of hydrogen by molecular oxygen, when both are available, has nevertheless been demonstrated. Hydrogen is usually made available by metabolic processes carried out by other organisms and may serve the dual role of synthetic reduction of carbon dioxide and exogenous electron donor in chemolithotrophic hydrogen bacteria because these organisms are usually autotrophic.

An oxidizing agent, or electron acceptor, must be provided for the derivation of energy from hydrogen metabolism. In reactions with molecular oxygen as oxidant, the dual role of H_2 in autotrophic metabolism may be illustrated by a two-step process. The first step is as follows:

$$2H_2 + O_2 \rightarrow 2H_2O$$
$$ADP \rightarrow ATP$$

This step produces energy to combine more hydrogen with carbon dioxide to form cell substance as follows:

$$2H_2 + CO_2 \rightarrow (CH_2O) + H_2O$$
$$ATP \rightarrow ADP$$

Cell substance produced by organisms carrying out this type of reaction, however, may be better represented by a polymerized form of beta-hydroxybutyric acid $(C_4H_6O_2)_n$. Other reactions involving the utilization of H_2 are described earlier under anaerobic respiration. In anaerobic respiration, the O_2 of CO_2 serves as the final electron acceptor, and reactions involving combinations of H_2 and CO_2 are best treated there. Nitrate and sulfate may also be involved in anaerobic respiration of hydrogen bacteria.

Hydrogen bacteria are capable of diversified modes of life and constitute a heterogenous group. Both autotrophic and heterotrophic modes of life can be found in the group. Some species can grow either by oxidation of hydrogen by CO_2 or on organic substrates. Different enzyme systems from those utilized in chemolithotrophic metabolism are involved in organic substrate oxidation. Technically, only organisms that utilize available molecular hydrogen as electron sources to reduce molecular oxygen can be considered hydrogen chemolithotrophs.

Colourless sulfur bacteria include both filamentous multicellular forms and unicellular morphological forms. Filamentous forms occur in sulfur springs and other water where sulfur is available. They are frequently present just above water levels in sewage channels. As was stated there, some microbiologists consider filamentous sulfur bacteria to be colourless blue-green algae forms. The extremely large unattached and unsheathed cells of *Beggiatoa alba* characteristically contain sulfur globules. The process of chemosynthetic autotrophy was discovered as a result of experiments with milky white-appearing *Beggiatoa* cultures.

Winogradsky's early experiments (1887) showed that cultures grown in a moist chamber on microscopic slides in an atmosphere of H_2S deposited sulfur globules intracellularly. Sulfate would not promote growth but when H_2S was not present, it accumulated in growth media. Later experiments have shown that *Beggiatoa* organisms oxidize H_2S. The oxidation of H_2S by these forms has been shown to proceed in two steps, although little more than this is known about their oxidation reduction. Phosphorylation occurs at each step. In the first step, electrons (H_2) are removed from H_2S, combine with $\frac{1}{2}O_2$ to produce energy by oxidative phosphorylation, and leave elemental sulfur deposited intracellularly.

$$\overset{ADP \rightarrow ATP}{H_2S + \tfrac{1}{2}O_2 \rightarrow S + H_2O}$$

Further oxidation in step two also produces energy by final oxidation of elemental sulfur by molecular oxygen as follows:

$$\overset{ADP \rightarrow ATP}{2S + 3O_2 + 2H_2O \rightarrow 2H_2SO_4}$$

The autotrophic character of *Beggiatoa* was not established in earlier experiments, but it has been confirmed by a number of workers since 1900. The other colourless sulfur form, *Thiothrix*, resembles *Beggiatoa* in structure, except that it is attached to a central mass at one end with radiating cell filaments. Metabolic and energy relationships of the two genera are also similar.

Unicellular colourless sulfur forms, with few exceptions, consist of small, motile, nonsporeforming, mesophilic, gram-negative rods. These forms do not deposit sulfur granules intracellularly, although all tested members of the genus oxidize H_2S, and some can oxidize thiosulfate or elemental sulfur. In addition to their chemolithotrophic character these organisms are also autotrophic.

The short gram-negative nonsporulating rods that obtain electrons from inorganic sulfur can be grouped into the genus *Thiobacillus*. Four important species will illustrate individual characteristics of these organisms. *T. thioparus* organisms grow best at a pH near neutrality and deposit elemental sulfur outside their cells. Sodium thiosulfate ($Na_2S_2O_3$) is converted into elemental sulfur and sulfuric acid. *T. novellus* organisms also oxidize sodium thiosulfate to sulfuric acid but do not deposit elemental sulfur in the process. *T. thiooxidans* organisms oxidize elemental sulfur to sulfuric acid rapidly and make their growth media very acid. The ability of these organisms to grow at extremely low pH levels has distinguished them from all others. The sulfuric acid they produce may be damaging to concrete and other materials when in areas favourable for their growth. *T. denitrificans* organisms are able to oxidize sulfur to sulfuric acid anaerobically and utilize nitrate as the electron acceptor. Nitrate is reduced to nitrogen gas, but neither nitrogen form can be utilized as a nitrogen source. Ferrous salts are oxidized by *T. ferrooxidans.* Reduced sulfur compounds may be oxidized by heterotrophic bacteria and other fungi, but energy obtained in the process does not appear to be the only energy available to these forms.

Two microbial genera derive energy from the oxidation of nitrogen compounds, but few details concerning the process are available. Ammonia is usually present in a vast array of environments and can be found in varying concentrations in the natural habitats of microorganisms because it is the nitrogen compound formed by the breakdown of protein by a large number of organisms. Ammonia does not accumulate in soil, however, because it is converted to nitrate by microorganisms. The conversion of ammonia to nitrate is carried out in two steps, and two genera of bacteria are involved in the process. The overall process is termed *nitrification.* The conversion of ammonia to nitrite is termed *nitrosification* and is carried out by the genus *Nitrosomonas* as follows:

$$2NH_3 + 3O_2 \rightarrow 2HNO_2 + H_2O$$
$$ADP \rightarrow ATP$$

Four other genera have been named in connection with nitrification, but the action of the *Nitrosomonas* genus is best established, and there is some doubt concerning the metabolism of other genera. The conversion of nitrite to nitrate is called *nitrification* and is carried out by the genus *Nitrobacter* as follows:

$$HNO_2 + \tfrac{1}{2}O_2 \rightarrow HNO_3$$
$$ADP \rightarrow ATP$$

The preceding two reactions, which together result in the conversion of ammonia to nitrate, not only. provide nitrifying organisms with energy to synthesize cell substances from carbon dioxide autotrophically but also furnish essential nitrates for the soil. In. the first reaction, ammonia,is the exogenous electron donor, and in the second, nitrite is oxidized to nitrate.

In early experiments with nitrifying bacteria, experimenters noted that media used for isolation became acid. In later work, acidity was attributed to a change of NH_4^+ (cations) to NO_3^- (anions) and H^+ (hydrogen) ions. In some cultures, oxidation proceeded only to nitrite, and in others ammonia was not converted to nitrite, but nitrite was changed to nitrate. The addition of fresh soil to cultures, however, usually promoted the oxidation of both steps. Isolation of the two bacterial forms involved was difficult and was finally accomplished by Winogradsky, who realized that each oxidation step produced energy and that oxidative energy could be utilized by the organism to reduce carbon dioxide to organic matter. Establishing that chemosynthetic energy was derived from oxidation of inorganic nitrogen compounds greatly facilitated the isolation and characterization of involved organisms.

The group of specific *iron-oxidizing bacteria* are thought to be autotrophic, but doubt exists concerning this point. There is doubtless variation in the obligatory carbon source of many of the included groups. Transformations in nature also take place in iron media by some nonspecific groups, which are not autotrophic. The outstanding feature of the true iron bacteria is deposition of ferric hydroxide around the cell, presenting a typical brown-to-rust colour. Iron incorporated into material surrounding the cells comes from water and not from the corrosion of pipes in which bacteria live in nature. The fouling of pipes, therefore, is from the accumulation of iron from water and not from pipes. The amount of ferric hydroxide in proportion to the cell is large and may reach a volume of several hundred times that of the cell that deposits it. It may be deposited on glass or on other materials beside iron surfaces.

Some specific iron bacteria oxidize ferrous to ferric iron. This process is illustrated by the reaction of the *Ferrobacillus ferrooxidans*, which is a simple rod:

$$4Fe^{++} + 4H^{+} + O_2 \rightarrow 4Fe^{+++} + 2H_2O$$
$$ADP \rightarrow ATP$$

This action proceeds rapidly at low pH (3.0 to 3.6) at 37 C. Other simple unicellular forms are small curved rods, but historically workers mistook excreted materials for the organisms themselves. Iron bacteria are widely distributed in nature in water that contains iron. In sheathed filamentous forms the sheath becomes impregnated with ferric hydroxide and sometimes manganese hydroxide. These organisms may prevent activated sludge from settling in sewage plants. Physiology of this type, represented by *Gallionella*, is unclear. Members of the genus are thought to oxidize ferric carbonate ($FeCO_3$) with the production of carbon dioxide. Organisms with sheaths that become impregnated with ferric and manganese hydroxides usually belong to the Chlamydobacteriales. Members of the genus *Sphaerotilus* produce polarly flagellated single cells at the ends of mature filaments; single cells then grow new filaments.

PHOTOSYNTHESIS

Photosynthesis may be Carried on by Both Procaryotic and Eucaryotic Microorganisms

Although energy for a large number of microorganisms and other forms of life can be derived from one or more of the previously described processes, eventually (actually quite soon) all available compounds would be fermented or oxidized to their end products, and CO_2 would be the final carbon compound. No more substrates would be available from which to derive energy, and no organic compounds of which carbon skeletons are built would be present. For the continuation of life in its present form, an entirely different process for energy must be available. Also, a process for the reformation of CO_2 into organic compounds is essential, and the essential processes are available in *photosynthesis*. Any type of photosynthesis is a metabolic process that converts radiant energy into chemical bond energy, which is generally expressed in the production of ATP from ADP. ATP generated may then be utilized for the synthesis of more cell materials. As the name photosynthesis implies, it denotes synthesis by energy derived from light. To the present world of life, photosynthesis is the most important of all processes, both for obtaining energy and synthesizing cell material. Cell material is usually synthesized from CO_2 by photosynthetic organisms, although we shall see that this is not always the case. The only form of life that seems not to depend on photosynthesis for energy is chemolithotrophic, in which inorganic nitrogen, sulfur, H_2, and iron are utilized as electron donors. This type of life, found only in a few microorganisms, also is usually autotrophic and can obtain carbon skeletons from CO_2. To be self-sustaining, however, a form of life must also possess the power of nitrogen fixation. Chemoautotrophic life would also soon cease in the absence of photosynthesis because all substances that they utilize for energy would soon be used up.

The equation given for photosynthesis gives no indication of the internal working of the system:

$$H_2O + CO_2 \xrightarrow[\text{Chlorophyll}]{\text{Light}} (C_6H_{12}O_6) + E$$

As is well known, one can mix CO_2, H_2O, and extracted chlorophyll and expose the mixture to light and obtain only a carbonic acid-chlorophyll mixture. For the process of photosynthesis to occur a number of conditions must be fulfilled. Photosynthesis usually requires living cells with intact mechanisms for carrying out the process. Some processes have been demonstrated by the use of cell particles, but there must still be organization in the particles. A classical experiment termed the *Hill reaction* shows that chloroplasts from plant leaves, when exposed to light, will evolve O_2, provided iron, quinone, or some other substance is present to act as a hydrogen acceptor. When live cells are absent, however, no CO_2 is changed. The *photo* reaction occurs, but not *synthesis*. The combination of a series of CO_2 molecules to form more complex organic molecules (usually sugars) is referred to as CO_2 fixation, or CO_2 assimilation. It is apparent that live cells are essential for CO_2 fixation but not for the splitting of H_2O and production of O_2. It is also apparent that photosynthesis proceeds in two steps. The first step results in the production of electrons (H_2) for reduction as well as phosphate bond energy. This reaction requires the presence of light. The second reaction involves the reduction of CO_2 by electrons (H_2) and energy produced in the first reaction and does not require light. The second process has been termed the *dark reaction*.

Experiments just described have been carried out with chloroplasts of *eucaryotic cells*. The process of photosynthesis in *procaryotic blue-green algae* seems to follow that found in other algae and higher plants; the same starter substances are essential, and the same end products evolve. The main difference appears to be that blue-green algae do not contain chloroplasts like those in eucaryotic photo-synthesizing cells. There is evidence, however, that small particles found in blue-green algae are similar in function to chloroplasts of eucaryotic plant cells.

Each group of organisms that contains chlorophyll possesses a particular type or combination of types. Each chlorophyll molecule is built around an atom of magnesium, which is enclosed by a tetrapyrrole ring. The tetrapyrrole ring, with the central magnesium atom, is sometimes referred to as the *head of the molecule*, and the chain of linked carbons that is attached to it, as the *tail*. The tail may contain twenty or more carbon atoms, and variations in its structure and minor variations in the head structure account for different chlorophylls. The tails of chlorophylls a and b, for example, consist of phytol structures of twenty carbon atoms, but differences occur in the head in relation to double bonding, and there are other minor variations.

Photolysis of water is apparently related to the presence of a particular type of chlorophyll, called chlorophyll a, which is apparently the only type present in blue-green algae; it is also found in eucaryotic algae and plants. Photosynthetic bacteria do not contain chlorophyll a, and as will be shown later, they do not photolyze H_2O. The process of photosynthesis in microorganisms other than bacteria, as well as that observed in green plants, seems to follow a pattern, although it should be remembered that many facts concerning the process are obscure or completely unknown. Furthermore, variations occur, and only an overall plan or process can be presented.

When molecules of chlorophyll a absorb light, they become *ionized* by the emission of electrons, with resulting positive charges. This ionization of chlorophyll and subsequent neutralization (reduction), combined with photolysis, account for two important phenomena: the generation of chemical energy (ADP → ATP) from light energy excitation, and the reduction of NADP to NADPH and H^+. There are two methods by which these processes occur. Cyclic electron transfer, which results in photophosphorylation, appears to be a process carried out by all photosynthetic forms. Electrons separated from chlorophyll by light energy may return either directly to neutralize positively charged chlorophyll from which they were separated or indirectly by means of one or more electron carriers. Direct return of electrons to chlorophyll results in radiation, which may produce fluorescence or heat, or it may be expressed in some other manner. If chlorophyll is reduced by emitted electrons that are returned by way of.cytochrome carriers, however, energy is recovered in the form of high energy phosphate bonds (ADP → ATP). This type of phosphorylation, produced by cyclic electron return by means of an electron transport system, to

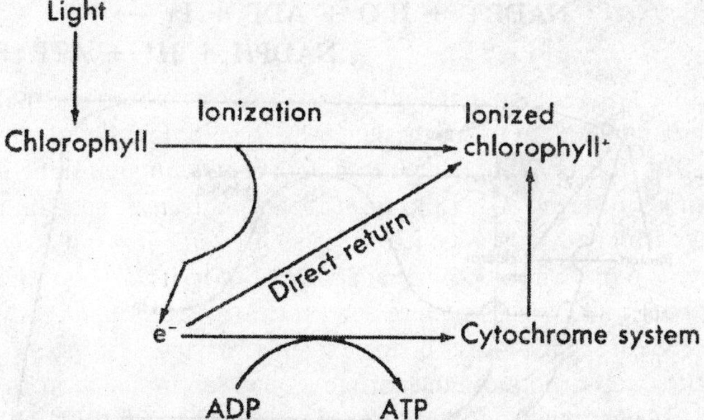

Fig. 6.4. Cyclic photophosphorylation, showing return of electrons directly to oxidized chlorophyll or by way of a cytochrome system.

the chlorophyll from which electrons were removed, is known as cyclic photophosphorylation. By returning to their origin, from which they were separated by light. energy, electrons generate ATP from inorganic phosphate. This type of photophosphorylation appears to be the only common process carried on in all photosynthetic reactions.

Photosynthetic processes in green plants, eucaryotic algae, and blue-green algae seem to follow similar photosynthetic patterns. Several patterns have been proposed, and the following description includes only one hypothesis. Electrons are expelled from chlorophyll at a high energy level by light. Electrons are also removed from the OH^- of H_2O and may neutralize positively charged chlorophyll ions. In this event, electrons that were separated from chlorophyll may reduce NAD nucleotides because chlorophyll has been reduced by other electrons. In this type of reaction, known as *noncyclic electron transfer*, light is responsible for chlorophyll excitation and also directly or indirectly for photolysis (dissociation) of H_2O into a proton (H^+) and an OH^- ion.

The process of chlorophyll neutralization by electrons from hydroxyl ions is not direct but is accomplished through an electron chain similar to that of cyclic photophosphorylation. Accordingly, ADP is changed to ATP by energy resulting from the movement of electrons through the electron (cytochrome) chain. This process is known as *noncyclic photophosphorylation*. Electrons separated from chlorophyll and H_2 ions derived from the photolysis of water, reduce $NADP^+$ to NADPH and H^+. These processes can be diagrammed, although many steps are poorly understood. The sum of these reactions is as follows:

$$NADP+ + H_2O + ADP + Pi \longrightarrow NADPH + H^+ + ATP + ½\ O_2$$

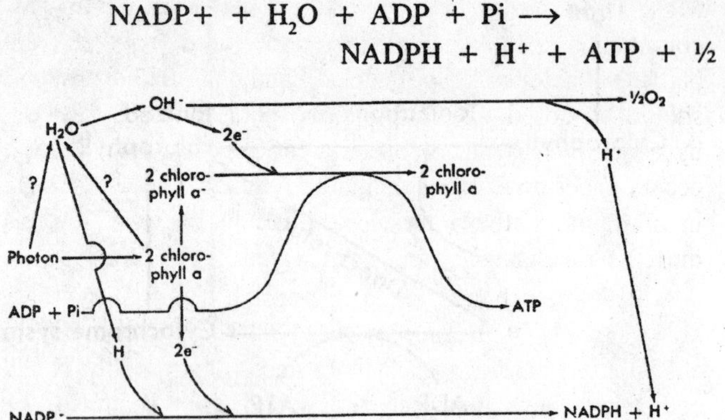

Fig. 6.5. Noncyclic photophosphorylation showing conversion of ADP to ATP and reduction of $NADP^+$.

The transfer of electrons from H_2O to chlorophyll probably generates the majority of energy responsible for the conversion of ADP to ATP, but it is not known whether this is the sole source of phosphate energy.

The photolysis of H_2O is coupled with reduction of chlorophyll by electrons carried through the electron transport system. Electrons from chlorophyll plus H_2 ions from H_2O reduce $NADP^+$ to NADH and H^+. Nucleotide reductants are later utilized in the reduction of CO_2, and the generated ATP is utilized for energy in CO_2 assimilation. Both *noncyclic photophosphorylation* and *coenzyme reduction* are involved in the preceding process.

Noncyclic electron transfer produces NADPH and H^+ and ATP on an equimolar basis. Reduction of CO_2 to cell substance requires more ATP than NADPH, and the additional amount necessary is supplied by cyclic photophosphorylation. The overall equation for the algae and plant type of photosynthesis may be written as follows:

$$CO_2 + 2H_2O \xrightarrow{\text{Light}} (CH_2O)_2 + H_2O + O_2$$

A further point in connection with this equation should be noted. Experiments with heavy isotopes of O_2 have shown that free O_2 evolved is derived from H_2O and that O_2 in cell material comes from CO_2.

Photosynthesis in bacteria differs from that in blue-green and other algae in that water is not utilized as an electron donor, oxygen is not given off in the process, chlorophyll a is absent, photolysis does not occur, and the process is anaerobic. Molecular hydrogen, reduced sulfur compounds, or even organic compounds may serve as electron donors. Most forms are autotrophic and grow well in the absence of organic compounds. Although not fully understood, another major difference between bacterial photosynthesis and that found elsewhere in nature should be noted. Cyclic photophosphorylation is the only mechanism by which photosynthetic bacteria can generate ATP because the process occurs under anaerobic conditions. A suitable reductant must be provided in all photosynthetic processes, and in bacterial photosynthesis this must be a substance other than water.

Structures thought to be responsible for photosynthesis in bacteria are known as *chromatophores*. Isolated chromatophores contain chlorophyll, carotenoids, cytochromes, flavoprotein, NAD, ferredoxin, coenzyme Q, and other protein and lipid particles. The origin of chromatophores is not known, but they possibly arise from cytoplasmic membranes by budding and are discrete particles analogous to particulate organelles in eucaryotic cells. The number present in bacterial cells can be altered by environment; high light intensity will reduce the number present. The main pigments involved in bacterial photosynthesis are *bacterial chlorophylls* and *carotenoids*. Pigments are contained within chromatophores and are layered so that photoenergy can be transferred from one layer to the other. Light absorbed by carotenoids can thus be utilized in photosynthesis by the transfer of absorbed energy to chlorophyll, but carotenoids are not essential for photosynthesis. Carotenoids help utilize a broader spectrum of radiant energy and protect cells from photodynamic destruction. They may also protect chlorophyll from bleaching. Absorption of different wavelengths of light by carotenoids and chlorophylls allows for photosynthetic utilization of a broader spectrum of light by bacteria. Bacterial photosynthesis can occur in the infrared region by utilizing radiation of 7500 to 9200 A in wavelength. Green bacteria, purple bacteria, and green plants respond photosynthetically to different wavelengths of light because of differences in the absorption bands of their different pigments. This difference in pigment allows photosynthetic bacteria to grow slightly below the surface of stagnant waters, where aerobic algae grow on the surface and absorb oxygen. Wavelengths of light that are not absorbed by the surface algae filter through and are utilized by photosynthetic anaerobic bacteria.

Photosynthetic bacteria are grouped under the order Pseudomonadales, and three families are recognized. The family Thiorhodaceae contains the so-called purple sulfur bacteria which are unicellular. Individual cells may be spherical, short rods, vibrioid, long rods, or spirals, and growth may occur in packets, tetrads, clumps, or singly. Although this group is named for the rose colour, combinations of green bacteriochlorophyll and yellow and red carotenoids produce colour variations in masses that may range from blue violet through purple and brownish to deep red. Photoaction or enzyme action on hydrogen sulfide usually produces elemental sulfur, which is stored as globules inside the cell. Thirteen genera are characterized primarily on their cell and aggregate morphologies.

When H_2S is utilized as an electron donor, it is first oxidized to elemental sulfur and later to sulfate if the amount of H_2S is limited. Conversion apparently proceeds in two steps as follows:

$$CO_2 + 2H_2S \xrightarrow{\text{Light}} (CH_2O) + 2H_2O + 2S$$

$$3CO_2 + 2S + 5H_2O \xrightarrow{\text{Light}} 3(CH_2O) + 2H_2SO_4$$

The overall result is eventual oxidation of H_2S to sulfate as follows:

$$2CO_2 + H_2S + 2H_2O \xrightarrow{\text{Light}} 2(CH_2O)_n + H_2SO_4$$

This is a case of oxidation reduction, where H_2S serves as the oxidant and CO_2, as the reductant. Light is not essential for the conversion of H_2S to S or S to $-SO_4$ by bacteria, provided energy is available from another source. The role of photosynthesis in the preceding reactions, therefore, is apparently only that of furnishing energy through cyclic photophosphorylation.

The Chlorobacteriaceae, or green sulfur bacteria, contain chlorophyll, which differs from that contained in either the algae or purple sulfur bacteria. Cells are usually small and may grow singly or in various types of cell masses. Growth is best in high concentrations of H_2S. Sulfur is usually not deposited intracellularly, but the deposition of elemental sulfur outside the cell is characteristic. Genera are characterized by colonial growth or by relationships with other microorganisms. For example, the genus *Chlorobium* consists of free-living forms, which are not united into well-defined colonies and are not associated with other microbes, but the genus *Chlorobacterium* consists of bacteria in aggregates with protozoa, and the genus *Cylindrogloea* consists of green sulfur bacteria, which grow in aggregates around central filamentous forms of unknown classification.

Photosynthesis in these forms appears simplified in the overall process as follows:

$$CO_2 + 2H_2S \xrightarrow{\text{Light}} (CH_2O)_n + H_2O + 2S$$

The difference between this equation and that for blue-green algae photosynthesis is as follows:

$$CO_2 + H_2O \xrightarrow{\text{Light}} (CH_2O)_n + H_2O + O_2$$

This difference appears to be only in the electron donor and the end product formed.

Lack of photolysis in the bacterial process and possible enzyme action on H_2S, however, are differences not shown by equations.

The Athiorhodaceae contain the nonsulfur purple and brown bacteria, which do not contain sulfur granules even in the presence of H_2S. Individual cells are small, and many morphological forms have been described. Members of both the rod-shaped and spherical *Rhodopseudomonas* and the spiral *Rhodospirillum* are gram negative and are motile by polar flagella. Cell masses may show various colourations, depending on the type of pigmentation and quantities of each pigment present. All specimens tested require organic growth factors. A few species may oxidize thiosulfate or sulfide. The line of demarcation between sulfur and nonsulfur forms is not clear-cut, but general differences are evident.

Photosynthesis in Athiorhodaceae apparently consists of a light-catalyzed assimilation of organic compounds. Acetate and butyrate are rapidly incorporated into poly-beta-hydroxybutyrate, which can be utilized in general biosynthesis by the organisms. Photometabolism of succinate, malate, and propionate produces a glycogen-like polysaccharide. An input of reductants is essential for photometabolism of organic substrates, and these essential reductants may be furnished from the tricarboxylic cycle. The production of poly-beta-hydroxybutyrate from acetate competes for H_2 with CO_2 fixation. The two processes are not complementary and probably do not occur as major metabolic patterns in the same organism. Photosynthesis of acetate may be coupled with molecular hydrogen uptake as follows:

$$2CH_3{-}COO^- + H_2 \longrightarrow (C_4H_6O_2) + 2H_2O$$

Reducing power of the photochemical process is unnecessary if enzyme action can remove electrons from reductants at the pyridine nucleotide level. The only role light plays in bacterial photosynthesis where H_2S, H_2, and sulfate are utilized as reductants is probably that of cyclic photophosphorylation because enzyme action can account for some processes previously associated with photosynthesis. There is some question regarding the participation of light energy in changing H_2S, but since nonphotosynthetic organisms also utilize the compound, there appears to be no reason to suppose light activation is essential for its utilization.

In cases where succinate and some other organic compounds act as electron (H_2) donors, however, NAD is probably not directly reduced by enzyme action because electron transport in nucleotides is at a higher potential than that of succinate. Additional energy to raise the succinate potential level is necessary and may be provided directly or indirectly by a light reaction that is independent of cyclic photophosphorylation. A simple primary shift in the electrons of chlorophyll probably occurs. The relationship of chlorophyll excitation to the raising of succinate potential is not understood. The primary electronic shift in chlorophyll is also probably closely associated with electron transfer, but details of the process are unknown. One postulated mechanism for the reduction of CO_2 and one for the reduction of organic compounds. Both systems have the production of ATP from ADP by cyclic

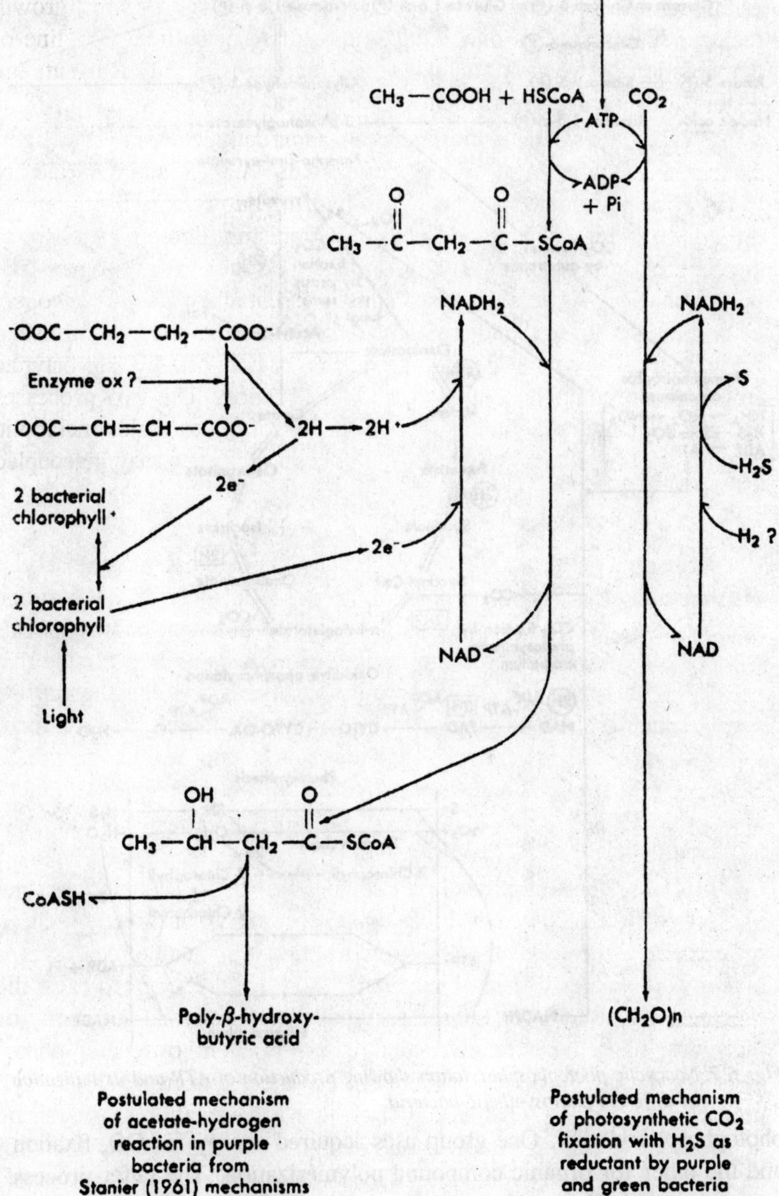

Fig. 6.6. Noncyclic photophosphorylation showing conversion of ADP to ATP and reduction of $NADP^+$.

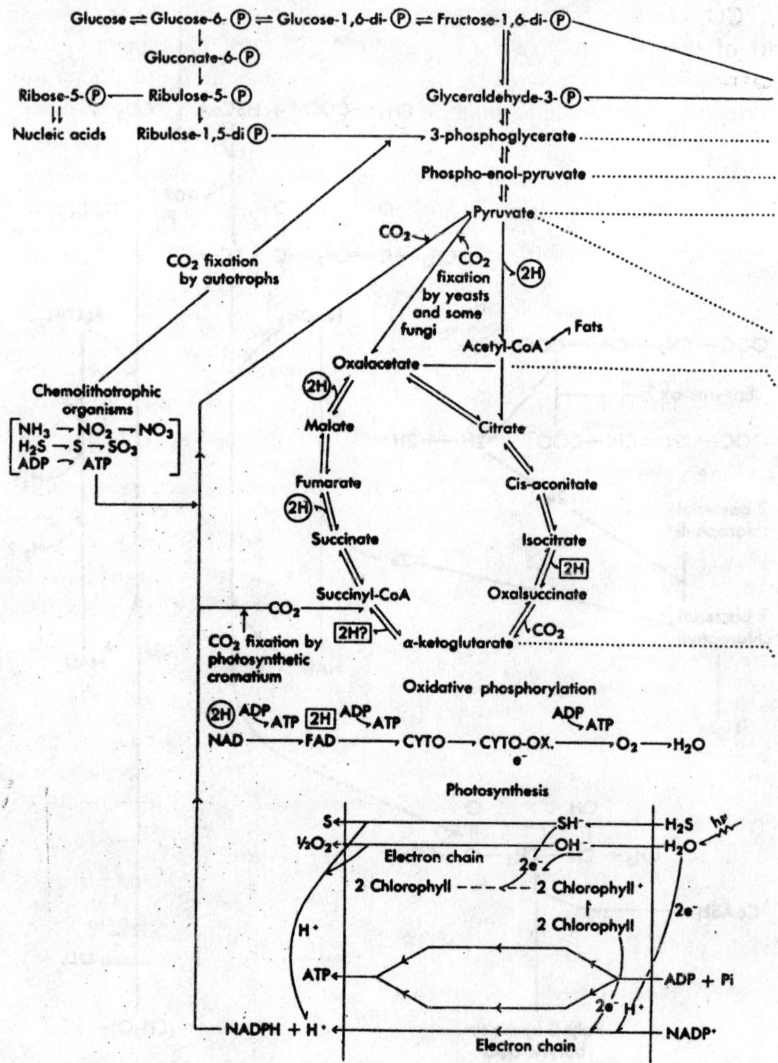

Fig. 6.7. Noncyclic photophosphorylation showing production of ATP and its utilization by different photosynthetic bacteria.

photophosphorylation. One group uses acquired energy for CO_2 fixation and the other for organic compound polymerization. The latter process occurs in nonsulfur purple bacteria, and it also occurs in purple sulfur bacteria under special conditions when they are growing in the presence of organic compounds.

CO_2 fixation by plants and probably by algae usually occurs as part of the pentose phosphate. According to most experiments CO_2 apparently combines with ribulose 1,5-diphosphate to form an intermediate carboxydiphosphopentose, which breaks into two molecules of phosphoglycerate.

An important observation in regard to CO_2 fixation by photosynthetic *Chromatium* has recently been made by Buchanan, Bachofen,

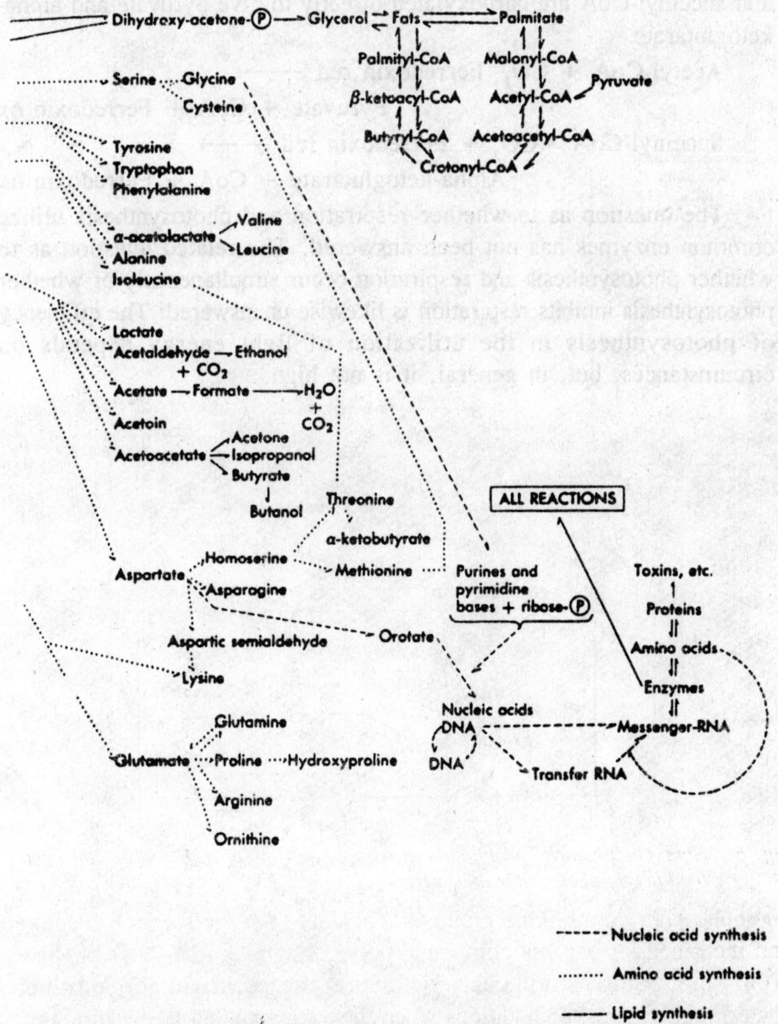

Fig. 6.8. Chart showing relations of some metabolic processes.

and Arnon (1964). This report shows the part played by ferredoxin on CO_2 assimilation by photosynthetic bacteria. The CO_2 fixation process, designated as the pyruvate synthesis system, does not involve the 1,5-diphosphocarboxylase, previously thought to occur in all autotrophic assimilations of CO_2. Ribulose 1,5-diphosphate is not carboxylated by the enzyme to form the six-carbon unit, which breaks to form two molecules of phosphoglycerate. In the ferredoxin system, acetyl-CoA and succinyl-CoA are carboxylated directly to give pyruvate and alpha-ketoglutarate.

Acetyl-CoA + CO_{2+} Ferredoxin red ⟶
Pyruvate + CoA + Ferredoxin ox

Succinyl-CoA + CO_2 + Ferredoxin red ⟶
Alpha-ketoglutarate + CoA + Ferredoxin ox

The question as to whether respiration and photosynthesis utilize common enzymes has not been answered. The related question as to whether photosynthesis and respiration occur simultaneously or whether photosynthesis inhibits respiration is likewise unanswered. The efficiency of photosynthesis in the utilization of light energy depends on circumstances, but, in general, it is not high.

7

BACTERIOPHAGES

Bacteriophages, or phages, have played an important role in the development of molecular biology. At the present time a few phages are the most completely understood of any organisms. Because of their lesser complexity than bacteria and higher cells and the availability of an enormous number of mutants, phages have been extraordinarily useful in the study of basic processes, such as replication, transcription, translation, and regulation. In a short book it is not possible to look into the molecular biology of phages in any depth. Instead, we shall examine some basic features of phage biology and provide a few examples of reproductive strategies by looking at particular stages of the life cycles of certain phages. For more information, consult MB.

A bacteriophage is, first of all, a bacterial parasite. By itself, it can persist, but a phage can neither grow nor replicate except within a bacterial cell. Most phages possess genes encoding a variety of proteins. However, all known phages use the ribosomes, protein-synthesizing factors, amino acids, and energy-generating systems of the host cell.

Each phage must perform some minimal functions for continued survival. These are the following:

1. To protect its nucleic acid from environmental chemicals that could alter the molecule.
2. To deliver its nucleic acid to the inside of bacterium.
3. To convert an infected bacterium to a phage-producing system which yields a large number of progeny phage.
4. To release phage progeny from an infected bacterium.

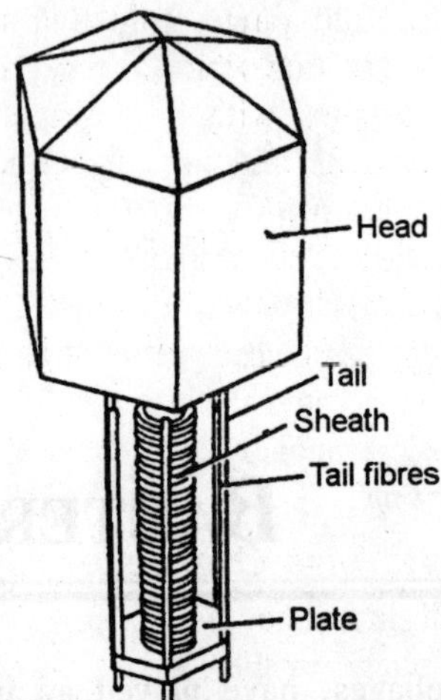

Fig. 7.1. A T-even bacteriophage in "untriggered" form.

These problems are solved in a variety of ways by different phage species.

Phage particles also differ in their physical structures from species to species and often certain features of their life cycles are correlated with their structure. The most common type of nucleic acid in phages is double stranded linear DNA; however, double-stranded circular DNA, single-stranded linear and circular DNA, and single- and double-stranded linear RNA are also found. The molecular weight of the nucleic acid

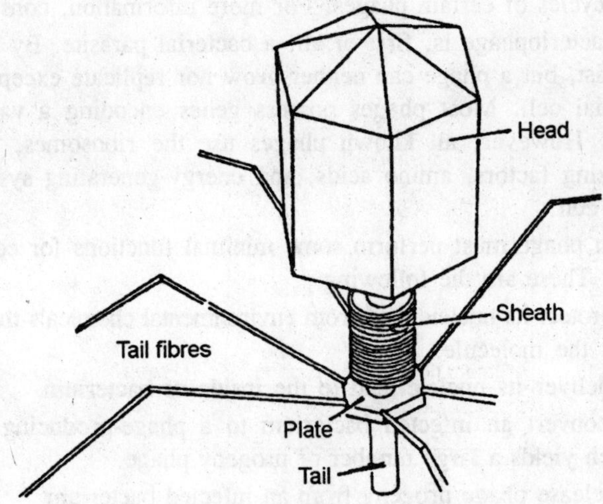

Fig. 7.2. A T-even bacteriophage in "triggered" form.

also varies over a hundredfold range from one species to the next, which contrasts with bacteria, for which the variation is rarely more than about ten percent. Phages with larger nucleic acid molecules have more complex life cycles and are less dependent on bacterial enzymes for their reproduction. An unusual feature of the DNA of some phages is the presence of bases other than the standard A, T, G, C. For example, T4 contains glucosylated 5-hydroxymethylcytosine instead of cytosine and SP01 has hydroxymethyluracil instead of thymine. The nucleic acid is a ways isolated from the environment, and thereby protected from harmful substances by an enclosing protein shell called either the *coat* or the *capsid*.

Stages in the Lytic Life Cycle of a Typical Phage

Phage life cycles fit into two distinct categories—the *lytic* and the *lysogenic* cycles. A phage in the lytic cycle converts an infected cell to a phage factory, and many phage progeny are produced. A phage capable only of lytic growth is called *virulent*. The lysogenic cycle, which has been observed only with phages containing double stranded DNA, is one in which no progeny particles are produced; the phage DNA usually becomes part of the bacterial chromosome. A phage capable of such a life cycle is called *temperate*. In this section only the lytic cycle is outlined. The lysogenic cycle will be described late. There are many variations in the details of the life cycles of different virulent phages. There is, however, what may be called a basic lytic cycle, which is the following.

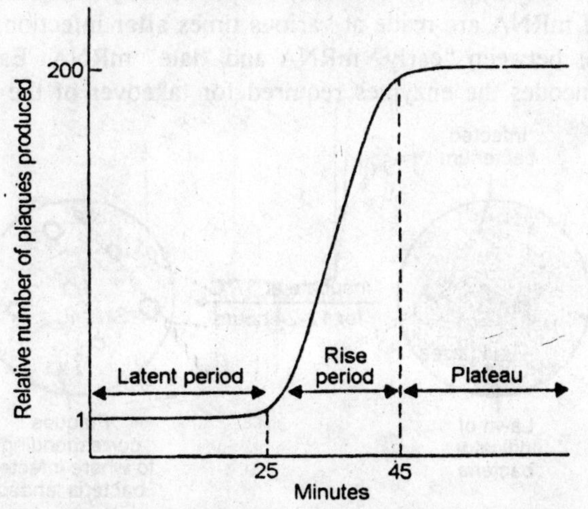

Fig. 7.3. One-step growth curve for phage T4 infection of E. coli.

Adsorption of the Phage to Specific Receptors on the Bacterial Surface

These receptors are very varied and serve the bacteria for purposes other than phage adsorption.

Passage of the DNA from the Phage through the Bacterial Cell Wall

Some types of tailed phages use an injection sequence shown schematically.

Conversion of the Infected Bacterium to a Phage-producing Cell

Following infection by most phages a bacterium loses the ability either to replicate or to transcribe its DNA; sometimes it loses both. This shutdown of host DNA or RNA synthesis is accomplished in many different ways depending on the phage species.

Production of Phage Nucleic Acid and Proteins

By several mechanisms the phage directs the synthesis of a replicative system that specifically makes copies of phage nucleic acid. This programming is accomplished either by synthesis of phage-specific polymerases or by addition of specificity elements to bacterial enzymes. Transcription is almost always initiated by the bacterial RNA polymerase but after the first transcription event either the bacterial polymerase is modified to recognize phage promoters or a phage-specific RNA polymerase is synthesized. Transcription is regulated and phage proteins are synthesized sequentially in time as they are needed. Distinct classes of mRNA are made at various times after infection; the major division is between "early" mRNA and "late" mRNA. Early mRNA usually encodes the enzymes required for takeover of the bacterium,

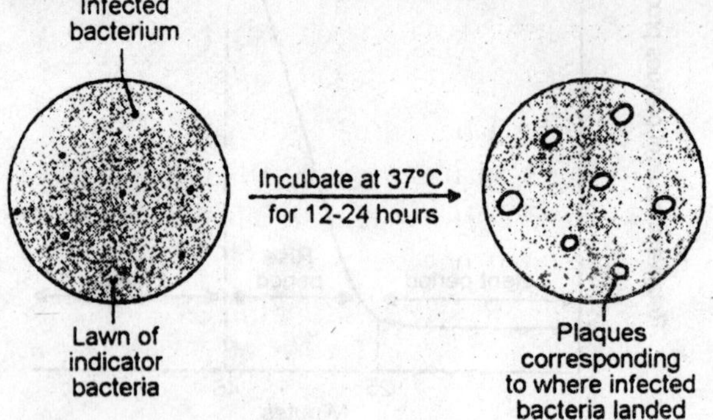

Fig. 7.4. Diagrammatic representation of the production of phage plaques on a bacterial lawn when phage-infected bacteria are spread on it.

for DNA replication, and for the synthesis of late mRNA. Late mRNA encodes the components of the phage particle, proteins needed for packaging of nucleic acid in the particles, and enzymes required to break open the bacterium. The more complex phage usually synthesize several classes of early mRNA, which are made in a particular time sequence.

Assembly of Phage Particles

This process is often called *morphogenesis*. Two types of proteins are needed for the assembly process; *structural proteins*, which are present in the phage particle, and *catalytic proteins*, which participate in the assembly process but do not become part of the phage particle. A subset of the latter class consists of the maturation proteins, which convert intracellular phage DNA to a form appropriate for packaging in the phage particle. Usually 50-1000 particles are produced, the number depending on the phage species.

Release of Newly Synthesized Phage

With most phages a phage protein called a *lysozyme* or an *endolysin* is synthesized late in the cycle of infection. This protein causes disruption of the cell wall. Other proteins called *membranases* dissolve the cell membrane and together with the lysozyme cause total destruction of the cell, so that phage are released to the surrounding medium. This disruption process is called *lysis*.

The events just described occur in an orderly sequence, as exemplified by the life cycle of *E. coli* phage T4 listed below (times in minutes at 37^{0}C):

- t=0 Phage adsorbs to bacterial cell wall. Injection of phage DNA probably occurs within seconds of adsorption.
- t=1 Synthesis of host DNA, RNA, and protein is totally turned off.
- t=2 Synthesis of first mRNA begins.
- t=3 Degradation of bacterial DNA begins.
- t=5 Phage DNA synthesis is initiated.
- t=9 Synthesis of "late" mRNA begins.
- t=12 completed heads and tails appear.
- t=15 First complete phage particle appears.
- t=22 Lysis of bacteria; release of about 300 progeny phage.

The total duration of the cycle is typical, most phages having a life cycle of 20-60 minutes, which is comparable to the generation

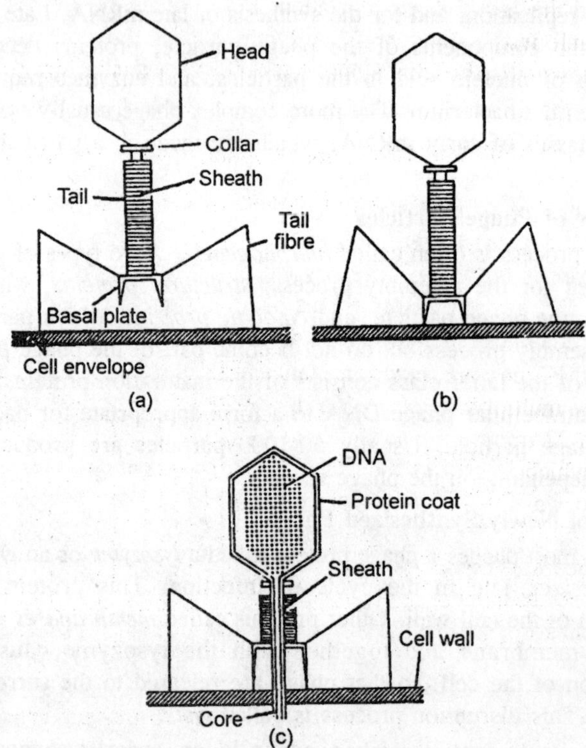

Fig. 7.5. Mechanism of attachment and introduction of DNA into host cell in a T-even bacteriophage.

times of most bacteria. The life cycles of animal viruses are considerably longer—24 to 48 hours—but this is again comparable to the life cycle of an animal cell.

Specific Phages

On a biological scale of complexity, a phage is a relatively simple form of life. However, phages are sufficiently complex that there is not a single phage type for which all of the molecular details of its life cycle are completely understood. The major progress has been with a small number of phage species that grow in *E. coli*, *Bacillus subtilis*, and *Salmonella typhimurium*. Special features of each phage have resulted in particular phages being more suited to the study of certain processes. For example, regulation of transcription is best understood in phages λ and T7, the study of morphogenesis has been more successful with phages T4 and λ than with other phages, phages P1 and P22 gave the first clue to understanding transduction, T5 has yielded the best information about DNA injection, T4 has been most profitable in the study of DNA synthesis, and the study of T4 and T7 has shown clearly how a phage takes over a bacterium.

In the following sections several well-understood features of a few phages will be described.

E. coli phage T4 is a well-studied phage with an unusual DNA that lacks cytosine; instead, there is a modified form of cytosine that is called *5-hydroxymethylcytosine* (HMC), which base-pairs with guanine. This base is further modified by glucosylation—that is, sugar is coupled to the OH group of HMC. This has the consequence that purified T4 DNA is somewhat resistant to a variety of DNases.

HMC has an important function in the T4 life cycle. T4 is a rapidly multiplying phage, producing about 500 progeny per infected cell in 22 minutes. The high rate of T4 DNA synthesis in infected *E. coli*, which is greater than that in an uninfected cell, is accomplished in two ways: (1) the number of enzymes needed to make the DNA precursor nucleoside triphosphates is increased by synthesizing phage-specified enzymes and (2) *E. coli* DNA is degraded by phage-encoded nucleases to mononucleotides, which can be converted to triphosphates. The phages-specified DNases could not distinguish T4 DNA from *E. coli* DNA were it not for the presence of the HMC in the phage DNA—HMC-containing DNA is resistant to the enzymes.

E. coli does not possess enzymes for forming HMC. This synthesis is instead accomplished by two phage enzymes, which convert dCMP to dHDP.

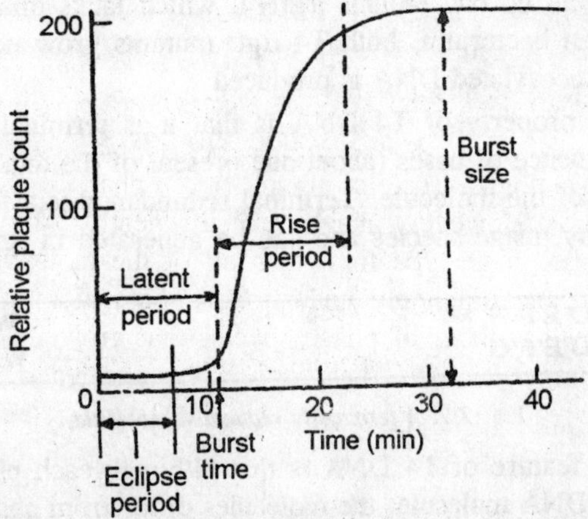

Fig. 7.6. One-step multiplication curve of a T-even phage.

$$\text{dCMP} \xrightarrow{\text{T4 hydroxymethylase}} \text{dHMP}$$

$$\text{dHMP} \xrightarrow{\text{HM kinase}} \text{dHMP}$$

The *E. coli* enzyme, nucleoside phosphate kinase, which forms all triphosphates in *E. coli*, then converts dHDP to dHTP, the immediate precursor of the HMC in the DNA.

When *E. coli* DNA is degraded, dCMP is produced. Whereas most of his is converted to dHMP, as shown above, some will be converted to dCTP, which could be incorporated into progeny T4 DNA. This would yield a T4 DNA molecule that could be attacked by the phage-encoded nucleases that degrade *E. coli* DNA. Thus, a mechanism exists for preventing such incorporation: a phage-specified enzyme, dCTPase, degrades any dCDP and dCTP that is formed, forming dCMP again. Ultimately, all of the dCMP is in the form of dHMP.

HMC creates another problem for T4, since *E. coli* possesses an endonuclease that attacks certain sequences of nucleotides containing HMC. To avoid this damage the HMC residues in T4 DNA are glucosylated. This is accomplished by two phage enzymes, α-glucosyl transferase (αgt) and β-glucosyl transferase (βgt), each of which transfers a glucose from uridine diphosphoglucose (UDPG) to HMC that is already in DNA. Thus glucosylation is a postreplicative modification. The *E. coli* endonuclease is inactive against glucosylated DNA; therefore, glucosylation is a protective device. A simple genetic experiment shows that this is the only essential function of glucosylation. A T4 αgt^- mutant cannot carry out glucosylation, so its newly synthesized DNA is destroyed by the *E. coli* HMC endonuclease. However, if an *E. coli* mutant ($rgIB^-$), which lacks this nuclease, is used as a host bacterium, both T4 αgt^- mutants grow normally even though nonglucosylated DNA is produced.

Another property of T4 DNA is that it is terminally redundant that is, a sequence of bases (about one present of the total) is repeated at both ends of the molecule. Terminal redundancy is a feature of the DNA of many phage species and can be generated in several ways.

Fig. 7.7. A terminally redundant molecule.

Another feature of T4 DNA is that although each phage particle contains one DNA molecule, the molecules differ from phage to phage, even if the population was produced by replication of a single phage particle and its progeny. A sample of T4 DNA molecules is *cyclically permuted*. The figure indicates schematically that the termini of the DNA molecules in the population can be found at many different bases within an overall sequence—in reality, probably at any point in the base sequence. Note that cyclic permutation is a property of a phage population, whereas terminal redundancy is a property of an individual phage DNA molecule. With T4 both terminal redundancy and cyclic permutation are a consequence of the mechanism by which DNA is packaged into the phage head as next described.

The mechanism of packaging of T4 DNA in a phage head is not yet elucidated, though the process can be carried out in vitro. The basic problem is how the long strand of DNA is tightly folded so that it will fit in the phage head. It is thought that packaging begins by the attachment of one end of a DNA molecule to a protein contained in the phage head. Then either a condensation protein or a small, very basic molecule (such as a polyamine) induces folding. The best-understood feature of the process is that the molecule is cut from long concatemers. The cuts are not made in unique base sequences in the DNA because, if they were, T4 DNA could not be cyclically permuted. Instead, the cuts are made at positions that are determined by the amount of DNA that can fit in a head. Presumably a free end of the DNA molecule enters the head and this continues until there is no more room; then the concatemer is cut. This is known as the *headful mechanism* and it explains how both terminal redundancy and cyclic permutation arise. The essential point is that the DNA content of a T4 particle is greater than the length of DNA required to encode the T4 proteins. Thus, when cutting a headful form a concatemeric molecule, the final segment of packaged first—that is, the packaged DNA is terminally redundant. The first segment of the second DNA molecule that is packaged is not the same as the first segment of the first phage. Furthermore, since the second phage must also be terminally redundant, a third phage-DNA molecule must begin with still another segment. Thus, the collection of DNA molecules in the phage produced by a single infected bacterium is a cyclically permuted set.

E. coli Phage T7

Phage T7 is a midsized phage. Its DNA has a molecular weight of about 26 x 10^6 and contains equal amounts of adenine, thymine, guanine, and cytosine, and has no unusual bases. It has a terminal redundancy of 160 base pairs but it is not cyclically permuted. The DNA molecule is encased in a head that is attached to a very short tail.

T7 has been of interest for many years because it has a somewhat small number (49-50) of genes, which simplifies analysis of its life cycle. Furthermore, most of the gene products have been identified by gel electrophoresis, 42 have been purified, and many have been sequenced by chemical means. Of great importance is the recent elucidation of the complete sequence of 39, 930 base pairs of T7 DNA; this information allows one to locate, by direct inspection, all of the promoters, termination sites, regulatory sites, spacers, leaders, initiation codons, termination codons, and genes.

A great deal is known about the molecular biology of T7. However, we shall only discuss how it regulates its transcription, and this only in barest essentials. The life cycle of T7 can be divided into three transcriptional stage. Transcription is temporally regulated with transcript I made first. A basic principle of transcription of phage DNA is that the *first transcript must be made by the host RNA polymerase*, as no other enzyme is available; this is true of transcript I. One of the gene products in transcript I is a protein kinase that phosphorylates and thereby inactivates *E. coli* RNA polymerase; this is the first stage in takeover of the bacterium by T7, though the inactivation is not complete. A second protein made from transcript I is T7 RNA polymerase; this enzyme is essential for further transcription since it alone (not *E. coli* RNA polymerase) recognizes the promoters for transcripts II and III. T7 RNA polymerase does not recognize the termination site for transcript I and hence allows polymerization to proceed past this site. Transcript II encodes the replication proteins, the major protein from which the phage head is constructed, and a second inhibitor of *E. coli* RNA polymerase. Synthesis of the latter protein completes the takeover of the bacterium and ensures that no energy is wasted making unnecessary bacterial RNA. A termination site for transcript II cause termination with 90 percent efficiency, meaning that 10 percent of the transcripts proceed on to the end of the DNA molecule. The extended region encodes the tail proteins. Since the tail is very short, less tail protein is needed than head protein; the extension of transcript II in 10 percent of the initiation events maintains the appropriate ratio of head and tail proteins. The lysis proteins are also included in the extended segment, but their concentration is not high enough to cause lysis during the period in which transcript II is being made. The DNA replication proteins are enzymes and hence are needed in very small amounts compared to the head and tail proteins. Therefore, late in the infection there is no need to make these enzymes; hence, at this time transcript III, which encodes all of the structural proteins and the lysis enzymes, but not the replication proteins, is made. Two proteins of T7 are responsible for the reduction in the synthesis of transcript II and the initiation of transcript III; synthesis of a general inhibitor of transcription, and slow injection of T7 DNA. The promoter for transcript II (*pII*) is a very weak promoter compared to that for transcript III (*pIII*). One of the proteins encoded in transcript II causes an overall reduction in all transcription (the mechanism is unknown), when this proteins is made, transcription is initiated only from the strong promoter. However, since T7 RNA polymerase can act at both *pII* and *pIII*, a mechanism is needed to prevent activation of *pIII* in the interval when the DNA replication enzymes must be synthesized. The mechanism is one that has not been observed very often with phages. Most phage species inject their DNA within about 30 seconds of adsorption, whereas with T7 it takes about 10 minutes. Thus, *pIII* is not injected until about seven minutes after infection, which delays synthesis of transcript III.

Let us now summarize how the three transcripts are made in an orderly way. At first, *E. coli* RNA polymerase makes transcript I. An inhibitor, translated from transcript I, prevents further synthesis of transcript I. T7 RNA polymerase, translated from transcript I, enables transcript II to be made. An inhibitor, translated from transcript II, prevents further synthesis of this transcript. Slow injection delays entry of *pIII* into the cell; however, once *pIII* is available, transcript III is made. The net result is early synthesis of the DNA replication proteins and late synthesis of the structural proteins of the phage particle.

E. coli Phage ϕX174

Study of *E. coli* phage ϕX174 has been important for many reasons: it was the first organism to be discovered containing single-stranded DNA as its genetic material and having DNA in a circular form. Its DNA was also the first to be replicated *in vitro* to yield biologically active DNA, and study of its mode of replication led to the hypothesis of rolling-circle replication. The phage is one of the smallest known, contains 11 genes, and has a fairly simple life cycle. Each particle contains one circular, single-stranded DNA molecule consisting of 5386 nucleotides, whose base sequence is known. This strand is known as the (+) strand. The mode of DNA replication of ϕX174 DNA is clearly different from that of double-stranded DNA, and it is that which will concern us here.

The mechanism by which a base sequence is transmitted to daughter molecules in complementary base-pairing; thus, at some stage of the replication cycle of ϕX174 a DNA single strand (a) (-) strand must be synthesized whose sequence is complementary to the (+) strand. There are two ways that such a molecule could be produced: (1) a complementary strand could be synthesized and removed from the (+) strand as it is being made, like RNA transcribed from DNA, or (2) the circular single strand could be converted to a circular double-stranded molecule. Mechanism 2 is the one used by ϕX174, though some phages containing linear single-stranded DNA use the first mechanism.

Following infection of *E. coli* with ϕX174 and entry of the DNA in the cell, the first stage of replication occurs: conversion of the phage DNA molecule to a double-stranded supercoil by the *E. coli* replication system. Conversion occurs in four steps: synthesis of an RNA primer, extension of the primer by DNA polymerase III until replication of the circle is complete, removal of the RNA nucleotides and replacement with DNA nucleotides by polymerase I, and final sealing by DNA ligase. This supercoiled intermediate is known as RFI (replicating form I). In the second stage of replication RFI replicates many times to produce about 60 circular double-stranded molecules per cell. The second stage of ϕX174 replication—synthesis of (+) strands for packaging in the phage cost—then occurs.

Table 7.1. Replication cycle of ϕX174

Stage	*Time, minutes at 33°C*	*Event*
ss→RF	0-1	Adsorption and penetration; viral single strand converted to parental RF, RF transcribed
RF→RF	1-20	Parental RF replicates and 60 progeny RF are formed.
	25	Replication of RF molecules and host DNA stops
RF→ss	20-30	About 35 rolling circles form from progeny RF; from these, about 500 ss molecules are produced: DNA is inserted into phage particles
	40	Lysis occurs

A variant of the rolling circle mode, called *looped rolling circle replication*, is used to generate a progeny single-stranded circle from a double-stranded circular template. For phage ϕX174 this is accomplished in the following way. A phage protein called the gene-A protein makes a nick at the origin and in the process becomes covalently linked to the newly formed 5'-P terminus. Using the *E. coli* proteins—Rep, ssb, and polymerase III-chain growth occurs from the 3'-OH group, displacing the broken parental strand (which we call the (+) strand). This strand becomes coated with ssb protein and is not replicated. Leading strand synthesis continues until the origin is reached. At this point, the *A*-gene product binds to the 3'-OH group of the (+) strand and, using the energy obtained from the original nicking event, it joins the 3'-OH and 5'-P groups of the (+) strand, dissociates, and attaches to the newly synthesized (+) strand. This process can continue indefinitely, generating numerous circular (+) strands. Note that in looped rolling-circle replication, the displaced strand never exceeds the length of the circle, in contrast with ordinary σ replication. A summary of the replication cycle, with the timing of each stage. All transcription of ϕX174 DNA is from the (-) strand and hence does not occur until the first RFI is made. Therefore, synthesis of RFI must utilize host enzymes only, as is the case.

E. coli Phage λ: the Lytic Cycle

E. coli phage λ may be the best understood of the double-stranded DNA phages, especially with respect to the elegant way in which transcription is regulated. The study of λ has made many important contributions to molecular biology, such as the discovery bidirectional replication, transcription termination and the Rho protein, antitermination proteins, DNA ligase, DNA gyrase, site-specific recombination, and SOS repair, to name just a few.

λ is of particular interest because it can engage in both a lytic cycle, in which progeny phage are produced, and a lysogenic cycle, in which phage DNA becomes incorporated into the bacterial chromosome. Special features of the regulatory systems of λ, not found with all phages, determine the pathway that is used in any given infection. In this section, we will discuss significant features of the lytic lysogenic cycle is described in the next section.

The DNA of λ has two components: a double-strand containing 48,489 nucleotide pairs of known sequence a nucleotide single-strand extensions at the 5'-P terminus of. The two terminal single strands have complementary base and are called *cohesive ends*. Immediately following injected DNA the cohesive ends base-pair, yielding a circle adjacent 5'-P and 3'-OH groups are quickly sealed by DNA then the circle is supercircled.

As is the case for many phages, the genes are clustered at a function. For example, the head, tail, replication, and red genes form four distinct clusters. Many λ proteins—form regulatory proteins and those responsible for DNA synthesis particular sites in the DNA. In general these proteins are adjacent to their sites of action (when there is a single instance, the origin of DNA replication lies within the coding for gene O, which encodes a DNA replicating-initiation protein.

Transcription of λ occurs in several stages, with genes allowing synthesis of related genes at the appropriate time T7, λ makes an early mRNA, which encodes regulatory proteins and DNA replication enzymes, and a late mRNA, which encodes the structural proteins of the phage particle. In contrast with T7, *E. coli* RNA polymerase is used for all transcription events. With most phages (and those discussed so far) transcription is regulated by controlling the availability of particular promoters. For example, T7 early mRNA is made by *E. coli* RNA polymerase, which does not recognize the late promoters, and T7 RNA polymerase functions at the late, but not the early, promoter. Other phages—for example, T4—use the host polymerase throughout the life cycle but modify the enzyme by the addition of accessory molecules that enable it to recognize other promoters. Control of promoter availability is used in the lysogenic cycle of λ, but in the lytic cycle a totally different system is used. With λ all promoters are recognized by the host enzyme, but modification of the enzyme determines whether particular transcription-*termination* sites are recognized. That is, unmodified RNA polymerase initiates at a promoter and stops at a particular termination site, but the modified enzyme-passes the termination site and stops at a later site. This mechanism produces a temporal delay in the production of particular proteins in the following way. Consider a DNA segment containing the sequence *p*—A—B—t1—C—t2, in which *p* is a promoter, ABC are genes, and t1 and t2 are termination sites. Initially, RNA polymerase will attach to *p* and transcribe A and B. Let us assume that B encodes a regulatory protein that can bind to RNA polymerase so that t1 is ignored. In that case, before the C product can be made, the B product must be made and joined with RNA polymerase, thus, the C product will be synthesized at a later time than the A product.

Figure shows a portion of the genetic map of λ, with three regulatory genes (*cro*, N, Q), three promoters (*p*L, *p*R, *p*R2) and five termination sites (*t*L1, *t*R2, *t*R3, *t*R4). Seven mRNA molecules are also shown as arrows, the L and R mRNA series are transcribed

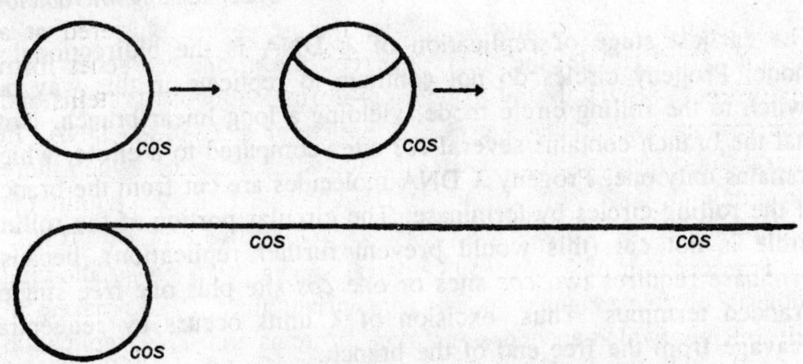

Fig. 7.8. The replication sequence of phage λ.

leftward and rightward, respectively from complementary DNA strands. The arrows indicate the site of action of the regulatory protein. The critical genes whose products modify RNA polymerase are N and Q; their products are called *antiterminators*. Synthesis of each of the mRNAs—L1, R1, and R4—begins at the same time. The terminator tR1 is a weak one, so very little of the O and P products are made Once the N product is translated from L1, it binds to the *nut*L and *nut*R sites on the DNA molecule. During the next passage of RNA polymerase along the DNA, the enzyme picks up the N protein from these sites and is thereby modified. The terminators *t*L1, *t*R1 and *t*R2 are then bypassed, leading to production of L2 (which we will not discuss further) and R3. Translation of R3 yields the DNA replication proteins, allowing replication to begin, and the product of gene Q. The Q protein then binds to the *site qut*, where it is picked up by RNA polymerase, which is thereby modified so that it ignores *t*R4. Extension of R4 yields R5, the late mRNA, from which heads, tails, other structural proteins, and the lysis enzyme are made. Note that the delay of production of the late proteins compared to the early proteins is simply a result of the time required to transcribe and translate the N and Q genes. Details of the regulation of transcription and also of the roles of the genes *cro* and *c*I can be found in MB.

It was mentioned earlier that λ DNA circularizes shortly after infection and that the cohesive ends base-pair to form the double stranded *cos* region. Clearly, at some stage of the life cycle the single-stranded termini must be regenerated, since the DNA is linear within a phage head. This is accomplished by a cutting system, called *Terminase* (or *Ter*), which cuts within the *cos* region of progeny DNA. However, the cuts are not made in progeny circles; instead, DNA replication and cutting are coordinated in a way that is found with many phage species.

The earliest stage of replication of λ DNA is the bidirectional θ mode. Progeny circles do not continue to replicate in this way but switch to the rolling circle mode, yielding a long linear branch. Note that the branch contains several *cos* sites compared to a circle, which contains only one. Progeny λ DNA molecules are cut from the branch of the rolling circles by terminase. The circular portion of the rolling circle is not cut (this would prevent further replication), because terminase requires two *cos* sites or one *cos* site plus one free single-stranded terminus. Thus, excision of λ units occurs by sequential cleavage from the free end of the branch.

E. coli Phage λ: the Lysogenic Life Cycle

There are two types of lysogenic cycles. The most common one, for which *E. coli* phage λ is the prototype, follows:

1. A DNA molecule is injected into a bacterium.
2. After a brief period of transcription, needed to synthesize an integration enzyme, transcription is turned off by a repressor.
3. A phage DNA molecule is inserted into the DNA of the bacterium, forming a *prophage*.
4. The bacterium continues to grow and multiply and the phage genes replicate as part of the bacterial chromosome.

The second and less common type, for which *E. coli* phage P1 is the prototype, differs from the preceding one in that there is no integration system and the phage DNA becomes a plasmid (an independently replicating circular DNA molecule) rather than a segment of the host chromosome.

Two important properties of lysogens are the following: (1) A lysogen cannot be reinfected by a phage of the type that first lysogenized the cell. (2) Even after many cell generations, a lysogen can initiate a lytic cycle, in this process, which is called *inducion*, the phage genes are excised as a single segment of DNA.

The molecular mechanism for immunity and the circumstances that give rise to induction will be described shortly.

The resistance of a λ lysogen to infection by λ is called *immunity*. The cause of the phenomenon is the following. Phage λ contains a repressor-operater system. The repressor gene is called *cI*; the repressor protein binds to two operators *o*L and *o*R, which are adjacent to two promoters *p*L and *p*R. The letter L and R means leftward and rightward and refer to the direction of synthesis of two early mRNA molecules, when the genetic map is drawn in a standard orientation.

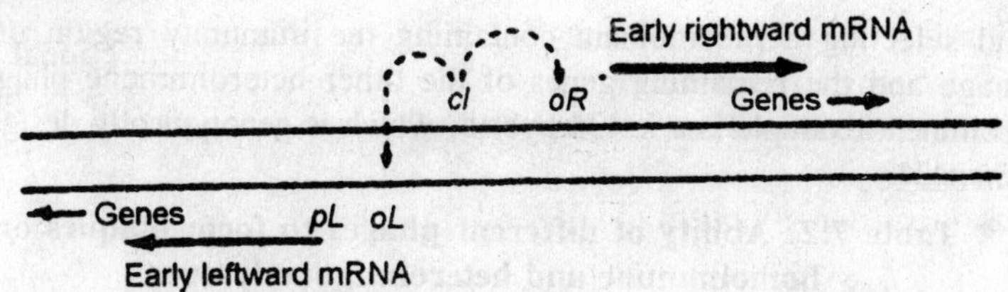

Fig. 7.9. The repressor-operator system of λ, showing the two early mRNA molecules. Symbol cI repressor gene: p, promoter; o, operator; L, left; R, right.

In a lysogen the cI repressor is synthesized continuously and in slight excess with respect to the operators. Both *o*L and *o*R sites contain bound repressor molecules so that *p*L *p*R are unavailable to RNA polymerase. Thus, in a lysogen, transcription from the two early promoters is prevented. It will be seen later that this is sufficient to keep the prophage in an "off" state thus the lysogen grows indefinitely. If a normal cell is infected by λ, the two operators of the incoming λ DNA molecule will be unoccupied, since phage repressor has not yet been made and transcription will occur. However, if a phage tries to infect a lysogen, the excess repressor molecules already present in the lysogen bind to the two operators on the infecting DNA molecule before as RNA polymerase can bind to the *p*L and *p*R sites. This operator-binding prevents to phage from proceeding into lytic development. This inhibition is referred to as *resistance to homoimmune superinfection*.

What happens to superinfecting DNA? It is able to form a supercoil—phage gene products are not needed for supercoiling—but it cannot replicate. The bacterium is unaffected by the presence of this DNA molecule and grows and divides normally, so that the superinfecting DNA is progressively diluted out.

There are many temperate phages of *E. coli* other than λ. Two related to λ are phages 21 and 434. Each of these phages has its own immune system—that is, its own repressor and repressor-specific operators. Thus, the 434 repressor cannot bind to a λ operator and a λ repressor cannot bind to a 434 operator. Such a pair of phages is said to be *heteroimmune* with respect to one another. A temperate phage can form a plaque on a heteroimmune lysogen because the repressor made in the lysogen does not bind to the operator of the superinfecting phage. The immunity region of the DNA, which includes the *cI* gene, operators, and promoters, is denoted by *imm*—specifically, *imm*λ, *imm*21 and *imm*434. Interesting hybrid phages, which have been very useful in the laboratory, have been created by crossing two heteroimmune phages and selecting a recombinant containing the immunity region of one phage and the remaining genes of the other heteroimmune phage. A prominent example is a λ-434 hybrid, which is genotypically designated λ*imm*434.

Table 7.2. Ability of different phages to form plaques on homoimmune and heteroimmune lysogens

Superinfecting phage	*Lysogen* B(λ)	B(434)	B(21)
λ	–	+	+
434	+	–	+
21	+	+	–

Induction is also based on the repressor-operator interaction. If, for any reason, the repressor in a lysogenic cell is inactivated, the λ operators will be free and transcription will begin. In addition to all of the gene products that are made in the lytic cycle and are necessary for phage production, an enzyme called *axcisionase* is made. This enzyme cuts the prophage DNA from the bacterial chromosome and forms a circle; thus the λ DNA is in the same state as in the beginning of a lytic cycle. The molecular mechanism of induction is not well understood. It is known that the repressor is cleaved by a protease, but the events leading to induction are unclear. The initial signal seems to be damage to DNA, in particular, single-stranded DNA produced by the repair of damage. If DNA is damaged, cell death might occur, so it is to the advantage of the prophage to get out of the bacterium. Precisely how the damaged DNA activates the protease that cleaves the repressor is not known, but is an active field of study. Ultraviolet radiation is a powerful inducing agent because of its ability to damage DNA.

In the lysogenic cycle of λ and other phages, a phage DNA molecule is inserted (or integrated, to an equivalent term) into the bacterial chromosome to form a prophage. In order to form lysogen, (1) repression must be established and (2) a prophage must be formed by insertion of a λ DNA molecule.

When λ DNA integrates, it is inserted at a preferred position in the *E. coli* chromosome, this site is between the *gal* operon and the *bio* (biotin) operon and is called the λ attachment site or, genotypically, *att*. For most temperate phages integration occurs at one preferred site, though there are a few phages that either have several preferred sites or appear to be able to insert their DNA anywhere in the *E. coli* chromosome. An important feature of the insertion mechanism was derived from the genetic observation that the gene order in the λ prophage is a permutation of the gene order of the DNA in the phage. This permutation was explained by a model (due to Allen Campbell and widely known as the *campbel model*) for the mechanism of integration. In this correct model, λ DNA is circularized first; and then prophage integration occurs as a physical breakage and rejoining of phage and host DNA—precisely, between the bacterial DNA attachment site and another attachment site in the phage DNA is located near the center of the phage DNA molecule. A phage protein, *integrase* (its gene designation is *int*), recognizes the phage DNA and bacterial DNA attachment sites and catalyzes the physical exchange. This results in integration of the λ DNA molecule in into the bacterial DNA. As a consequence of the circularization and integration, the linear order of the genes in the infecting λ DNA molecule is permuted. A question that arose after this model was proposed was: why doesn't the integrase excise the prophage shortly after integration occurs, since

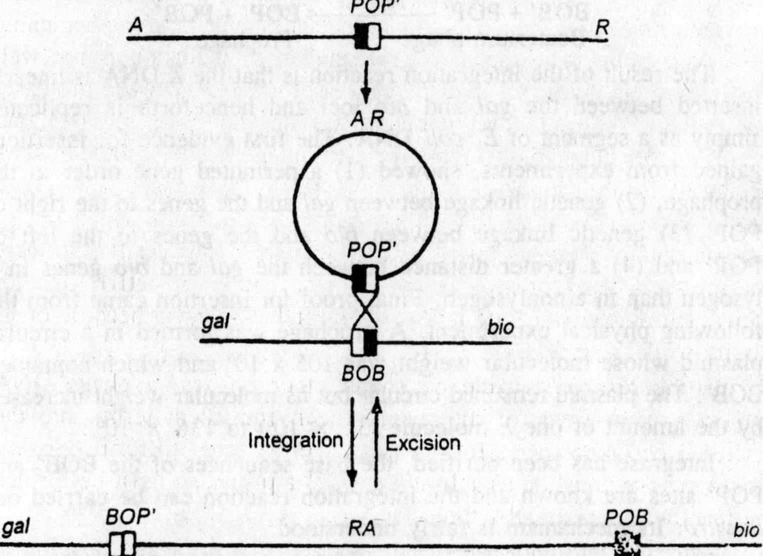

Fig. 7.10. The Campell model for the mechanism of prophage integration and excision of phage λ. The phage attachment site has been denoted POP' in accord with subsequent findings. The bacterial attachment site is BOB. The prophage is flanked by two new attachment sites denoted BOP' and POB'.

the prophage is flanked by two attachment sites? Furthermore, excision would seem to be the preferred reaction since, kinetically, two attachment sites in the same DNA molecule (that is, the host chromosome) ought to interact with each other more rapidly than two attachment sites in different DNA molecules (phage and bacterial DNA). Why this does not happen became clear when it was discovered that, inasmuch as the DNA attachment sites in the phage and the bacterium are not the same, they recombine to form prophage attachment sites that, as a consequence, also differ from the sites joining the bacterial and phage DNA.

All of the λ attachment sites have three different components. One of these is common to all sites and is denoted by the letter O. The phage attachment site is written POP' (P for phage) and the bacterial attachment site is written BOB' (B for bacteria). Thus, in the integration reaction two new attachment sites, BOP' and POB', are generated. There are often written *att*L and *att*R to designate the left and right prophage-attachment sites, respectively. Integrase cannot catalyze a reaction between BOP' and POB', so that the reaction is irreversible if integrase is the only enzyme present.

$$\underset{\text{Bacterium phage}}{\text{BOB}' + \text{POP}'} \xrightarrow{\text{integrase}} \underset{\text{Prophage}}{\text{BOP}' + \text{POB}'}$$

The result of the integration reaction is that the λ DNA is linearly inserted between the *gal* and *bio* loci and henceforth is replicated simply as a segment of *E. coli* DNA. The first evidence for insertion, gained from experiments, showed (1) a permuted gene order in the prophage, (2) genetic linkage between *gal* and the genes to the right of POP' (3) genetic linkage between *bio* and the genes to the left of POP' and (4) a greater distance between the *gal* and *bio* genes in a lysogen than in a nonlysogen. Final proof for insertion came from the following physical experiment. A prophage was formed in a circular plasmid whose molecular weight was 105 x 10^6 and which contained BOB'. The plasmid remained circular but its molecular weight increased by the amount of one λ molecule (31 × 10^6) to 136 × 10^6.

Integrase has been purified, the base sequences of the BOB' and POP' sites are known and the integration reaction can be carried out *in vitro*. Its mechanism is fairly understood.

The synthesis of integrase is coupled to the synthesis of the c. repressor. This is efficient because integrase and the repressor are both needed in the lysogenic cycle and neither is needed in the lytic cycle. In the absence of a positive regulatory element (the product of

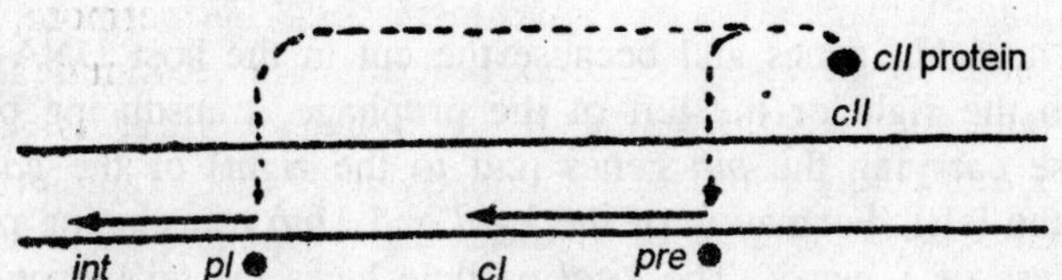

Fig. 7.11. The role of the cII protein in simulating synthesis of cI and int mRNA from pre and pI respectively.

the λ gene *cII*) the promoters for both the *int* and *cI* genes are unavailable to RNA polymerase. Shortly after infection, the *cII* gene product is translated from R3 mRNA. If the concentration of the protein is high enough, the *cII* product binds to sites near the promoters for the *cI* and *int* genes (designated *pre* and *cI*, respectively) and thereby renders them accessible to RNA polymerase. (In this respect the *cII* protein is similar to the cAMP-CAP complex in the *lac* operon). RNA polymerase then transcribes both genes and the gene products are made. Synthesis of the *cII* protein is also the key step in the decision between a lytic and lysogenic pathway in a particular infection; however, details of its role are not yet understood.

Transducing Phages

Some phase species, known as *transducing phases*, are able to package bacterial DNA as well as phage DNA. Particles containing bacterial DNA, called *transducing particles*, are not always formed and usually constitute only a small fraction of a phage population. Such particles can arise by two mechanisms, aberrant excision of a prophage and packaging of bacterial DNA fragments. We begin with a discussion of the first mechanism. One mechanism by which transducing particles form depicts the formation of the galactose and biotin-transducing forms of phage λ namely, λ *gal* λ *bio*.

When λ prophage is induced, an orderly sequence of events ensues in which the prophage DNA is precisely excised from the host DNA. In the case of phage λ this is accomplished by the combined efforts of the *int* and *xis* genes acting on the left and right prophage attachment sites. At a very low frequency—namely, in about one cell per 10^6—10^- cells—an excision error is made: two incorrect cuts are made—one within the prophage and the other cut in the bacterial DNA. The pair of abnormal cuts will not always yield a length of DNA that can fit in a λ phage head—it may be too large or too small. However, it the spacing between the cuts produces a molecule between 79 percent and 106 percent of the length of a normal λ phage-DNA molecule, packaging can occur. Since the prophage is located between the *E.*

coli gal and *bio* genes and because the cut in the host DNA can be either to the right or the left of the prophage, transducing particles can arise carrying the *bio* genes (cut to the right) or the *gal* genes (cut to the left). Formation of the λ*gal* and λ*bio* transducing particles entails less of λ genes. The λ*gal* particle lacks the tail genes, which are located at the right end of the prophage; the λ*bio* particle lacks the *int xis*,... genes from the left end of the prophage. The number of missing phage genes of course depends on the position of the cuts that generated the particle and thus correlates with the amount of bacterial DNA in the particle. The missing phages genes come from the prophage ends, but because of the permutation of the gene order in the prophage and the phage particle, the deleted phage genes are always form the central region of the phage DNA.

Since these transducing particles lack phage genes, they might be expected to be nonviable. This is indeed true of the λ*gal* type, which lack genes essential for synthesis of the phage tail and, in the case of the larger deletion-substitutions, the genes the phage head also. Thus, these particles are incapable of producing progeny; they are defective and this is denoted by the symbol d—thus, a *gal*-transducing particle is written λd*gal* (and sometimes λdg). The λd*gal* particle contains the cohesive ends and all of the information for DNA replication and transcription; it therefore goes through a normal life cycle, including bacterial lysis. In fact, if the head genes are not deleted, the concatemeric branch produced by rolling circle replication is cleaved by the Ter system. However, tails are not added to the filled head, no viable particles are produced in the lysate, and plaques are not formed on a host lawn. Note that there is no discrepancy between the formation of a λd*gal* transducing particle and its ability to reproduce itself. A λd*gal* particle fails to reproduce only because it lacks the tail genes, but such a particle arises from a normal prophage having a full set of genes.

The situation is quite different with λ*bio* particles for these usually lack only nonessential genes—*int*, *xis* etc. These genes are needed for the lysogenic cycle but not for the lytic cycle, so these particles are able to replicate and to form plaques. To denote thus, the letter p, for plaque-forming, is added; this is, the particle is called λphio.

Transducing phages, like λ, able to form transducing particles containing bacterial genes from selected regions of the bacterial chromosome, are called *specialized transducing phage*. Other phage species are able to package fragments of bacterial DNA from all parts of the chromosomes. These are called *generalized transducing phages*. The specialized transducing phages producing transducing particles only come from a lysogen. In contrast, the generalized transducing phages form transducing particles from infected cells. The mechanism by which these particles are produced is straightforward. Some phage species, of which T4 is an example, degrade bacterial DNA early in infection. With T4 the degradation is complete, but in other species, of which *E. coli* phage P1 and *Salmonella typhinurium* phage P22 are examples, fragments of chromosomal DNA comparable in size to phage DNA are present at the time phage DNA is packaged. The packaging system of λ is a site-specific one, requiring *cos* sites. However, some phages, like T4, either merely fill a head, or like P1 and P22, are able to package fragments having a range of sizes. The latter phages cannot distinguish bacterial DNA from phage DNA and occasionally package a piece of bacterial DNA. In a P1 population about one percent of the particles contain only a fragment of bacterial DNA.

Transducing particles are quite useful in genetic experiments. For example, they can be used to transfer bacterial as means of isolating particular segments of a bacterial chromosome.

Genetic Maps in the Bacteriophages

Early maps were constructed by standard complementation and recombination techniques using simple plaque morphology mutants. A larger range of mutants was obtained when conditional lethal mutants were used, e.g., temperature-sensitive mutants and amber mutants. It should be realized that recombination in phages is effectively as population phenomenon, and repeated rounds of replication are accompanied by rounds of recombination. Consequently, multiplicity of infection and the ratio of the two parental phages has a significant effect on recombination. Consequently, multiplicity of infection and the ratio of the two parental phages has a significant effect on recombination values obtained. A theory for phage recombination was proposed by Visconti and Delbruck (1953) which is discussed in detail by Lewin (1977). Recombination data following mixed infection of bacterias built up linkage maps which for many phages were circular, e.g., T4, although some maps were linear, e.g., λ. It must not be assumed that, because the genetic map is circular, the physical map is also circular. T4 molecules have been shown to be linear in phage heads, and the linear molecules in the heads of λ can be made to circularize in the test-tube. The maps of the small bacteriophages of

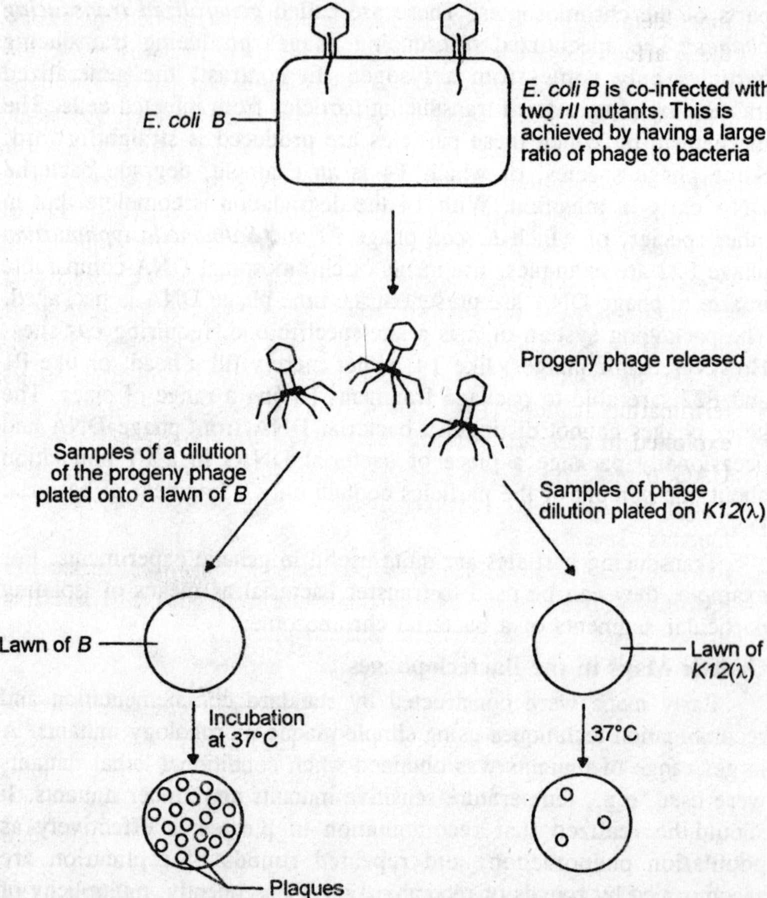

Fig. 7.12. Protocol for determining the number of wild-type recombinants (= one-half the total number of recombinants) produced in a cross of two rII mutants of T4.

ϕX174 (9 genes), M13 (8 genes) and MS2 (4 genes) provide fascinating insights into the minimum number of genes required to maintain an infectious virus particle. MS2 is particularly interesting as it has a single stranded RNA molecule which acts both as a mRNA and as a template for replication. Recombination does not occur and maps have been constructed using other techniques.

Genetic Analysis of T4

Detailed maps have been built up using complementation, recombination analysis, heteroduplex analysis and deletion data. The most detailed maps were constructed using a amber mutations, as these affected a wide variety of proteins. The proteins of T4 can be broadly divided into two groups: the early proteins and the late proteins. The early proteins are those involved in eliminating bacterial functions and replicating phage DNA. An impression of often given that viruses can 'take over' the biosynthetic machinery of the host cell and exploit this for their own replication. In T4, however, at least 24 genes are involved in early functions which range from a phage-coded DNA polymerase to enzymes involved in synthesizing precursors for DNA replication. T4 contains 5-hydroxymethyl cytosine in place of cytosine in its DNA, and an enzyme is synthesized by the phage to destroy dCTP; the phage also synthesizes dCMP hydroxymethylase. Chain terminating triplets UAG, UAA and UGA have been exten-sively exploited in the analysis of bacteriophage T4 and other systems. The UAG triplet has been called an *amber* mutant after Bernstein who discovered it, and by extension UAA mutants are known as *ochre* mutants. Their use depends on host bacterial strains which can suppress the effects of the chain termination. In non-suppressor strains chain termination occurs to give a shortened form of, for example, the head protein subunit. Lysates of the infected bacteria can be seen under the electron microscope to have all the phage components except the phage heads. The normal phage head protein has a serine residue at the site of the amber mutation. In the suppressor strain C there is a mutant tyrosine transfer RNA which has an anticodon CUA instead of the normal GUA. The mutant tRNA can recognize UAG as a codon and tyrosine is inserted at the serine position to give a functional head protein. The use of these mutants has increased the range of mutants which can be studied as in theory any protein, even essential proteins, could be affected. They have also been useful as a safety device in gene cloning vectors to reduce the likelihood of escape of recombinant DNA. Suppressio strains are rare in nature and grow rather poorly, therefore the probability of a recombinant amber vector escaping from the laboratory and finding such a strain is low. Amber mutations have been extensively used to analyze late proteins. Growth of on amber mutation on a non-suppressor strain often resulted in the production of incomplete phage particles which had tails but no heads, or heads but no tails. Analysis of the details of the process has enabled a morphogenetic sequence to be constructed. More than 20 genes have been implicated in the assembly of the head, although only gene 23 actually codes for the major head protein. More than 30 genes are involved in tail assembly, so that this phage has a complex organization. The structure of the head has been implicated in packaging DNA. The basic idea is that the volume of the head dictates the length of the chromosome packaged. This is called the 'headful hypothesis' and it has important genetic implications. Polyhead mutations result in heads which are up to 20 times longer than a normal head, and these heads contain much larger DNA molecules than wild-type phages.

Table 7.3. A selection of enzymes and proteins encoded in the T4 genome.

Gene	*Enzyme or protein*	*Function*
1	DNA-kinase	DNA synthesis
30	Ligase	Joining single-strand breaks in DNA
32	DNA unwinding protein	DNA synthesis
42	dCMP hydroxymethylase	Production of hydroxymethyl cytosine
43	DNA polymerase exonuclease	DNA synthesis
46	Nuclease	Host DNA degradation
56	dCTP and dCDP ases	Degradation of cytosine compounds
e	Lysozyme (endolysin)	Destruction of cell wall

Chromosome Structure of T4

The structure of the T4 chromosome caused considerable confusion for a number of years. The genetic map was clearly circular, and yet the molecules from phages were linear. The first clue towards solving this paradox was the occurrence of heterozygosity at a frequency of about 2% per market. A cross between an r^+ and an *rII* mutant gave about 2% mottled plaques which had properties of both plaque morphologies. Eventually these plaques were shown to derive from two types of phage. One type was a genuine heteroduplex r^+/r, the hybrid DNA discussed earlier, which segregated on DNA replication to give r^+/r^+ and r/r duplexes. The second type was shown to have two sets of the same genes, one at each end of the chromosome. This was described as *terminal redundancy*. The genes present in double dose varied from phage to phage, but each gene would be a duplicate in 2% of the phages. The origin of this terminal redundancy was obscure until it was realized that the chromosome were cut out from a long molecule consisting of repeating unit genomes called *concatenates*.

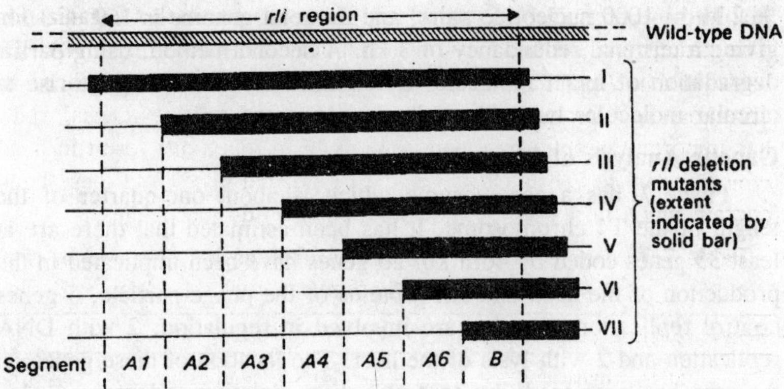

Fig. 7.13. Division of rII region into segments by overlapping deletion mutants.

unit genome

1 2 3 4 5 6 1 2 3 4 5 6 1 2 3 4 5 6

chromosome 1 chromosome 2

The chromosome is longer than the unit genome, and consequently each molecule will contain different duplicated segments. Each molecule is a circular permutation of the gene order. The length of the chromosome is dictated by the volume of the head, and evidence for this is that a deletion of genes from the chromosome increases the length of the terminal redundancy. Concatenates can be generated either by a rolling-circle DNA replication mechanism or by recombination. In T4 they are generated by recombination between two terminally redundant molecules:

1 2 3 4 5 6 1 2 × 1 2 3 4 5 6 1 2 → 1 2 3 4 5 6 1 2 3 4 5 6 1 2

Direct proof of these proposals for chromosome structure has been obtained by examination of DNA molecules in the electron microscope. Two methods were used; firstly, phage DNA molecules were denatured to give single strands and were then allowed to reanneal. A variety of circular molecules resulted, depending on which single strands reannealed. For molecules with different terminally redundant sequences, only one sequence will be able to find a partner, leaving two single strands unpaired. This permits measurements to be made of the unit genome, the circle, and the terminally redundant regions. For T4, Kim and Davidson (1974) have estimated that the unit genome is 166

± 2 kb = 1000 nucleotide pairs, and the total genome is 169 ± 3 kb, giving a terminal redundancy of 3 kb. A second method, using partial degradation of intact molecules by exonuclease III, also gave rise to circular molecules but without the single-strand tails.

Genetic Analysis of Phage Lambda, λ

Phage λ has a chromosome which is about one-quarter of the length of the T4 chromosome. It has been estimated that there are at least 35 genes coded by 46.5 kb. 20 genes have been implicated in the production of the head and tail proteins of the phage particle, 5 genes control replication, 6 genes are involved in regulation, 2 with DNA replication and 2 with lysis of the host. The location of these genes on the map has been made by recombination, deletion and heteroduplex analysis.

The genetic map is highly structured with genes for head formation, tail formation, recombination, regulation, immunity and lysis all being to a large extent grouped in blocks along the chromosome. The genetic map is also linear, and a combination of biochemical, biophysical and genetic techniques has shown that the chromosome is a linear duplex which has single-stranded regions of 12 bases at each end. These single-stranded regions can pair with each other to give rise to a circular molecule which is finally sealed by a ligase. The lambda chromosome has a number of unique properties which have allowed it to be manipulated extensively *in vivo* and *in vitro*. When the molecule is stirred vigorously, it can be broken into two halves which can be separated by their different buoyant densities. The leaf half has 55% G-C pairs and has a greater density than the right half, which has 45% G-C content. Mapping of the genes on these fragments can be made using a marker rescue technique. Helper phages with mutant genes allow the free wild-type DNA to infect the host cell. Wild-type plaques are obtained when the free DNA carries the wild-type allele corresponding to the mutant gene in the helper phage. The single strands of DNA can also be separated by centrifugation into the left strand (*l*) which is transcribed to the left, and the right strand (*r*) which is transcribed to the right.

Integration of the λ Chromosome

When λ DNA is not replicated in the lytic cycle, it is reduced to the prophage state. The model for the formation of the prophage was proposed by Campbell (1962). Figure summarizes the process, which assumes that the phage chromosome circularizes after entering the host cell, and that a cross-over between the two circular molecules results in integration, a process very similar in principle to the insertion of *F* to produce *Hfr* strain. However, the region of homology between the phage and bacterial chromosomes is a sequence of only 15 bases. Apart from these 15 bases, the attachment site *att* on the phage chromosome has a different base sequence from the corresponding attachment site in the bacterial chromosome. Integration of the phage chromosome requires the protein coded by the *int* gene whereas the reverse process of excision is catalyzed by the *int* + *xis* gene products. Excision occurs at low frequency and results in the lytic cycle. This can also be induced by ultraviolet light. The accuracy of this model has been confirmed in detail, and some of the evidence will be mentioned

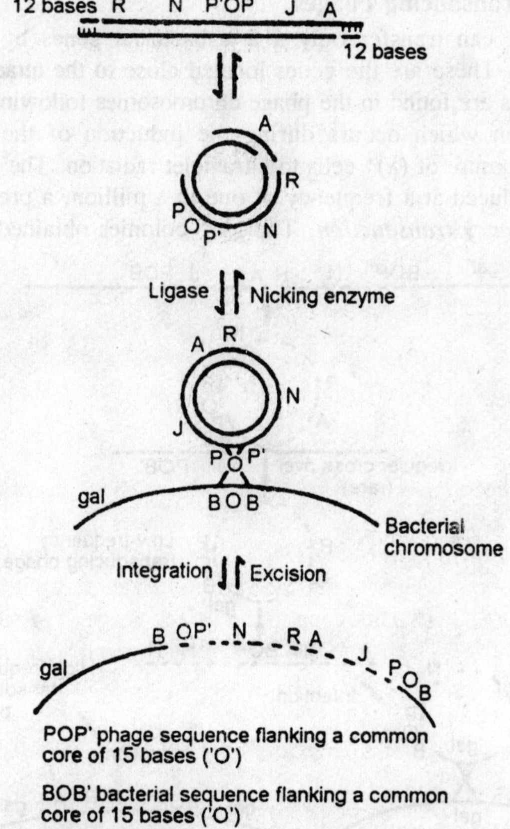

Fig. 7.14. Model for the circularization of the lambda chromosome and its insertion into the bacterial chromosome.

briefly. The prophage behaves as a bacterial gene and can be transferred during conjugation to give free phage by a process called *zygotic induction*. Crosses between two lysogenic bacteria with different phage mutations allow the prophage map to be constructed. The gene order is different from that of the phage map and is exactly as predicted by the model. Deletion mapping has been carried out by using bacterial deletions which extended into the prophage. Finally, heteroduplex analysis of bacterial plasmids carrying the prophage has also confirmed the model. An F-factor containing a λ^+ prophage was used to produce a heteroduplex with λ containing two deletions, *b*2 and *b*5. The position of the deletions relative to the insertion points showed that the prophage markers were a permutation of the linear phage genome.

Origin of Transducing Phages

Phage λ can transfer only a few bacterial genes by specialized transduction. These are the genes located close to the attachment site, *Gal*$^+$ markers are found in the phage chromosomes following inaccurate recombination which occurs during the induction of the lytic cycle following exposure of $(\lambda)^+$ cells to ultraviolet radiation. The *gal*$^+$ marker can be transduced at a frequency of one in a million, a process known as *low-frequency transduction*. The *gal*$^+$ colonies obtained are in fact

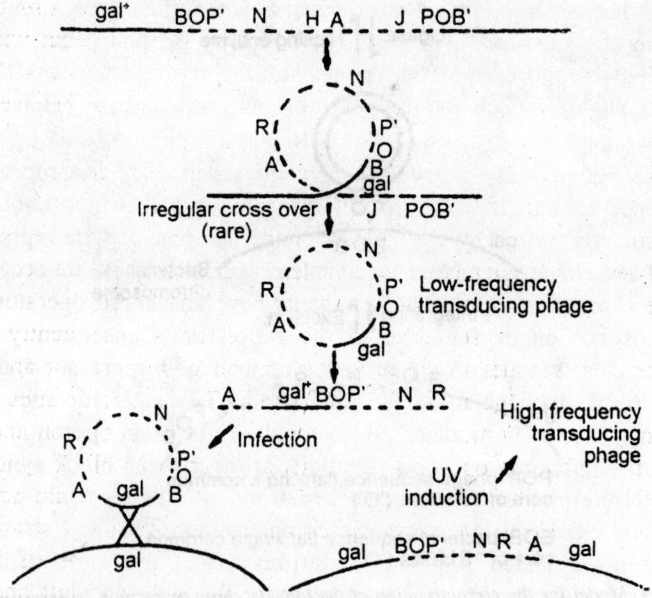

Fig. 7.15. Mechanism of transduction of gal$^+$ *genes by lambda.*

heterozygous and genotypically are *gal*$^-$/λ-*gal*$^+$. When these bacteria re induced with ultraviolet radiation, phage stocks are produced which transduce at a much higher frequency (*high-frequency transduction*). When these lysates are used to infect cells at low multiplicities of infection, only one phage infects a bacterium and some *gal*$^+$ recombinants are immune to infection by further phage but cannot be induced to lyse with ultraviolet light. This can be explained by assuming that the incorporation of the *gal*$^+$ genes has resulted in the loss of essential gene functions necessary for the lytic cycle of the phage. These phages are described as λ*dgal* or λ*dg* for defective *gal* transducing phages. These phages very much resemble *F'* prime factors.

Genetic Basis of Lysogeny

The normal plaques of lambda are turbid due to the establishment of lysogenic bacteria which are immune to further phage infection. Mutant phages can be detected which have *clear plaques*. These have been used to analyze the genetic control of lysogeny. When the clear-plaque mutants were tested on lysogenic and non-lysogenic strains, they were found to fall into two groups. *c* mutants produced clear plaques on the non-lysogenic strain, but failed to lyse the lysogenic strain. The second-class *v* mutants produced clear plaques on both strains. The explanation for the *c* mutants is that they have a mutation affecting the synthesis of a repressor coded by the phage but with an unaltered operator site. The absence of the repressor means that *c* mutants cannot switch themselves off, and consequently behave like virulent phages.

The operator site, however, remains sensitive to the repressor, and hence *c* mutants fail to infect lysogenic strains which contain a repressor synthesized by the resident phage. *cI* codes for the repressor, and *cII* and *cIII*, distinguished by complementation analysis, are necessary for the expression of *cI*. The *v* mutants have an altered operator site which is no longer sensitive to the repressor. Consequently they produce clear plaques in the presence of their own repressor and that of the resident phage in a lysogenic strain. Two operator sites have been identified by *v* mutants, *v*1 *v*3 for the rightwards operon and *sex* *v*2 for the leftwards operon. Mutations at the *x* locus block synthesis of mRNA complementary to the *r* strand of DNA, while *sex v*2 mutations block mRNA synthesis complementary to the *l* strand. It has been suggested that the mutations affect the sites of RNA polymerase attachment, the *promoters*, for the right and left operons respectively.

Gene Expression in the Lytic Cycle

When lambda infects a non-lysogenic cell, no repressor is present and therefore the early genes are transcribed and translated. The leftwards operon (O_LP_L) give the *N* gene product and the rightward operon (P_RO_R) gives the *cro* gene product. Termination signals stop mRNA synthesis immediately after these genes. However, in the presence of the *N* gene product the host RNA polymerase can transcribe past the termination signals, tL1 and tR1, and synthesis continues into *cII*, O, P and Q in the right operon and to *cIII* in the left operon. The N-protein acts in cooperation with the host *nus* proteins at the nut_R and nut_L regions in transcribing past these stop signals, *nus* stands for N-utilization substance and *nut* for N-utilization. It will be seen from figure that, although *t* must be downstream for *nut*, there can be a variable distance between the two regions. The N-protein is short-lived and therefore needs to be synthesized continuously to exert its effect. The fate of the cell then depends on the balance of control of other gene products in the lytic or lysogenic cycles. The Q gene product is also an anti-terminator which acts at the *qut* region in much the same way as the N-protein acts at *nut* regions. In the absence of the protein a short transcript is produced from P'_R due to the presence of a termination region but, when the Q-product is present, transcription occurs through this region to give expression of the late proteins in a cascade fashion.

This results in the synthesis of the tail and head proteins. However, the *cII* and *cIII* products have also been synthesized and these can stimulate the synthesis of the *cI* repressor, resulting in a complete shutdown of both operons. The gene product for *cro*, however, can reduce the synthesis of *cII* and *cIII* activities, thereby preventing *cI* expression. Consequently, the decision between the lytic and the lysogenic pathways depends on the balance between the amount of *cro* and N gene products. Excess *cro* product will prevent repressor synthesis giving the lytic cycle while N synthesis followed by *cII* and *cIII* synthesis will result in repressor synthesis and lysogeny. The *cII* product interacts with the DNA at *Pre* which allows high level synthesis of the *cI* repressor while the *cIII*-gene product protects the *cII*-protein from the proteolytic activity of a bacterial protease coded by the *hfl* locus.

M13 and Mu1: Phage Important in Recombinant DNA Techniques

M13 is a single-stranded DNA filamentous phage related to fd, and Mu is a strange phage which combines the properties of a temperate phage and a transposable genetic element. M13 is useful because it

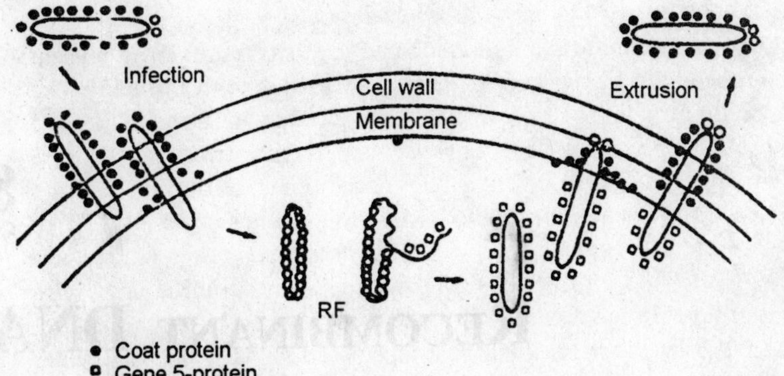

Fig. 7.16. Model for the infection and release of the filamentous single-stranded DNA phage M13.

can be used to clone, amplify and separate single strands of DNA which can then be used for heteroduplex analysis or for base sequencing. There is also a double-stranded replicative intermediate which can be handled with the same techniques as for other DNA molecules. Mu phage has been used to promote mutation by insertion, to produce deletions and inversion, and also to produce fusions between *replicons* (units for replication). M13 is unusual in its mode of infection and release. Protein coded by gene-3 are required for adsorption and release, both of which occur by the filamentous phage passing through wall and the membrane without lysing the cell. The gene 5-protein protects the single-stranded DNA prior to its packaging into the coat protein in the membrane of the cell. M13 produce plaques as infection shows the growth of the host cells giving thinner areas of bacterial lawn.

8

Recombinant DNA Technology

The basic techniques of genetics have relied heavily on spontaneous or induced mutations, followed by standard recombination methods using sexual or asexual processes. There are natural barriers to limit recombination between different organisms, although some of these can be circumvented. Much interest has recently centered around techniques which avoid natural barriers to recombination. A wide variety of techniques has been used, including production of hybrids by cell or protoplast fusion, transfer of genes between different genera of bacteria by promiscuous plasmid, and extracellular manipulation of DNA from widely different sources.

The latter techniques have opened up a completely new range of experiments and possibilities. It is now possible to produce hybrid organisms with genes from bacteria and man, and this has allowed the isolation of the specific base sequences of e.g. human genes, a technique known as *cloning DNA*. New methods are now available for mapping chromosomes, sequencing genes, and preparing genes and gene products in quantities previously thought impossible. The potential applications range from vaccines against malaria to improved crop plants and from insulin to diagnostic probes for 'genetic fingerprinting'. There has been a considerable increase in the patenting of discoveries made in this area and the most notable is the patenting in the United States of America of the basic cloning techniques. This chapter aims to provide an introduction to these techniques with particular emphasis on aspects which are relevant to microbial genetics.

As with many major advances in science, progress was possible only when a number of different lines of research came together. The first was the discovery of very specific enzymes (restriction endonucleases) which could cut DNA in a very specific way. Next methods for purifying and analyzing large quantities of DNA were required and here electrophoresis of supercoiled DNA was important. Further, enzymes were required to link DNA molecules together and to label nucleic acids so that highly specific hybridization could be detected. Finally, delivery systems were required to insert the recombinant DNA into suitable host cells to produce more DNA or to obtain high levels of synthesis of particular proteins. Some of these techniques have been mentioned in earlier chapters but the next few sections will discuss the genetic basis on which these techniques depend. This chapter aims to review the basic techniques and recent progress made.

Basis of Restriction and Modification

In 1953 Luria reported a strange phenomenon in which the plating efficiency of a bacteriophage was seen to depend on the host on which it had been grown. This was nothing more than a genetic oddity until the mechanism of the system was better understood. The inability of certain phages to grow on a particular strain was found to be due to the presence of highly specific endonucleases, nuclease which cut within a polynucleotide. These could be viewed as a bacterial defence system which degraded incoming foreign DNA (*restriction*). The cell protected its own DNA by highly specific methylation of particular bases (*modification*). Consequently the system is often described as a restriction/modification phenomenon. A genetic analysis of the system revealed that there were three genes: ***hsdS*** responsible for site recognition, *hsdR* coding for the restriction endonuclease, and *hsdM* coding for the modifying methylase. Systems like this have been found encoded in a variety of plasmids, phages and bacteria, and they appear to be of very general occurrence.

An examination of a large range of restriction endo-nucleases has revealed a variety of types. Type I endo-nucleases recognize a sequence of 7 nucleotide pairs, but these enzymes cleave the DNA at random. Type II endonucleases recognize a sequence of 4, 5, 6 or more nucleotides, and can make two staggered breaks at various positions or two breaks immediately opposite. The staggered breaks produce single-stranded complementary strands called 'sticky' ends or cohesive tails which permit rejoining of molecules. The enzymes have been named after the bacterium from which they were isolated, and therefore *Eco*RI was obtained from *Escherichia coli*, carrying a resistance transfer factor RI, while *Hind*III was isolated from *Haemophilus influenzae*, strain dIII. At the time of writing about 180 of these enzymes have been isolated, 40 of which have been characterized for the sequence recognized. A further 30 enzymes have been identified but not isolated, and it is very likely that the list will grow. An interesting point about some of the sequences is that they form a *palindrome*, i.e. they have a centre of symmetry and the sequence of bases on one strand is the same as the sequence on the complementary strand when they are read in the same direction of polarity. Thus *Eco*RI recognizes 5'GAÅTTC' with a complementary strand of 3'CTTÅAG5' which read backwards is the same sequence. It has been suggested that this may relate to the recognition of two sites by a dimer enzyme involved in restricting the DNA. The bases marked can be methylated, in which case the DNA is protected from restriction.

Interest in these enzymes increased in 1973 (Cohen et al.), when it was realized that restriction endonucleases could be used to fragment DNA into pieces with complementary single-stranded ends which allowed the fragments to be rejoined. This effectively meant that recombination could be carried out almost at will in the test-tube. Further, there was no limitation on the source of the DNA. Bacterial DNA could be made to recombine with human DNA, opening up the possibility of cloning human genes or isolating human proteins from bacterial cultures.

RESTRICTION ENDONUCLEASES

The Restriction-modification Phenomenon

One reason for the rapid development of recombinant DNA technology was discovery of a variety of enzymes that catalyze the cleavage of DNA at a small number of reproducible sites. These enzymes are called *restriction endonucleases* or *restriction enzymes*. (Recall from previous discussions that an endonuclease results in a cut within a nucleic acid chain.)

In the 1950s, S. Luria and Colleagues found that the progeny phage produced after infection of a particular strain of *E. coli* (B/4) by phage T2 could no longer reproduce in the normal host strain of *E. coli*. They reasoned that the phage released from B/4 had become modified in some way such that the normal host was no longer permissive for growth of T2. This phenomenon was called *host-controlled modification* and *restriction*.

In the 1960s W. Arber and D. Dussoix shed some light on the restriction phenomenon. They grew phage λ on *E. coli K* and used the progeny to infect *E. coli B*, with the result that most of the DNA was broken into pieces; that is, the DNA was *restricted*. Apparently a few DNA strands remained intact since some progeny phages were produced from the injection of strain *B*. These could now infect *E. coli B* normally and about 2% of them could also infect *E. coli K*.

To investigate this in more detail, Arber's group grew phage λ in *E. coli K* that was growing in a culture medium containing the heavy isotopes ^{15}N and ^{2}H (deuterium) so that the DNA of the resulting progeny phage was more dense than normal phage DAN. The heavy phages were then used to infect *E. coli B* growing in normal light medium and the densities of the progeny phage's DNA were determined by cesium chloride density gradient centrifugation. Most of the progeny phages contained normal light DNA, which has been entirely synthesized from new material. These phages could not grow in *E. coli K*. A few of the progeny phage were heavy, and these retained the ability to grow in *E. coli K*. They reasoned, therefore, that phages grown in particular bacterial strains become modified in a way specific for that particular strain. Further, the modification occurred on the DNA itself and involved a covalent bond since it was not lost by passage through the bacterial host. It now appears that in many bacteria the modification involves the addition of methyl groups on certain bases in particular DNA sequences in the genome.

In 1970 a major breakthrough occurred. This was the identification of restriction enzymes, which play a role in the restriction phenomenon. These enzymes recognize a specific nucleotide sequence in DNA and this may then result in cleavage. The way this relates to the modification-restriction phenomenon is as follows. A bacterial cell has two enzymes (or sets of two enzymes, depending on the number of modification of which it is capable) both of which have a recognition site for a particular nucleotide-pair sequence. One of them, the *modification enzyme*, catalyzes the addition of a modifying group (often a methyl group) to one or more nucleotides in the sequence. The other enzyme, the *restriction enzyme*, can only recognize the unmodified nucleotide sequence and hence serves to digest invading DNA unless it too is modified like the host DNA. To date a number of restriction-modification systems have been identified in bacteria. However, not all of the enzymes that are being used in recombinant DNA experiments have been shown to have a role in a restriction modification system in the bacterium from which they have been isolated, even though they are called restriction enzymes.

Properties of Restriction Enzymes

Over 100 restriction enzymes have been isolated so far. All of these enzymes have been found in prokaryotes; no similar enzymes have been identified in the few eukaryotic organisms that have been examined.

The restriction enzymes fall into two classes with regard to the way they cleave DNA. Class I enzymes recognize a specific nucleotide-

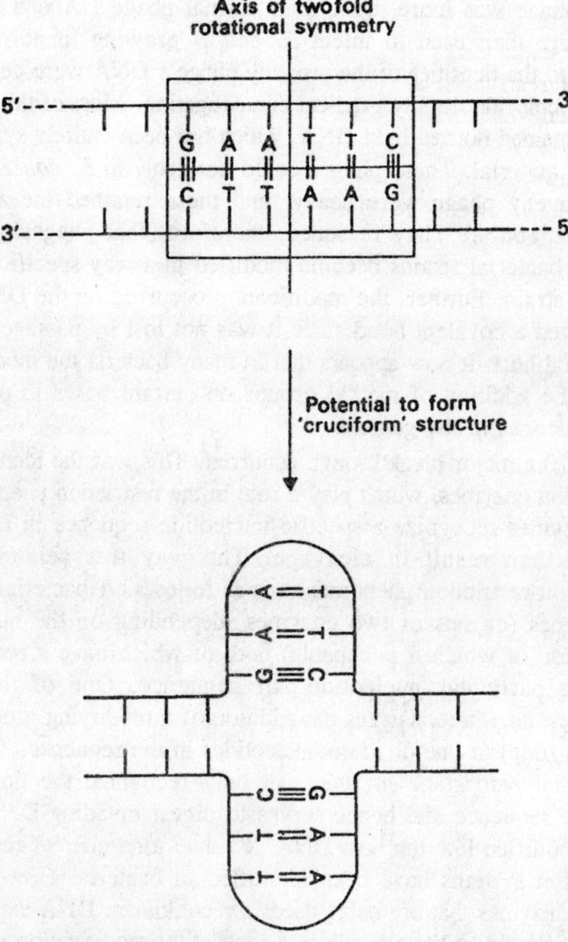

Fig. 8.1. Region of DNA showing two-fold rotational symmetry of nucleotide pair-sequence.

pair sequence and then cleave the DNA at a nonspecific site away from that recognition site. The enzymes involved in the *E. coli K* and *B* restriction system belong to this class. Class II enzymes, on the other hand, cleave the DNA at the specific recognition site. For that reason, and the fact that the recognition sites have two fold rotational symmetry, these enzymes are valuable for constructing recombinant DNA molecules, as we shall see.

Class II restriction endonucleases have been isolated from a large number of microorganisms. All of the enzymes are sequence specific, and thus the number of cuts they make in a particular DNA molecule or population of molecules is dependent upon the number of times the particular sequence is present in the DNA. The cleavage sites for a number of restriction enzymes are shown along with the number of cuts each enzyme makes in the commonly tested DNAs from phage λ, adenovirus-2 (Ad2), and simian virus 40 (SV 40). This information can be used as a guideline for choosing an enzyme for a particular application. As can be seen, some enzymes cut both strands of DNA between adjacent nucleotide pairs, whereas others make staggered cuts in the symmetrical nucleotide pair sequence.

Construction of Recombinant DNA

As briefly mentioned, the new techniques permitted the construction of recombinant DNA molecules. Three basic methods are available for joining DNA fragments. The first method makes use of the complementary single-strand tails, as these automatically ensure rejoining. Ligase treatment is necessary to close up the single strand gaps. The disadvantage of this method is that the fragments will join up at random, and special selective methods have to be adopted to distinguish the recombinant molecules from the original reestablished molecules. The second method can be applied to sheared DNA, as it adds on polyadenine tails to one set of fragments and polythymine tails to the other set. On reannealing, this automatically means that hybrid molecules will result. The third method utilizes a special T4 ligase which can join DNA molecules without single-strand projections (*flush end ligation*).

Basic Procedure Involved in a Recombinant DNA Experiment

Figure summarizes the steps required to clone DNA, from any source, in *E. coli*. The first step is the treatment of a plasmid and target DNA with a restriction enzyme to give fragments with complementary ends. The two sets of fragments are mixed, allowed to anneal and then joined together with a ligase. The mixture of regenerated plasmid, recombinant plasmid and other fragments is used to transform an *E. coli* cell and selection is made for ampicillin resistance. The cells selected can contain either the original plasmid or one in which target DNA has been inserted into the plasmid. A variety of methods are then used to identify the cells containing the required piece of DNA.

Although this is the basic method for cloning DNA, there are many variations which lead to increased efficiency of the technique. The source of DNA fragments can be:

(i) Restriction enzyme treatment, already described
(ii) Mechanical shear
(iii) mRNA used to produce copy DNA (cDNA) which is further used to produce ds-DNA.
(iv) Direct chemical synthesis.

Methods for joining DNA to the vector have already been discussed and the recombinant DNA can then be used to transform or transfect cells directly or it can be packaged *in vitro* into a phage coat. Great ingenuity has been used to recognize the recombinant cells produced. These methods include genetic, immunochemical and hybridization methods. One genetic method involves the insertion of DNA into a tetracycline resistant gene resulting in loss of resistance *insertional inactivation*). Other methods select for the recombinant clones by having, for example, a normal gene for leucine synthesis leu^+ on the vector and *leu*$^-$ in the host organism. Immunochemical methods require an antibody to a protein produced by the recombinant clone. The antibody is tagged with a fluorescent or radioactive label and colonies are treated to see which ones take up the label. A similar principle can be applied to labelling DNA in colonies with a radioactively-labelled DNA or RNA probe. Cells are denatured *in situ* on a plate and then flooded with the radiolabelled probe. Radioactive colonies are then detected using autoradiographic techniques.

Cloning Vehicles

To clone a piece of DNA for study, it must first be attached to a *cloning vehicle* (*vector*). One type of cloning vehicle used for these experiments is *plasmids*. Plasmids are extra-chromosomal genetic elements that replicate autonomously within bacterial cells. Their DNA is circular and double-stranded and they carry the genes required for replication of the plasmid as well as for the other functions that plasmids have. The plasmids most extensively used in recombinant DNA experiments carry antibiotic-resistance genes that play an important role in identification and selection of recombinant DNA molecules. In general the plasmid vehicles have a different buoyant density than that of the host DNA and thus they can easily be purified. Some plasmids have the ability to integrate into the host's chromosome, and these are called *episomes*. The F factor that is involved with conjugation of *E. coli* is an example of an episome. Further, only some of the plasmids are able to promote conjugation between bacteria. Thus to clone a piece of DNA the DNA is spliced into the plasmid DNA, and the chimeric molecule is used to transform a host bacterium such as *E. coli*. The plasmid then replicates within the host as the host grows and divides, and the piece of DNA of interest becomes clones. A plasmid that has been used extensively for molecular cloning is pBR322. This plasmid is of the non-conjugal type; it will not promote conjugation, and each *E. coli* cell transformed with it will have six to eight copies per host chromosome.

pBR322 is a plasmid that has been "engineered" in the laboratory from natural plasmids so that it has features useful for molecular cloning experiments. Its replication in the *E. coli* cell is dependent upon the presence of *rep*, the origin of replication sequence (about "7

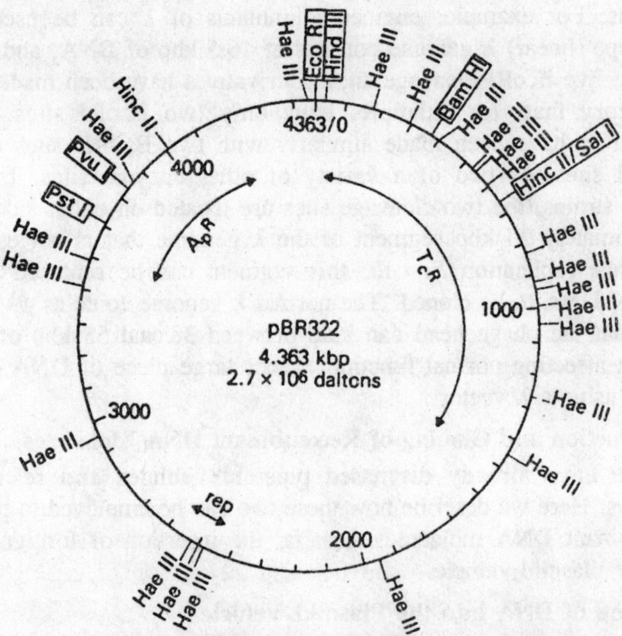

Fig. 8.2. Physical map of the pBR322 cloning vehicle.

o'clock"). pBR322 is 4363 base pairs (4.363 kilobase pairs [kbp]) long, weighing 2.7 x 10^6 daltons. This compares with 3600 kbp (2500 x 10^6 daltons) for the *E. coli* chromosome. pBR322 is cleaved once by any one of the following enzymes: EcoRI, HindIII, BamHI, SalI, PstI, PuvII, PruII, AvaI, and ClaI. The sites for HindIII, BamHI, and SalI all lie within the Tc^R gene (which confers tetracycline resistance upon cells with the plasmid), and the sites for PstI and PvuI are within the Ap^R gene (which confers ampicillin resistance upon cells with the plasmid). Such cleavage produces linear plasmid molecules that are suitable for use in the construction of recombinant DNA molecules as will be described later. None of the aforementioned enzymes cut with the *rep* sequence so that any recombinant DNA molecules generated are capable of replication within the bacterial host. A number of other restriction enzymes cut the plasmid but at two or more sites. These enzymes are not useful in cloning experiments, however, since the plasmid is cleaved into two or more pieces by their action and hence it is not possible simply to insert a piece of "foreign" DNA into the plasmid sequence.

Bacteriophages or their derivatives are also used as cloning vehicles. For example, engineered mutants of λ can be used. The wild-type (linear) λ genome consists of 46.5 kbp of DNA, and within this are five EcoRI cleavage sites. Derivatives have been made in the laboratory that, for example, have only two EcoRI sites. Other derivatives have been made similarly with two BamHI sites or two HindIII sites, or two of a variety of other enzyme sites. In these special strains, the two cleavage sites are located on either side of an approximately 25 kbp segment of the λ genome that is not essential for phage replication *E. coli*; this segment can be replaced with a piece of DNA to be cloned. The normal λ genome contains 49 kbp of DNA but the phage head can hold between 38 and 53 kbp of DNA without affecting normal function. Thus, large piece of DNA can be cloned using a λ vector.

Construction and Cloning of Recombinant DNA Molecules

We have already discussed plasmid vehicles and restriction enzymes. Here we describe how these two can be employed to prepare recombinant DNA molecules, that is, the insertion of foreign DNA into the plasmid vehicle.

Insertion of DNA into the Plasmid Vehicle

It can be seen that a number of restriction endonucleases make *staggered cuts* at specific recognition sites. Thus, for example, foreign

DNA can be cut into segments by a restriction enzyme that also makes one cut in the plasmid vehicle converting the latter to a linear molecule. By the very fact the EcoRI cleaves at a specific site in this way, the single-stranded ends of the linear plasmid vehicle are complementary to the single-stranded ends of the EcoRI-generated segments of foreign DNA. In solution the two DNAs can come together to produce a large circular DNA held together by hydrogen bonding of the complementary ends. In the presence of the enzyme polynucleotide ligase, the single-stranded gaps in the sugar-phosphate backbones are sealed and the structure is stabilized. The result is a recombinant DNA molecule. Since the ends of the DNA pieces produced by EcoRI

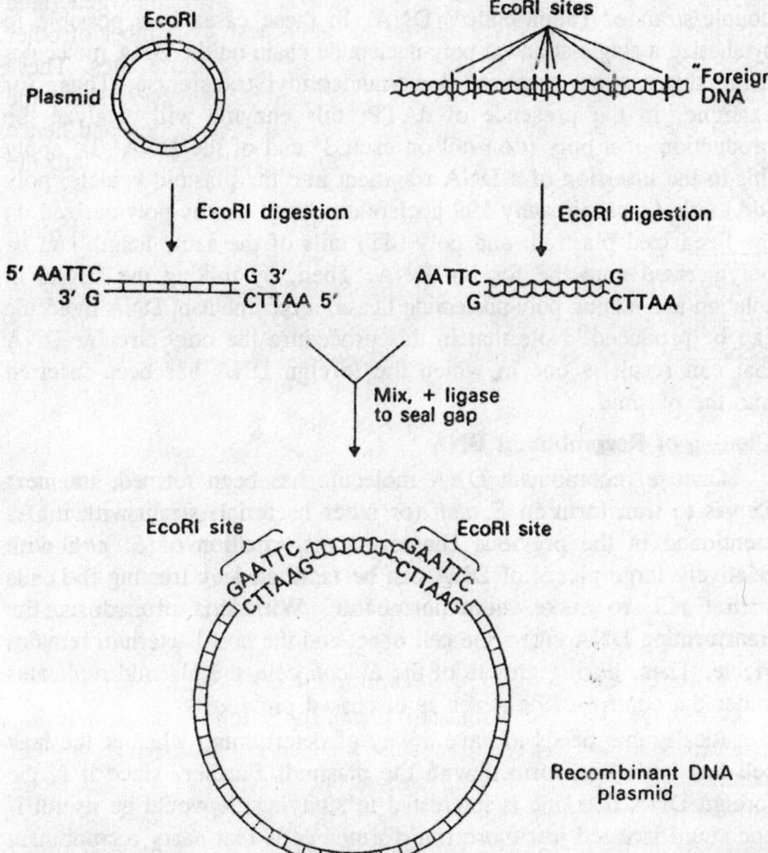

Fig. 8.3. Construction of a recombinant DNA plasmid through the use of the restriction enzyme EcoRI.

digestion are all identical, the foreign DNA can insert into the plasmid vehicle in two orientations, and this will occur in a random way. This orientation may have effects on transcription of the genes or gene fragments on the foreign DNA since the initiation of transcription is likely to depend on the location of promoter and other controlling sites on the vehicle.

Practically speaking, the number of single-stranded nucleotides on the DNA after digestion with a restriction enzyme that makes a staggered cut in small, and thus the probability of the complementary sequences finding one another in solution is relatively small. Further, some restriction enzymes do not make staggered cuts, and some methods for producing the DNA fragment to be cloned result in completely double-stranded (blunt-ended) DNA. In these cases it is possible to synthesize a single-stranded poly-nucleotide chain on the DNA molecules using the enzyme terminal deoxynucleotidyl transferase. Thus, for example, in the presence of dATP, this enzyme will catalyze the production of a poly (dA) tail on each 3' end of the DNA. To apply this to the insertion of a DNA fragment into the plasmid vehicle, poly (dA) tails (approximately 199 nucleotides long) can by polymerized on the linearized plasmid, and poly (dT) tails of the same length can be polymerized onto the foreign DNA. Then, by mixing the DNAs in solution and adding poly-nucleotide ligase, a recombinant DNA molecule can be produced. Note that in this procedure the only circular DNA that can result is one in which the foreign DNA has been inserted into the plasmid.

Cloning of Recombinant DNA

Once a recombinant DNA molecule has been formed, the next step is to transform an *E. coli* (or other bacterial) strain with it. As mentioned in the previous chapter, transformation of *E. coli* with relatively large pieces of DNA can be facilitated by treating the cells with $CaCl_2$ to make them permeable. With this procedure, the transforming DNA enters the cell intact and the host bacterium remains viable. Then, during growth of the *E. coli* cell, the plasmid replicates under the control of its genes as discussed previously.

Ideally one needs to have a way of determining whether the host cell has been transformed with the plasmid. Further, since it is the foreign DNA that one is interested in studying, it would be useful if one could isolated just those transformed cells that carry recombinant plasmids. The presence of a plasmid in an *E. coli* cell can be shown by the fact that it will confer antibiotic resistance on the bacterium

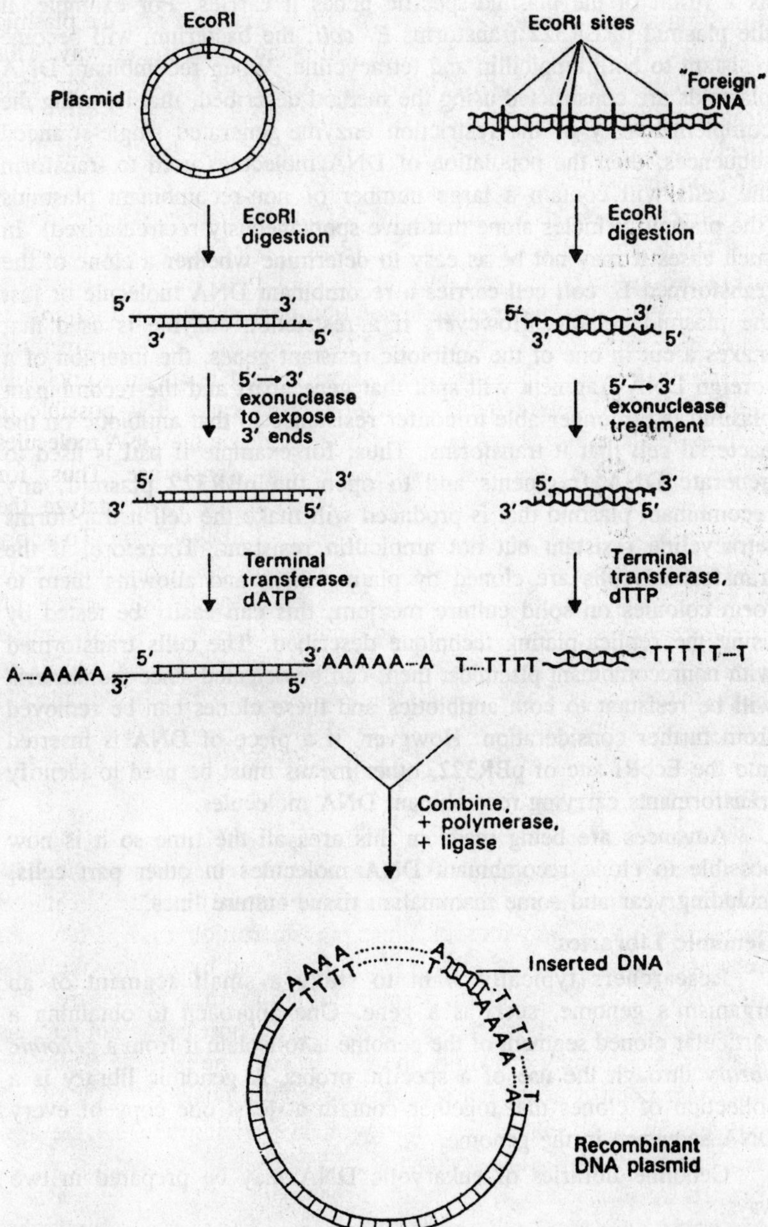

Fig. 8.4. Construction of a recombinant DNA plasmid using the enzyme terminal transferase.

as a result of the plasmid-specific genes it carries. For example, if the plasmid pRBR322 transforms *E. coli*, the bacterium will become resistant to both ampicillin and tetracycline. When recombinant DNA plasmids are constructed using the method described, that is, using the complementarity of the restriction enzyme-generated single-stranded sequences, then the population of DNA molecules used to transform the cells will contain a large number of non-recombinant plasmids (the plasmid vehicles alone that have spontaneously recircularized). In such cases it may not be as easy to determine whether a clone of the transformed E. coli cell carries a recombinant DNA molecule or just the plasmid vehicle. However, if a restriction enzyme is used that makes a cut in one of the antibiotic-resistant genes, the insertion of a foreign DNA fragment will split that gene apart and the recombinant plasmid is no longer able to confer resistance to that antibiotic on the bacterial cell that it transforms. Thus, for example if pstI is used to generate DNA fragments and to open the pBR322 plasmid, any recombinant plasmid that is produced will make the cell it transforms tetracycline resistant but not ampicillin resistant. Therefore, if the transformed cells are cloned by plating them and allowing them to form colonies on solid culture medium, this can easily be tested by using the replica-plating technique described. The cells transformed with nonrecombinant plasmids, then, can be detected since the bacteria will be resistant to both antibiotics and these clones can be removed from further consideration. However, if a piece of DNA is inserted into the EcoRI site of pBR322, other means must be used to identify transformants carrying recombinant DNA molecules.

Advances are being made in this area all the time so it is now possible to clone recombinant DNA molecules in other part cells, including year and some mammalian tissue culture lines.

Genomic Libraries

Researchers typically want to study a small segment of an organism's genome, such as a gene. One approach to obtaining a particular cloned segment of the genome is to isolate it from a *genomic library* through the use of a specific probe. A genomic library is a collection of clones that together contain at least one copy of every DNA sequence in the genome.

Genomic libraries of eukaryotic DNA may be prepared in two ways:

1. Genomic DNA is digested to completion with a restriction enzyme and the fragments are inserted into a suitable vector, usually λ.

One drawback of this method is that if the sequence of interest contains recognition site(s) for the restriction enzyme used, the sequence will be cloned in two or more pieces. Another drawback is that the average size of the fragment produced by digestion of eukaryotic DNA with restriction enzymes that have six base-pair recognition sequences is relatively small (about 4 kbp). Thus, an entire library would of necessity contain a very large number of recombinant phages, and screening by hybridization would be laborious.

2. Both of the problems of the first method can be avoided by cloning large about 20 (kbp) DNA fragments that are generated by random shearing of eukaryotic DNA. This method ensures that sequences are not excluded from the cloned library simply because of the distribution of restriction sites. The procedure is done as follows : High-molecular-weight eukaryotic DNA is fragmented randomly so that a population of molecules is produced with an average size of 20 kbp. Random fragmentation can only be achieved by mechanical shearing, but DNA prepared in this way must be subjected to several additional enzymatic manipulations (e.g. the addition of restriction enzyme sites) so that the fragments can be cloned. So, instead, the method typically used is partial digestion of the DNA with restriction enzymes that recognize frequently occurring four base-pair recognition sequences. Sucrose density gradient centrifugation or agarose gel electrophoresis can then be used to collect fragments of the size desired. The result is a population of overlapping fragments that is close to random and that can be cloned directly, since the ends of the fragments were produced by restriction enzyme digestion. For example, if the DNA is digested with the enzyme sau3A (which produces fragments with ends that are complementary to ends produced by BamHI digestion), the fragments generated can be cloned into a particular λ vector that has a central BamHI fragment which is not essential for phage replication in *E. coli*. This is done by digesting the λ DNA with BamHI, thereby producing a λ left arm, a λ right arm, and the disposable central segment. If the Sau3A digested eukaryotic DNA is mixed with the cut λ DNA, recombinant DNA molecules can be produced that can replicate in *E. coli*; these contain a left arm and a right arm on either side of a partially digested Sau3A fragment.

The aim of the method just described is to produce a library of recombinant molecules that is as complete as possible. However, not all sequences of the eukaryotic genome are equally represented in such a library. For example, if the restriction sites are very far apart or extremely close together in a particular region, the chances of obtaining a fragment of clonable size are small. Additionally, some regions of eukaryotic chromosomes may contain sequences that affect the ability of λ clones containing them to replicate in *E. coli*; these sequences would then be lost from the library. There is also evidence that thc presence of tandemly repeated sequences in the cloned eukaryotic DNA segment can lead to deletion of some of the sequence by recombination during phage propagation in *E. coli*.

Finally the probability of having any DNA sequence represented in the genomic library can be calculated from the formula

$$N = \frac{\ln(1 - P)}{\ln(1 - f)}$$

where N is the necessary number of recombinants P is the probability desired, and f is the fractional proportion, of the genome is a single recombinant *i.e.* f is the average size, in base pairs, of the fragments used to make the library divided by the size of the genome, in base pairs).

Selection of Specific Recombinant DNA Clones

In many instances the genome of an organism is digested with a restriction enzyme and the fragments produced are then cloned by the procedures described. By using different restriction enzymes, the genome can be cut up in a number of ways, and this results in the formation of an extremely large number of recombinant DNA clones. However, in general an investigator wishes to study a clone carrying a particular DNA segment. In some cases the clones or interest can be identified in a relatively easy way.

As an example, we shall discuss the cloning of a piece of the ribosomal DNA repeat unit of *Neurospora crassa*, which contains the 175 rRNA gene *Neurospora* rDNA. As discussed, eukaryotic ribosomes consist of two dissimilar subunits, each containing RNA and proteins. There are four different RNA species in the ribosome, which are named according to their sedimentation in density gradients: 17S, 25S, 5S, and 5.8S for the molecules from *Neurospora* ribosomes. The genes for these RNAs are found in the moderately repetitive DNA since they are present in many copies in the chromosomes. In the case of *Neurospora* there are about 200 copies each of the 17S, 25S, 5, 8S rRNA genes and 100 copies of the 5S gene. The genes for 17, 5.8 and 25S rRNAs are found in a tandem away with the general organization.

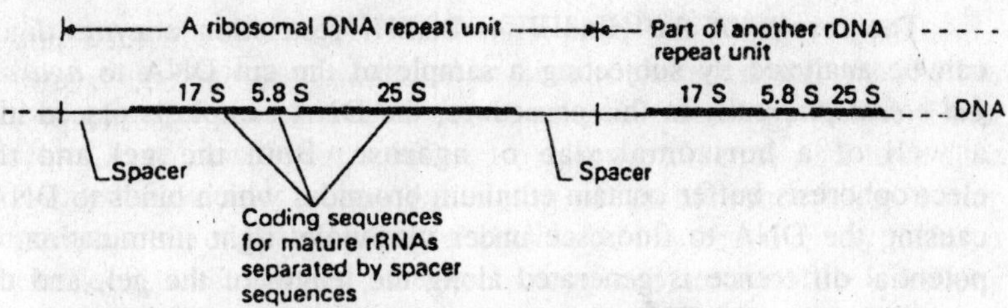

Fig. 8.5. Organization of the ribosomal DNA repeat unit of Neurospora crassa.

In *Neurospora*, the repeat units are homogeneous; that is, there is no variation in the length or organization of the rDNA repeat units. Shows the locations of the known restriction enzyme cleavage sites within the rDNA repeat units.

Preparation and analysis of the DNA fragment for cloning

The experiment discussed in this section has been reported in the literature and involves the cloning of approximately the left the half of the ribosomal DNA repeat unit.

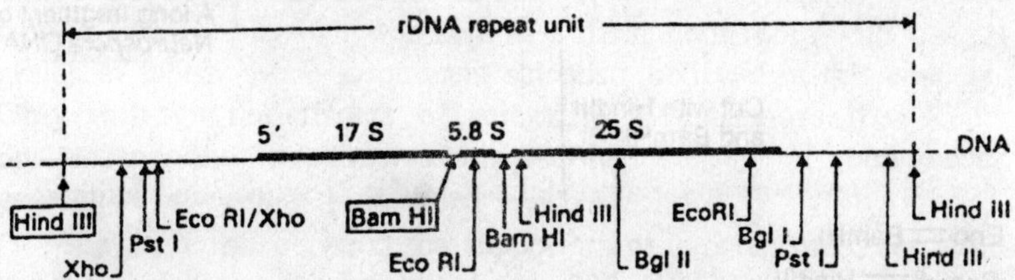

Fig. 8.6. Restriction endonuclease map of the ribosomal DNA repeat unit of Neurospora.

Suppose that the part of the DNA we wish to clone extends from the leftmost HindIII site to the BamHI site that is near the 3' end of the 17S rRNA coding sequence; these enzyme sites are boxed in. If total genomic DNA isolated from *Neurospora* is treated with both HindIII and BamHI, a large number of DNA fragments will be produced, and these will have a broad range of size owing to the distribution of the restriction sites over the genome. Because of the relative order of the HindIII and BamHI restriction sites along the DNA, various fragment types will be produced some fragments will have HindIII sites at each end, some will have BamIII sites at each end, and some will have a HindIII site at one end and a BamHI site at the other end. Among this last class will be the HindIII → BamHI fragment from the rDNA repeat unit that we want to clone. Since there are 200 copies of the repeat unit in the genome, there are 200 times as many of these restriction fragments as any other HindIII → BamHI fragment cut from unique sequence DNA.

The results of the HindIII + BamHI restriction enzyme digest can be analyzed by subjecting a sample of the cut DNA to *agarose gel electrophoresis*. In this procedure, the DNA sample is placed into a well of a horizontal slab of agarose. Both the gel and the electrophoresis buffer contain ethidium bromide, which binds to DNA, causing the DNA to fluoresce under ultraviolet light illumination. A potential difference is generated along the length of the gel, and the negatively charged DNA fragments migrate toward the anode. They do so as an approximately linear function of the logarithm of their length in base pairs, with the smallest fragments moving fastest and the largest fragments moving slowest. It is usual to have DNA size standards alongside the DNA sample so that the sizes of the DNA fragments in the sample can be calculated.

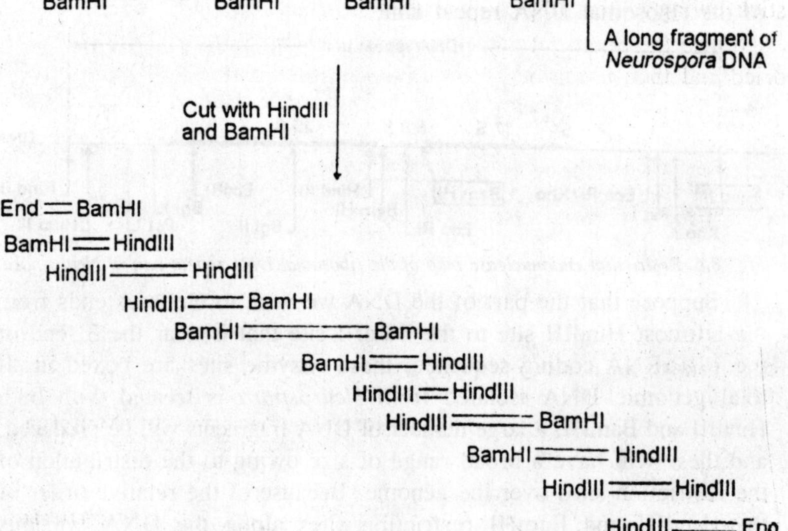

Fig. 8.7. The result of cleaving theoretical section of DNA with both HindIII and BamHI

After electrophoresis, the gel is examined under ultraviolet light to determine the distribution of the DNA fragments in the cut sample. The expected result for the HindIII + BamHI digest would be a smear of fluorescence down the gel lane. The reason for this is as follows; the genome of *Neurospora* contains a large number of restriction sites for the two enzymes used. Treatment with the two enzymes therefore produces an extremely heterogeneous collection of fragments with respect to size and this is reflected in the smear on the gel.

The smear on the gel is not wasted information. We can *probe* the smear in a very powerful way and identify the fragment of DNA that we want to clone. To do this, the DNA fragments are fist transferred from the gel to a sheet of nitrocellulose filter as they were in the gel. This is done by what has become known as the *Southern blot technique*, named after its developer, E.M. Southern. In brief, the gel is treated with alkali to denature the DNA to single strands. The gel is neutralized with buffer and is then placed on filter paper that acts as a wick to pull buffer from a reservoir. A piece of nitrocellulose filter is placed on top of the gel, and on top of the filter is placed a stock of absorbent filter paper. The latter wicks up buffer through the gel, then through the nitrocellulose filter, and onto the stack. By the movement of the buffer, DNA fragments are transferred out of the gel and onto the nitrocellulose to which they stick owing to the properties of the nitrocellulose itself.

Once the Souther transfer is completed, the nitrocellulose filter is dried and then it can be probed, meaning that a (usually) radio·ctive

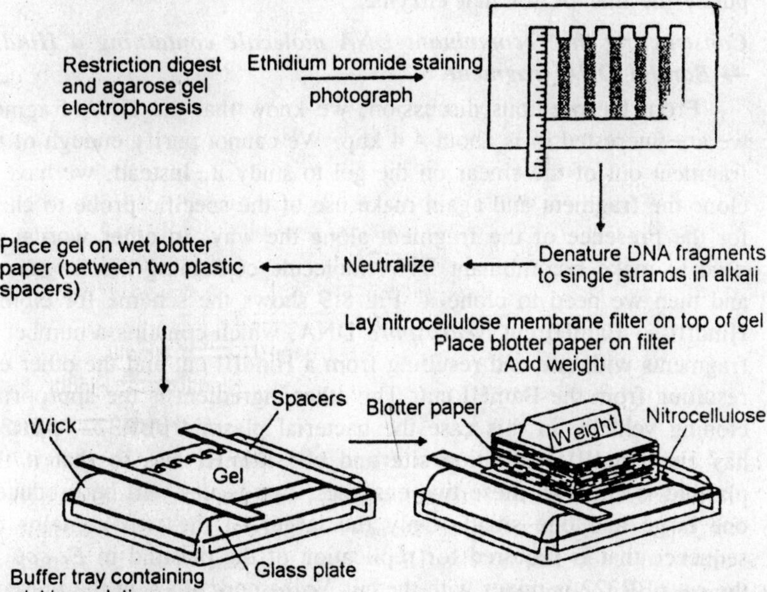

Fig. 8.8. Diagram of the Southern blot technique for the transfer of DNA fragments from gel onto nitrocellulose filters than can be used in radioactive probe experiments.

nucleic acid is added. At this stage single-stranded DNA fragment are fixed on the nitrocellulose filter, and thus any radioactive nucleic acid (probe) added will bind to any DNA fragment to which it is complementary. In this example, we want to find the DNA fragment that contains the 17S rRNA coding sequence.

Therefore the most appropriate probe to use is ^{32}P-labeled 17S rRNA. (Note: In general RNA probes are inferior to DNA probes.) If we hybridize with that probe, the 17S rRNA will hydrogen bond (hybridize) to the HindIII → BamHI fragment in which we are interested. If the filter is now washed free of any remaining unbound radioactive molecules and an autoradiograph prepared of the filter. As can be seen, there is only one band of silver grains on the autoradiograph, and this corresponds to a DNA fragment of about 4.4. kbp. We were able to pick out this fragment from all the rest by using the appropriate radioactive probe. This procedure is limited, therefore, only by the availability of the appropriate probe for doing the hybridization. Thus, as another example, we could probe for the DNA fragment or fragments for any enzyme, if we could isolate and purify the mRNA for that enzyme.

Constructing the recombinant DNA molecule containing a HindIII → BamHI DNA fragment

From the previous discussion, we know that the DNA fragment we are interested in is about 4.4 kbp. We cannot purify enough of the fragment out of the smear on the gel to study it. Instead, we have to clone the fragment and again make use of the specific probe to check for the presence of the fragment along the way. In other words, we need to make recombinant DNA molecule containing the fragments, and then we need to clone it. Fig 8.9 shows the scheme for cloning HindIII → BamHI cut *Neurospora* DNA, which contains a number of fragments with one end resulting from a HindIII cut and the other end resulting from the BamHI cut. The other ingredient is the appropriate cloning vehicle, in this case the bacterial plasmid pBR322. pBR322 has one HindIII restriction site and one BamHI site so that if this plasmid is cut with these two enzymes, two pieces will be produced, one large and one small. Only the larger of the two contains the sequence that is required for replication of the plasmid in *E. coli*. If the cut pBR322 is mixes with the cut *Neurospora* DNA in the presence of the enzyme, DNA ligase, recombinant DNA molecules will be produced. However, any one of a number of HindHIII → BamHI fragments can be inserted into the cut plasmid, including the plasmid

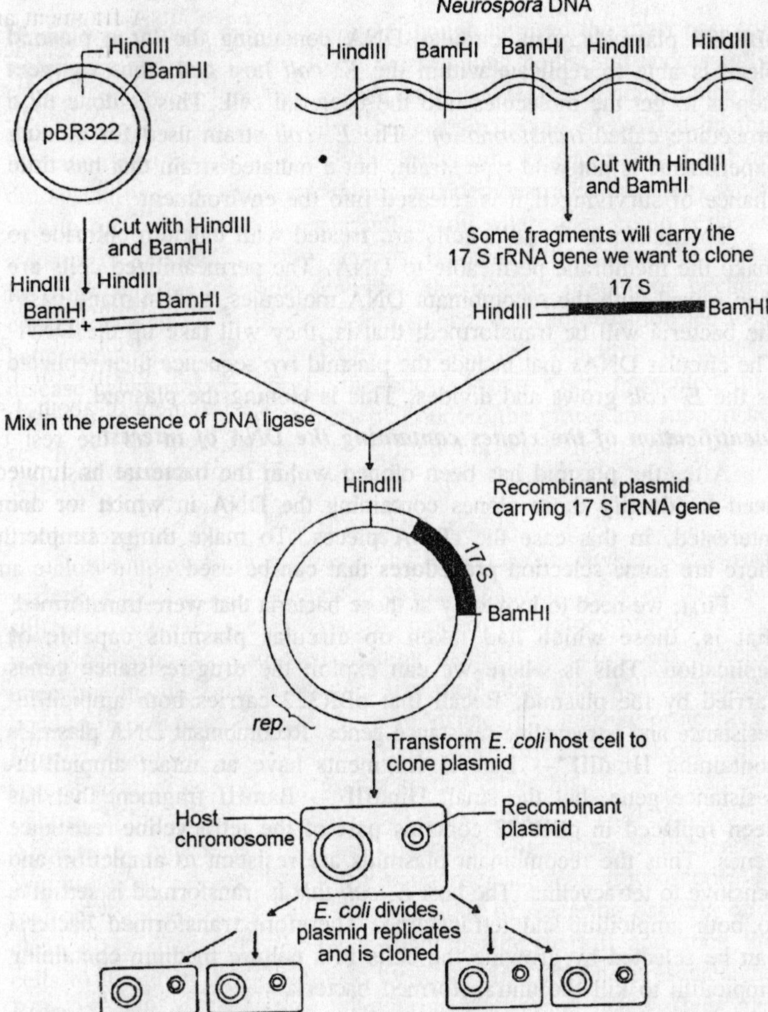

Fig. 8.9. Scheme for the cloning of HindIII → BamHI fragments of Neurospora using the pBR322 plasmid cloning vehicle.

piece that was originally cut back. It remains, then to find the plasmid carrying the piece of *Neurospora* rDNA that we want. To do that, the recombinant plasmids are first cloned.

Cloning the recombination DNA plasmids

At this point we have in the test tube a mix of plasmid + *Neurospora* recombinant DNA molecules and some reconstructed pBR322 plasmids. Any circular DNA containing the large plasmid piece is able to replicate within the *E. coli* host cell; thus the next step is to get the molecules into the bacterial cell. This is done by a procedure called *transformation*. The *E. coli* strain used for cloning experiments is not wild-type strain, but a mutated strain that has little chance of surviving if it is released into the environment.

To transform *E. coli*, cells are treated with calcium chloride to make the membrane permeable to DNA. The permeabilized cells are then mixed with the recombinant DNA molecules, and in many cases the bacteria will be transformed; that is, they will take up the DNA. The circular DNAs that include the plasmid *rep* sequence then replicate as the *E. coli* grows and divides. This is cloning the plasmid.

Identification of the clones containing the DNA of interest

After the plasmid has been cloned within the bacterial host, we need to identify those clones containing the DNA in which we are interested, in this case the rDNA pieces. To make things simpler, there are some selection procedures that can be used.

First, we need to look only at those bacteria that were transformed, that is, those which had taken up circular plasmids capable or replication. This is where we can exploit the drug-resistance genes carried by the plasmid. Recall that pBR322 carries both ampicillin-resistance and tetracycline-resistance genes. Recombinant DNA plasmids containing HindIII → BamHI fragments have an intact ampicillin-resistance gene, but the small HindIII → BamHI fragment that has been replaced in pBR322 contains part of the tetracycline-resistance genes. Thus the recombinant plasmids are resistant to ampicillin and sensitive to tetracycline. The host *E. coli* that is transformed is sensitive to both ampicillin and tetracycline. Therefore transformed bacteria can be selected by growing the cells in a culture medium containing ampicillin to kill the untransformed bacteria.

Once the transformed cells have grown, they are spread onto plates of solid medium containing ampicillin. On the plate, each bacterium will give rise to a colony, and each colony can picked to a microtiter dish containing wells of ampicillin-containing medium. This produces a series of clones in liquid culture. Replicas of each culture are printed onto a nitrocellulose filter that is on a plate of ampicillin-containing medium. Incubation of the plate gives rise to matrix of colonies on the filter. The filter is peeled from the plate, and it is treated to lyse the cells, to denature the DNA, and to cause the DNA to stick to the filter. The filter is then processed more or less as for the Southern blot. It is dried, and an appropriate radioactive probe is added. In our case, the probe is ^{32}P-labeled 17S rRNA. Each colony that contained a recombinant DNA plasmid carrying the HindIII → BamHI fragment with the 17S rRNA coding sequence is identified as a dark spot on the resulting autoradiograph. We can then go back to the microtiter well and grow large quantities of that clone for whatever purposes we have.

The radioactive probing here is highly selective. It will not pick up reconstructed pBR322 plasmids. Thus, in the few steps we have described, we can obtain the desired recombinant DNA clone.

Genes coding for proteins have often been cloned using a different approach. This approach involves the isolation of the mRNA for a protein. This approach is simplest for those mRNAs that are produced in large quantities by a cell (e.g. globin mRNA from rabbit reticulocytes). Once the mRNA has been isolated, a complementary DNA (cDNA) copy of it can be made in reactions involving, first, RNA-dependent DNA polymerase and, then *E. coli* DNA polymerase I. The former enzyme was first identified separately by D. Baltimore and H. Temin in 1970 as a component of the RNA tumor viruses, Rous' sarcoma virus (RSV) and mouse leukemia virus (MLV), respectively. The genetic material of these viruses is RNA, and they are able to transform a cell that they infect to the tumor state. This process apparently involves the production of a DNA copy of the viral genome by the RNA dependent DNA polymerase (also called *reverse transcriptase*, since the process it catalyzes, *reverse transcription*, is the opposite of transcription).

The use of reverse transcriptase in molecular cloning is illustrated Fig. 8.10. The resulting point is a polyadenylated mRNA, such as globin mRNA. The steps are as follows:

1. An oligo (dT) primer (a short, synthetic sequence of Ts) is annealed to the poly (A) tail of the mRNA.
2. The primer is extended in a reaction catalyzed by reverse transcriptase and using the four deoxyribonucleoside triphosphates as precursors. The result is a DNA-mRNA double-stranded molecule.
3. The RNA strand is destroyed by alkaline hydrolysis, and continued DNA synthesis by DNA polymerase I forms a short double-stranded, hairpin loop structure.
4. The 3' end of the hairpin loop acts as a primer for DNA polymerase I to catalyze the synthesis of the second DNA strand.

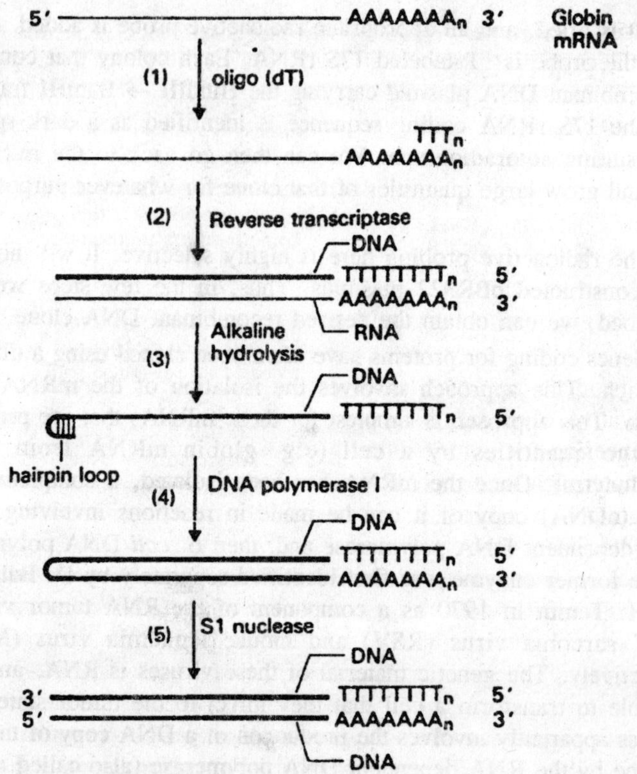

Fig. 8.10. The synthesis of double-stranded cDNA from a polyadenylated mRNA using reverse transcriptase.

5. S1 nuclease cuts the loop, resulting in a linear, double-stranded cDNA copy of the original poly(A)-mRNA. The cDNA can be inserted into a linearized plasmid and cloned.

The cloned cDNA may then be purified and used as a probe in any one of the several interesting applications such as the following: (a) to obtain sequence information about the mRNA (its coding and noncoding sequences); (b) to identify precursors to the mRNA (such as heterogeneous nuclear RNA); (c) to use as a hybridization probe to identify those recombinant DNA clones made from shared total DNA that contains part or all of the DNA region coding for globin mRNA; and (d) to attempt to get the gene produced synthesized in a bacterial or other simple organism host. In view of the discovery of intervening sequences, this can allow regions of interest, such as controlled sites or the intervening sequences themselves, to be analyzed by DNA sequencing.

Measuring the Activity of Fused Genes

Gene expression is best studied when the genes concerned code for products that can be assayed quickly and accurately with minimal expense. That the lactose operon of *E. coli* is still the best understood of all genetic segments is no accident. There is a simple colorimetric test for its principal product, the enzyme β-galactosidase, that unambiguously detects the enzyme's presence either in permeabilized cells or in cell-free extracts. Were β-galactosidase with its 1021 amino acids an unstable, hard-to-measure protein, we would never have concentrated on its underlying gene.

Yeasts, however, do not metabolize lactose and so normally do not produce any β-galactosidase-like enzyme. However, a close derivative of β-galactosidase can be made in yeasts using genetic engineering tricks to fuse the 5' upstream control elements and short aminoterminal sequences of selected yeast genes to the very long DNA fragments, coding the carboxylterminal 990 amino acids of β-galactosidase. Since the first 30 amino acids at the amino-terminal end of β-galactosidase are not required for enzyme activity, the resulting fusion proteins have full β-galactosidase activity and effective stability. Such fused genomes allow the activity of mutated yeast 5' control elements to be monitored through the amount of β-galactosidase activity that is synthesized, rather than by the much more laborious methods needed, say, to quantitate the amount of the uracil biosynthetic enzyme coded by the URA3 gene or the amount of cytochrome cl, the product of the CYCl gene

Detecting Related DNA Sequences

Once a gene has been cloned, its DNA can be used to make a restriction map and to determine whether similar DNA sequences exist elsewhere. The degree of relatedness between two DNA segments is best approached by measuring the ability of their component single strands to hybridize under appropriate conditions to form complementary double helices. A given cloned gene (restriction fragment) actually becomes a "probe" to search out other DNA segments of identical or nearly identical sequences. Now there are simple procedures in which such probes are used for the mass screening of many thousands of plaques or bacterial colonies. Such procedures always involve the production of replicas of the plaques or bacterial colonies on nitrocellulose paper. The DNA within these replicas is first fixed on the paper and then exposed to radioactively labeled DNA probes to pinpoint those plaques or colonies that contain related DNA sequences.

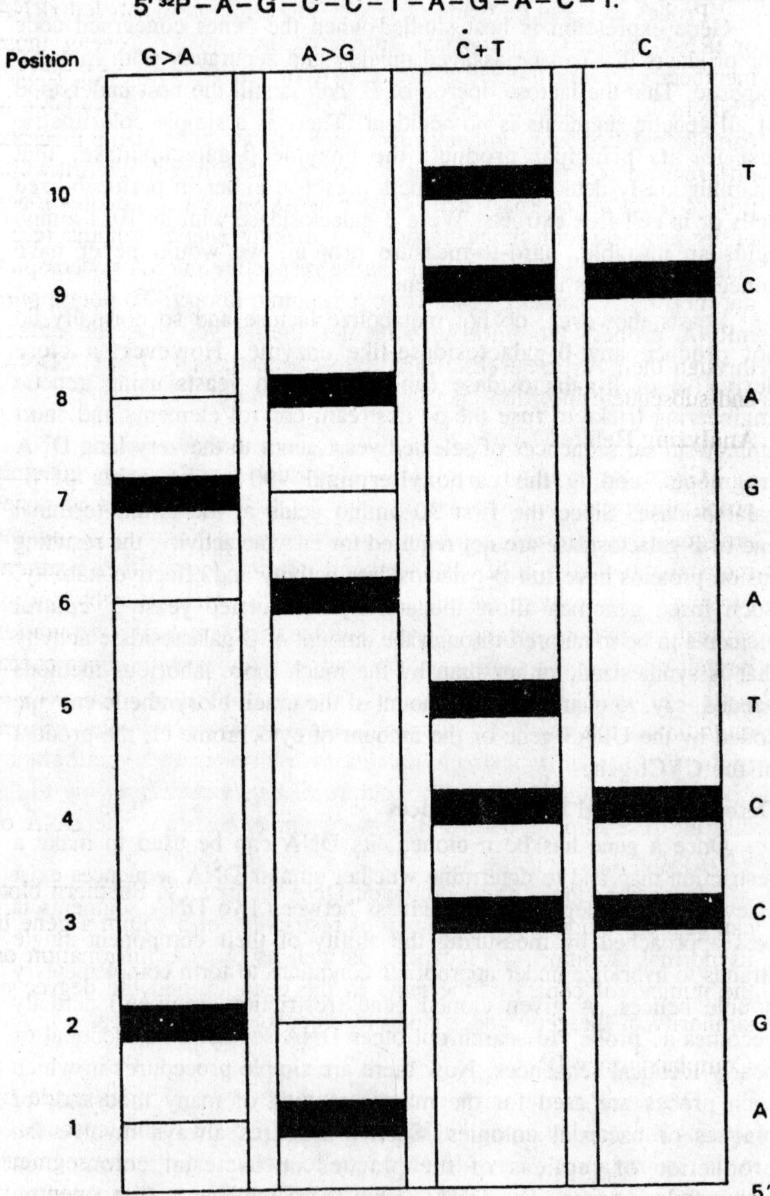

Fig. 8.11. Diagram of autoradiogram expected after the Maxam and Gilbert rapid DNA-sequencing technique is applied to the DNA fragment: 5'³²P-A-G-C-C-T-A-G-A-C-T.

Probes can also be made of RNA. The addition of labeled rRNA or tRNA chains to appropriate library replicas quickly reveals those members carrying the respective rDNA or tDNA genes. Messenger RNA molecules also can be used as probes when they are available in partially purified form. But since most cells contain more than a thousand different mRNA molecules, it is seldom possible to obtain worthwhile probes from the ordinary cell. Only when a single cell devotes much of its protein-synthesizing capacity to making single proteins (e.g., hemoglobin and immunoglobin) can be respective mRNA probes pick out their corresponding genes from a genomic library. To obtain pure mRNA probes, the mRNA molecules themselves must be cloned through their reverse transcription into complementary DNA (cDNA) and subsequent insertion into appropriate vectors.

Analyzing Related Genes

Given the availability of a suitable mRNA, cDNA, or genomic probe, important preliminary data on the structure of related genes are generated through a technique developed in Edinburgh by the molecular biologist E.M. Southern and which is now called *Southern blotting*. It starts with the digestion of a DNA population by one or several restriction enzymes. The resulting fragments, after separation on gels according to their size, are transferred to nitrocellulose sheets so that an exact replica of the DNA fragments in the gel is present on the nitrocellulose filter. A specific radioactive probe is then applied to the filter under hybridizing conditions. Subsequent autoradiography of the filter results in a specific pattern of bands corresponding to the discrete restriction fragments that are complementary to the DNA or RNA probe.

In conjunction with recombinant DNA technology, Southern blots can easily provide a physical map of restriction sites within a gene in its normal chromosomal location, as well as provide information on the number of copies of the gene in the genome, and the degree of similarity of the gene under study to other known homologs.

Complementary DNA

There are many important uses for the gene libraries made by cloning the population of mRNA molecules present within specialized cells. In particular, they afford a way to isolate those gene segments that code for exons away from their noncoding, intron components. In this procedure, the information in mRNA is copied into cDNA (complementary DNA) replicas using the retrovirus enzyme *reverse transcriptase*, which normally functions to convert infecting retroviral

RNA chains into DNA copies. Reverse transcriptase, like all other DNA polymerases, can initiate DNA synthesis only by adding onto primers. Fortunately, it is easy to prime cDNA synthesis by utilizing the stretches of poly A (called a *poly A tail*) at the 3' ends of most mRNA molecules. Added oligo dT binds to these poly A tails and serves as the necessary primer. Synthesis of a single-stranded complementary DNA chain can run to the 5' end of its template, where it often makes a hairpin loop turn, which serves as a primer for the synthesis of a complementary DNA strand. The loop of the hairpin can subsequently be dissected away by a nuclease specific for single-stranded DNA.

The resulting double-stranded cDNA can be inserted into a plasmid like pBR322 either by adding complementary tails (e.g., poly G and poly C) using the enzyme terminal transferase or by attaching chemically synthesized restriction enzyme sites (liners) using the enzyme DNA ligase. The cDNA-containing plasmids so constructed can then be introduced into *E. coli* and amplified.

Complementary DNA libraries will of necessity be heavily biased toward members representing the more abundant classed of mRNA found in given cells. It should not be taken for granted, however, that they must always contain the member we expect. The reverse transcription of long mRNA molecules into complete cDNA copies frequently aborts, producing short fragments representing only part of the gene. The marking of a highly representative large-copy-number *cDNA library* is thus a much harder task than the construction of its equivalent *genomic library*.

Identifying the Products of cDNA Clones

One of the most direct ways to screen a cDNA library is to use individual members as traps for the collection of specific mRNA molecules that in turn are added to *in vitro* systems for protein synthesis. With luck, the mRNA class selected out by its hybridization to the denatured strands of a given cDNA library member will direct the synthesis of a protein that can be identified by its position on a two dimensional protein gel. Such procedures work best when the proteins concerned have been previously well characterized and are major cellular constituents (e.g., the hemoglobin molecules synthesized by red blood cells, or one of four major histones). But with organisms like yeast, whose proteins have not yet been well characterized, this form of screening is not a practical way to find most desired cDNA clones. Once a cDNA clone has been so identified, however, it is very straightforward to pick out the corresponding genomic clone using DNA-DNA hybridization procedures.

Those cDNA made off less abundant mRNAs will obviously be much harder to identify. Directly looking for the protein products necessarily involves the testing of many hundred or thousands of clones, often a costly task necessarily limited to industries. If, however, the proteins of interest have been characterized by partial or full amino acid sequencing, then an important shortcut may be used. Given an amino acid sequence, intelligent guesses can be made as to its corresponding mRNA (DNA) sequence. Because all amino acids but one are specified by more than one codon, it is not possible to go from an amino acid sequence to a DNA sequence unambiguously. By focusing, however, on sequences that mainly contain the less common amino acids, it is usually possible to define a small collection of oligonucleotides, one of which should be exactly complementary to the segment of interest. Such a restricted collection can then be used as probes to identify the complementary cDNA clones by hybridization. Already this approach has been used to isolated a number of important cDNA clones, and while it never is as simple as described here, it nevertheless is practical technique that is bound to be increasingly employed.

Identifying cDNA Clones in *E. coli*

It is increasingly possible to find desired eukaryotic cDNA clones through techniques that allow their coding regions to function in *E. coli* strains that have been especially tailored to express foreign genes. Such strains are in fact called *expression strains*. In these techniques, the eukaryotic sequences are usually attached just downstream from a strong bacterial (or phage) promoter-ribosome binding site complex like those that govern the tryptophan and lactose operons or the expression the phage λ genome. These gene fusions can easily be made to yield fusion proteins containing short bacterial amino terminal segments linked to virtually all the amino acids of the eukaryotic protein. Often these hybrid proteins can be detected because they possess almost the total functional potential (e.g., enzymatic activity) of their pure eukaryotic equivalent. However, the hybrid proteins can usually be found more quickly by screening very large numbers of *expression vectors* for their possession of fusion products that specifically bind to appropriate antibodies. In such procedures, bacterial colonies that have been treated with chloroform to make them permeable to antibodies are exposed to radioactively labeled antibodies of the correct specificity.

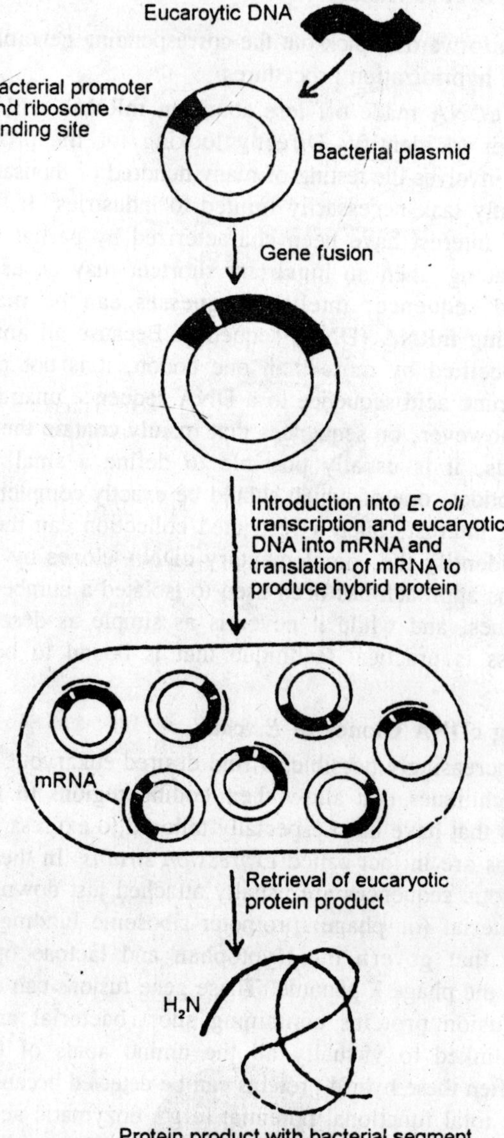

Fig. 8.12. Expression of eucaryotic DNA in E. coli allows high-level expression and identification of its protein product.

The usually rare colonies that are making immunologically cross reacting proteins are easily picked out by autoradiography. To replac

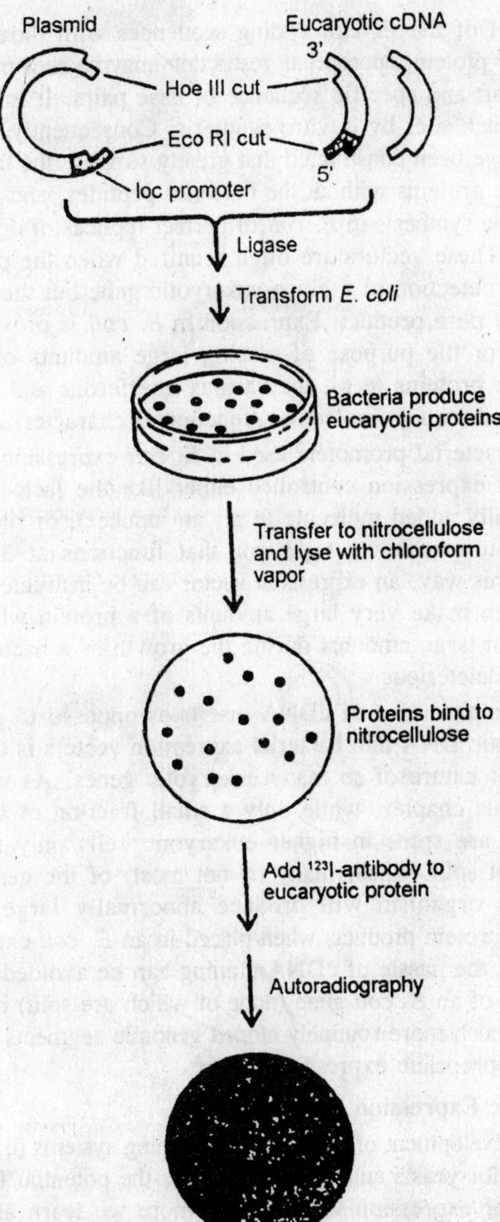

Fig. 8.13. Immunological screening of expression vector to detect eucaryotic-bacterial hybrid protein.

cleanly all of the *E. coli* coding sequences with those coding for an eukaryotic protein, appropriate restriction enzyme sites must occur within a very short and specific sequence of base pairs. It is now possible to produce such sites by in vitro synthesis. Consequently, new expression vectors have been constructed that greatly simplify the task of producing eukaryotic proteins without the bacterial peptides, and thus often make possible the synthesis in *E. coli* of perfect replicas of desired eukaryotic proteins. These vectors are often required when the primary purpose is not the detection of a given eukaryotic gene but the synthesis of its effectively pure product. Expression in *E. coli* is providing especially valuable for the purpose of making large amounts of normally rare eukaryotic proteins (e.g., the various interferons and growth factors) for their subsequent molecular (functional) characterization.

The bacterial promoters used in *E. coli* expression vectors usually have their expression controlled either like the lactose promoter, by an externally added molecule (e.g., an inducer), or like mutant λ, by a temperature-sensitive repressor that functions at 31°C but not at 37°C. In this way, an expression vector can be instructed at appropriate moments to make very large amounts of a protein whose continuous presence in large amounts during the growth of a bacterial population would be deleterious.

The routine used of cDNA inserts as opposed to genomic inserts of eukaryotic DNA into bacterial expression vectors is the consequence of the split nature of so many eukaryotic genes. As we explained in the previous chapter, while only a small fraction of the genes of *S. cerevisiae* are split, in higher eukaryotic cells only the exceptional gene is not split. Thus, many (if not most) of the genes of a higher eukaryotic organism will produce abnormally large, usually non-functional protein products when placed in an *E. coli* expression vector. Of course, the hassle of cDNA cloning can be avoided when the over expression of an *E. coli* gene (none of which are split) itself is desired. Then the much more routinely cloned genomic segments can be inserted into the appropriate expression vector.

Eukaryotic Expression Vectors

The development of multiple transforming systems (e.g., YEp, YRp, and YCp) for yeasts automatically creates the potential for using yeasts as hosts for expression vectors. The more we learn about how gene expression is controlled in eucaryotes, the more intelligently we can develop expression vector systems that can rival those already in use in *E. coli*. Several factors already encourage the development of eukaryotic systems for the expression of eukaryotic genes. First, such systems may more readily allow the passage through their external plasma membrane of proteins that normally are exported outside the cells in which they are translated. Such secreted proteins are frequently held together by large numbers of internal disulfide bonds (S—S) that are formed only after the passage of their polypeptide chains through cell membranes. Many of these exported eukaryotic proteins, for reasons that are still unclear, cannot pass through the *E. coli* membrane and, as a consequence, never form the correct sets of S—S bonds. The functional expression of many eukaryotic proteins thus demands their synthesis within the proper eukaryotic (as opposed to prokaryotic cells.

Also favouring the use of eukaryotic cells for the expression of eukaryotic genes is the ability of eukaryotic cells to process (splice) the newly synthesized pre-mRNA products of split genes. There is no reason why a yeast gene on a yeast vector ever needs to be converted into cDNA. In contrast, when such genomic insertions are placed in *E. coli*, only those genes lacking non-coding inserts (called introns) are likely to produce functional products. Thus, it is hoped that expression vectors for use in all genetically well-characterized organisms will soon be developed. When it comes to choosing which vector to use for a stated purpose, the fact that eukaryotic micro-organisms like yeast can be grown both faster and cheaper will favour their use whenever possible.

Long Stretches of Eukaryotic DNA

Eventually, the structure of very long stretches of eukaryotic DNA, if not complete eukaryotic chromosomes, will be worked out. In accomplishing this task, DNA hybridization techniques will provide the easiest way to pick out clones from gene libraries that have overlapping sequences. Thanks to such "chromosome walking;" (also called "overlap hybridization") the first centromeric sequences on a yeast chromosomes has already been isolated. Two cloned genes CDC10 and LEU2, that had been mapped to opposite sides of the centromere were used as starting points for two walks that finally reached each other in a total of 4 steps. Each of the clones so identified were then tested to see which one(s) ensured regular plasmid segregation during mitosis and meiosis. Walks have also been used to establish the structure of some very large *Drosophila* genes that are much too long even to be encompassed on cosmid clones.

Unfortunately, chromosome walks are almost impossible to carry out in higher plants and animals, since their chromosomes contains large numbers of nearly identical repetitive sequences. Their presence adjacent to so many genes leads to large numbers of cross-hybridizing clones, whose related sequences reflect the presence of common repetitive elements rather than overlapping unique chromosomal DNA. Hence, an easy and effective method to link up restriction enzyme-generated fragments along the chromosomes of higher organisms is not yet available.

Applications of Recombinant DNA Technology

In the years since recombinant DNA technology has been developed, it has been used in a large number of studies of both prokaryotic and eukaryotic genomes, and it is impossible to deal with them all adequately in this brief discussion. To give but three examples, recombinant DNA cloning has made it possible to obtain large amounts of particular genes, to determine the functions of segments of nuclear and organellar DNA, and to map genomes (e.g., ϕX174 and SV40 were mapped this way). Particularly useful applications of the techniques are studies in which the controlling regions for prokaryotic and eukaryotic genes are cloned with an eye towards understanding the regulation of gene expression at a very basic level.

The application of recombinant DNA cloning and the associated techniques has the potential to benefit mankind. For example, the genes for a number of human genes have been cloned in appropriately engineered vehicles so that the host bacterial or yeast cell synthesize the gene products. Thus, human insulin is being made using recombinant DNA methodologies and the hormone can be used to treat diabetics, particularly those who are sensitive to the porcine insulin now used. Similarly, to name but a few other examples, human growth hormone is being made from cloned genes to test a antiviral and antitumor agents, and bovine growth hormone is being made and used to increase milk production. Another application is to "construct" organisms to carry out specific functions. Thus, organisms have been engineered to "eat" oil spills, and experiments are underway to produce other organisms that can deal with toxic wastes or spill.

In medical diagnosis, progress is being made to use recombinant DNA methodologies to detect-genetic diseases. For example, at least some genetic mutations affect the sites for restriction enzymes. It then becomes possible to distinguish wild type and mutant genes by digesting a sample of genomic DNA with the appropriate restriction enzyme(s), separating the fragments on a gel, transferring the fragments to filter, and hybridizing with an appropriate labeled probe. Since fetal cells can be obtained by taking a sample of the amniotic fluid through a syringe, this approach can be used to screen fetuse for any genetic diseases that show a restriction enzyme digestion pattern difference.

Finally, there is a great potential for the use of recombinant DNA methodologies in plant genetics, particularly for improving crop yields. On exciting area of research concerns nitrogen fixation. This is the process whereby leguminous (e.g., peas, soybeans) capture atmospheric nitrogen and reduce it to a form that can be utilized directly by the plant. Special bacteria in the root modules called rhizobia are responsible for this nitrogen conversion. The many genes involved are called *nif* genes, and these are found in plasmids in the rhizobia. If these can be manipulated, it should be possible to increase the amount of nitrogen fixation in the plants. Experiments are also being done with cloned *nif* genes to try to induce nonleguminous plants to fix atmospheric nitrogen. As of this writing, a corn strain has been produced that get 1% of its nitrogen from nitrogen fixation.

In summary, great strides are being made in the application of recombinant DNA methodologies to projects related to human welfare. There is no doubt that significant achievement will continue to occur.

Recombination in Bacteria

In the late 1930s, no definite sexual differentiation of bacterial cells had been observed, no nuclei had been detected, and the chromosome had not been stained. However, the advent of selective techniques using nutritional mutants enabled a screen to be made for rare recombinant colonies. In addition, improvements in cytological techniques revealed the presence of darkly staining bodies. Lederberg and Tatum (1946) devised experiments in which two parental bacterial strains of *Escherichia coli* were each genetically labelled with three different nutritional mutations, i.e., $abcd^+e^+f^+$ x $a^+b^+c^+def$. Washed suspensions of the two types were plated on minimal medium and a+b+c+d+e+f+ colonies were detected at a frequency of about 1 in 106 bacteria plated. Reversion at three loci simultaneously could be ruled out as very improbable, and cross-feeding was ruled out by showing that the colonies could be cloned on to fresh medium. At first, transformation was thought to be explanation of these results, but this was eliminated by showing that cell-free extracts or filtrates could not substitute for direct cell-to-cell contact. Initially it was assumed that the recombination shown by these experiments resulted from processes already discovered, such as fusion of nuclei, followed

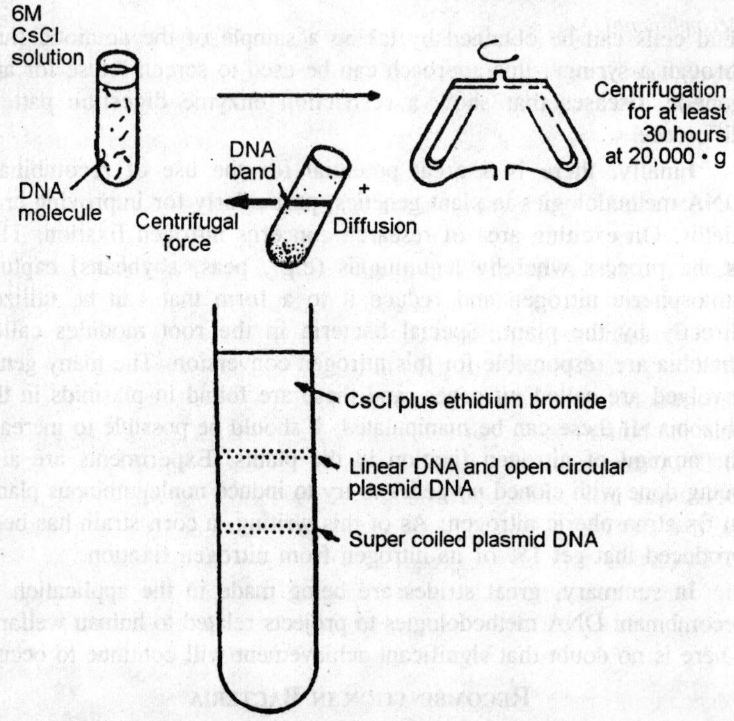

Fig. 8.14. Centrifugation techniques in the presence of ethidium bromide to give lower band of supercoiled plasmid DNA.

by meiosis or some other chromosomal segregation. However, it soon became clear that, although linkage maps could be drawn up, a number of results were puzzling. The explanation of these results was that transfer of the chromosome occurred in one direction only, and the transfer was frequently incomplete. Furthermore, the cells which could mate were differentiated into donor cells (males) and recipient cells (female). This was shown by Hayes (1953) who demonstrated differences between reciprocal crosses. Streptomycin-sensitive males (*str*^{s}) were mixed with str^r females and plated on a selective medium containing streptomycin. The reciprocal cross was *str*^{r} males plated with *str*^{s} females. Only first cross gave recombinants, showing that the viability of the recipient or female was essential conjugation. Evidently the chromosome was transferred from the male to the female and the female cell incorporated a fragment of the donor chromosome. Conjugation is therefore a one-way transfer of the chromosome which requires cell contact.

Discovery of the Sex Factor

Much of the confusion over the early experiments was removed by the discovery of the sex factor (F). This is an example of a plasmid, i.e., a *replicon* (unit of replication) which is stably inherited in an extra-chromosomal state. Naturally-occurring plasmids are normally non-essential and can be lost without death of the cell. Techniques are not available for the isolation of the sex factor and other plasmids and the most definitive technique relies on centrifugation in caesium chloride in the presence of ethidium bromide. Ethidium bromide can intercalate into DNA and reduce its buoyant density. The large peak contained chromosomal DNA and broken plasmid DNA, whereas the small peak, near the bottom of the tube, contained pure plasmid DNA. Examination of DNA samples under the electron microscope showed that the plasmid DNA was in a supercoiled form which was not readily accessible to the dye and was therefore more dense than chromosomal DNA. Plasmid DNA can be in three states: a supercoiled from (known as a covalently closed circle or *ccc*), a circular form, and a linear form. Some of the char-acteristic features of the F plasmids, the *E. coli* chromosome, and a variety of plasmids, which will be referred to later.

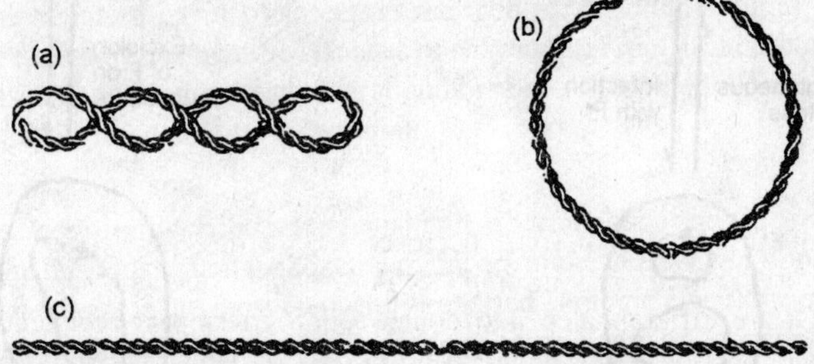

Fig. 8.15. Structure of (a) supercoiled, (b) circular, and (c) linear molecules of a plasmid (a) is a covalently closed circle (ccc), (b) has a single-strand nick and is an open circle, (c) has a double-strand break to give a linear molecule.

Type of *E. coli* Strains

It soon became apparent that there were several different strains involved in conjugation. The original strains were F^+, containing the sex factor, and F^-, lacking the sex factor. When F^+ and F^- strains were mixed, the F^- strains were rapidly converted into F^+ by an epidemic spread of the sex factor, later shown to be accompanied by DNA replication. No transfer to chromosomal genes occurred during the transfer. Present in the F^+ population, however, were rare strains in which the sex factor had integrated into the bacterial chromosome by a single crossover event. These strains were responsible for the low level of chromosomal transfer observed in the original experiments. When they were purified by cloning, they gave recombination frequencies of up to 100% per donor cell and were called *high-frequency mating strains* (H *fr*). In this strain the sex factor was assumed to transfer as in F^+ x F^- crosses, but this time the sex factor, being attached to the

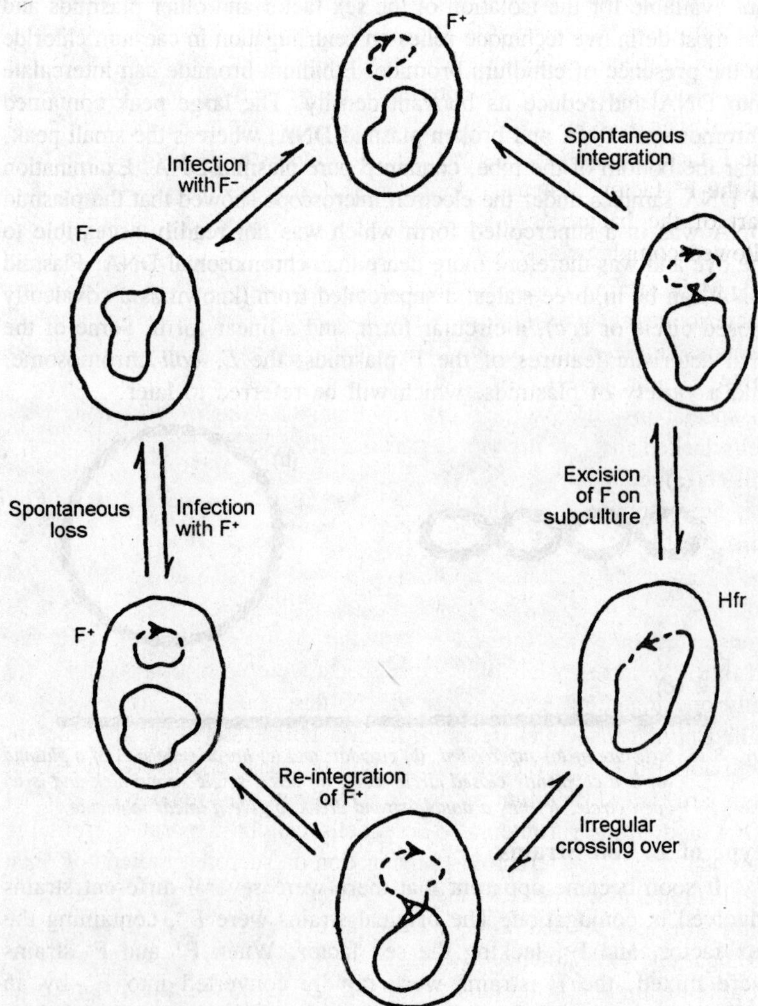

Fig. 8.16. Diagrammatic representation of the relationship between different strains of Escherichia coli.

bacterial chromosome, transferred this as well. Spontaneous breakage of the chromosome can occur, resulting in a gradient of gene transfer. Most of the recombinant strains formed were still female, as part of the sex factor remained in the H *fr* strain, except for rare situations when the whole chromosome was transferred. H *fr* strains could revert to the F^+ state by a crossover in the region which produced the H *fr* strain originally. Errors at this stage could produce a fourth strain which had a stretch of bacterial DNA incorporated into the sex factor. Crossing-over occurred at the wrong position, giving a plasmid containing bacterial DNA. The strain was called an F-*prime* or F'. This strain had properties intermediate between an F^+ and an H *fr* strain. The F' factor transferred at a frequency approaching that of the F-factor, but chromosome transfer can also occur due to integration of the F' factor. When the F' factor was transferred to a new strain, part of the bacterial genome was now present in duplicate, which allowed complementation tests to be done.

Male strains (F^+, H *fr* and F') were all found to sensitive to bacteriophages MS2 and *fd*, known as male specific phages. MS2 contains single-stranded linear RNA and *fd* has single stranded circular DNA. These were observed under the electron microscope to attach to special projections, the *sex pili*, on the bacterial cell surface. F^- cells lacked the sex pilus and were resistant to the phages. Ordinary pili and flagella, however, were present in both strains. The sex pilus has been implicated in the mating process, probably as a method of bringing the cells together prior to formation of a conjugation bridge.

Convincing evidence has been presented that DNA synthesis, by a rolling-circle method, occurs during conjugation. A single strand is transferred and this integrates into the double helix of the female strain. A number of labelling experiments were used to prove this, but only one ingenious experiment will be described. This relied on so-called *mini cells* which resulted from a temperature-sensitive mutant with abnormal cell division. At the non-permissive temperature, small cells are cut off without receiving DNA but with otherwise normal RNA and protein contents. These cells can be separated from the parental cells by differential centrifugation on sucrose gradients. Cohen et al. (1968) found that H *fr* strains mated with mini cells transferred only single-stranded DNA, even after mating for 90 minutes. Comparable experiments with an F' gave some single-stranded some double-stranded molecules, suggesting that a single strand was transferred and that some DNA synthesis was possible as a complete functional F factor had been transferred.

Mapping Chromosomes by Conjugation

The clearest evidence for the mechanism of chromosome transfer came from kinetic studies made by Wollman et. al. (1956). This gave a very useful technique for chromosome mapping and therefore will be described in some detail. The basic principle was to separate mating cells at particular times by a high-speed blender and then to plate an various selective media. Each marker from the donor is found to enter the female at a particular time and to reach a plateau level. The later the gene enters, the lower is the plateau level reached due to the high probability of spontaneous chromosome breakage. This time scale can be translated into a chromosome map if it is assumed that the chromosome carrying the genetic markers is transferred in a linear fashion with time.

—gal—lac—ton—azi →
25 16 10 9

Experiments with other independently-isolated H *fr* strains showed different orders of genes, and the results were finally explained by assuming that the sex factor could integrate at different positions and in different orientations into a circular bacterial chromosome. Transfer will therefore occur from different points and in different directions. Homology exists between the sex factor and the bacterial chromosome, as both contain insertion sequences, 1S2 and 1S3. These strange elements are between 800 and 1500 base pairs (0.8-1.5 kilobases) in length, and can move from one region of the chromosome to another. On insertion, a gene function is eliminated and a further method of detection is that of heteroduplex analysis in which DNA from two sources is allowed to hybridize. Single-stranded regions show sequences present in one strand but not the other. The sources of different H *fr* strains can be explained by the different location of 1S2 and 1S3 in the bacterial chromosomes of the original strains.

A second mapping technique is to select for one marker and then to classify for the unselected markers by use of different media. Results differ according to whether as early (proximal) or late (distal) marker is selected. Selection for a proximal marker gives results very similar to the interrupted mating experiment, as there will be decreasing gradient for the later markers due to chromosome breakage. Selection for a distal marker means that a long fragment of chromosome must be received by the recipient cell, and that the frequency of genotypes will depend on the probability that crossing-over can occur in a particular interval. This is necessary for the mapping of genes such as *thr*$^+$ and *leu*$^+$ (threonine and leucine-requiring mutations) which are so close that they cannot be separated by interrupted mating experiments.

Genetic Map of the Sex Factor

The sex factor is likely to have about 100 genes and a map of some of these genes is shown. The phenotypes which have been used are resistance to male specific phages, inability of an F' *lac*$^+$ to transfer and defective DNA synthesis. Complementation tests have been made by infecting the same cell with two F factors carrying different mutations, e.g., *tra*A$^-$ and *tra*B$^-$, and looking for restoration of transfer function. Cell with two sex factors are unstable but persist long enough to allow transfer of plasmids to recipient cells. By this method 16 *tra* loci have been identified. Mapping of the genes has been done using deletion techniques with an F *gal* system. The *tra* genes are arranged in an operon which means that a single polycistronic mRNA is synthesized. This is under the control of an operator region O, the region where a repressor can interact. Mutations in *tra* genes have polar effects, i.e., a mutation in *tar*K will reduce the expression of genes distally positioned relative to O, e.g., *tra*B and *tra*C. This also enables a map to be constructed. Recent techniques using restriction enzymes have allowed the sex factor to be divided up, recombined with other plasmids, and tested for activity with known defective sex factors. Thus EcoR1 splits F DNA into 19 fragments, which can be separated on gels, and individual fragments can be attached to another plasmid. This is then tested for its ability to complement known F-factors mutants.

Features of other Plasmids

A wide range of plasmids varying a length of DNA molecule from 2 μm to 60 μm or more have now been detected in a variety of bacteria and fungi. Plasmids are responsible for a variety of phenotypes from antibiotic resistance to the ability of *Pseudomonas* to degrade oil. Two groups have received most study—the *resistance transfer factors* (RTFs or *R-factors*) and the *colicin factors* (*Col*). RTFs, which resemble F' factors, are plasmids carrying antibiotic resistance genes; colicin factors code for species-specific bacteriocidal protein called *colicins*.

RTFs were first observed in Japan in 1957 by Watanable (1963). During an outbreak of dysentery, strains of *Shigella sonnei* were isolated which were resistant to more than one antibiotic. By 1964, 40% of all strains isolated were resistant to four to more antibiotics such as streptomycin, sulphonilamide, chloramphenicol and tetracycline. It was found that R-factors, plasmids which resembled the F-factor, could spread promiscuously between a variety of genera including *E. coli* and pathogenic strains of *Salmonella* and *Shigella*. This was recognized as a serious public-health hazard, and much work has gone into tracing the origin of R-factors. Two possible sources of selective pressure exist: the over-prescription of antibiotics to promote the growth of animals such as pigs and chickens. The situation was serious enough to require a Government inquiry which was later published. This banned the non-clinical use of chloramphenicol and recommended that antibiotics used for growth promotion should be different from those used to treat diseases of humans. The problem still remains in the world as a whole, due to the availability of antibiotics without prescription in certain countries. In Mexico in 1972, more than 10,000 people were infected with a chloramphenicol-resistant *Salmonella typhi* (typhoid fever) and 1400 died from the disease Antibiotic-resistant *Haemophilus influenzae* (causing meningitis) has been detected in the United States, and *Salmonella typhimurium* (food poisoning) resistant to six antibiotics has recently been detected in Britain.

When the F-factor and R-factor are compared, one major difference is apparent. F-factors are fully expressed in all cells, whereas R-factors are normally *repressed*. This means that there is a repressor/operator system which controls the expression of R-factor genes such that sex pili are present in very few cells and that only about $1/10^4$ cells can actually transfer the R-factor. Cells freshly infected with an R-factor transfer at a much higher rate, suggesting that the repressor has not been synthesized. Derepressed mutants *drd* can be isolated which show a 300-fold increase in R-factor transfer. Strains carrying R-factors are normally resistant to male specific phages, but R*drd* strains are susceptible due to the synthesis of sex pili. There are several different R-factors but details of these will be left until later. A genetic map of the R100 factor, based on heteroduplex analysis and mapping by transduction analysis with a phage vector. R-factors have been reported to integrate into the chromosomes, but this is not as frequent an event as with the F-factor.

The colicin factors can be compared with defective phages in which the DNA, the colicin factors, is not longer packaged into the 'phage coat, the proteinaceous colicin. The presence of a colicin factor (*Col*) in strain can be demonstrated by inoculating a Col^+ strain into nutrient agar in a glass Petri dish and growing overnight. The colony is killed by exposing to chloroform vapour for a few hours, and then the colony is overlaid with a colicin-sensitive strain in 0.6% agar.

Further incubation reveals a colony-centred area of clearing like a phage plaque. The area of clearing is due to inhibition of growth, and no phage can be detected. Colicins kill cells by a variety of mechanisms, e.g., colicin E1 inhibits active transport, and colicin *Ia* and *Ib* inhibit ATP production and protein and nucleic acid synthesis. Strains carrying *Col* factors are *immune* to their own colicins, and sensitive strains many become *resistant* to colicin by a different mechanism involving loss of the colicin receptors. This parallels similar effects with phage receptors and resistance of bacteria to phages. Colicin factors vary in size and the smaller *Col* factors such as *Col E1* are non-infectious, although they can be transmitted if a conjugative plasmid is also present. *Collb*, comparable in size with the F-factor, is infectious and may be *relaxed* or *stringent*. In relaxed plasmids, such as *Col E1*, between 5—20 copies are present per chromosome, whereas for stringent plasmids such as F and R100 there is strict relationship between the number of copies per chromosome, usually between 1 and 3.

Relationship between Plasmids

A sex factor is unable to infect a cell with a resident sex factor, a phenomenon known as *superinfection immunity*. In fact, this can be divided into two separate mechanisms, *surface exclusion*, which prevents the plasmid from entering at the cell surface, and *plasmid incompatibility*, which prevents the plasmid from becoming established inside the cell. Plasmids have been classified on the basis of incompatibility into a large number of groups, and Table shows examples of a few of these groups. Those R-factors which are compatible with the F-factor can be classified further into those which inhibit F-fertility fi^+ (fertility inhibition) and those which fail to inhibit F-fertility fi^-. Thus an $F^+R^+_{fi}$ cell would have the repressed fertility of an R-factor cell, whereas $F^+R^+_{fi}$ would transmit the F-factor at a high frequency. This implies that some F and R plasmids are related, and that their control mechanisms interact. A model for repression has been proposed which is based on two genes, *finO* (fertility inhibition) and $finP_R$, located in the R-factor and one gene $finPE_F$ located on the sex factor. The *finO* gene product can interact with the $finP_F$ protein to repress the transfer operon and switch off F-functions. Early models of this system assumed a simple repressor-operator system but more recent work has demonstrated complex interactions. The $finP_F$ and $finO_R$ complex exerts a negative effect on the o_J operator which prevents the synthesis of the *traJ* gene product. In the absence of the repressor complex, the *J* product normally exerts a positive control (i.e stimulation) over the main operator for the transfer operon, resulting in expression of *tra*-functions. Evidence for this system has been obtained form experiments in which a study was made of the transfer of plasmids from cells which transiently had two or more plasmids. These experiments demonstrated that $finP_F$ mutants were recessive whereas o_J mutations were dominant. However it should be noted that *Flac* o_J *Fhis*, R100 cells can transfer all three plasmids. The o_J operator is not cis-dominant because it controls the synthesis of a *trans*-active *J* proteins which acts on both of the F and R transfer operons. The situation is more complex than this as the J protein is also a positive regulator for *traM* which is not in the *tra* operon.

Table 8.1. Examples of compatibility groups in bacterial plasmids.

Properties	*Incompatibility group*	*Examples*
Like the sex factor F	IncFI	F, R386
	IncFII	R 1, R100
	IncFIII	ColB-K98
	IncFIV	R124
Unrelated to sex factor F	IncA	RA1
	IncI	ColIb-P9, R144
	IncN	R46

Detailed examination of strains of bacterial carrying plasmids has shown much variation in the molecular weight of plasmids, even within the same strain. Thus determinants for transfer of the plasmid (described as Δ) can separate from resistance genes, and plasmids may be in a state of constant change unless a selective pressure maintains a particular plasmid in an equilibrium state. Interchanges between plasmids and the host chromosomes also occur, and the ubiquitous occurrence of transposons and insertion sequences presumably facilitates this process. *Cryptic plasmids* with no known function have been detected, and these may be poised to pick up genes when selective pressures change. Cohen (1976) has reviewed the possible role of transposons in plasmid evolution, and Richmond and Wiedman (1974) have discussed the role of plasmids in bacterial evolution.

Analogies have been drawn between phages and plasmids, and the *episome concept* compared λ with F. Episomes were considered to be dispensable elements which could replicate either autonomously in the cytoplasm or integrated with the chromosome. There are some grounds for thinking of some plasmids as degenerate phages, e.g. the colicins, but on the whole the term *episome* is now considered to be an artificial group. Essentially the nonconjugative plasmids, conjugative plasmids and phages are three separate groups which may interact but can be distinguished on molecular weight and the presence or absence of a protein coat. Interchange between groups is possible by the agency of *transposons*, which are insertion sequences containing resistance genes, *e.g. Tn9* (*cam*) resistant to chloramphenicol. Each transposon has inverted repeat sequences flanking the resistant genes, as well as repeated bacterial sequences at the ends of the transposon. These elements are thought to transfer to other regions by replication.

9

MICROBIOLOGY OF FUNGI

Dematiaceous fungi are characterized by the development of a brown-to-olive-to-black color in the cell walls of their vegetative cells, conidia, or both. This cell coloring results in colonies that are olive to black. These ubiquitous and cosmopolitan opportunistic pathogens are normally associated with soil and plants, but occasionally they may cause infections in humans and animals. In medical mycology, dematiaceous fungi often are thought of as being exclusively hyphomycetes. This idea is in error because some ascomycetes, basidiomycetes, coelomycetes, and zygomycetes may be dematiaceous.

Mycotic infections caused by dematiaceous fungi include chromoblastomycosis, mycetoma, phaeohyphomycosis, and sporotrichosis. In this chapter, only chromoblastomycosis, phaeohyphomycosis, and sporotrichosis will be considered. Sporotrichosis is treated here because the etiologic agent is dematiaceous in culture, even though the yeast form in tissue is hyaline. Deciding whether a particular dematiaceous fungus is involved in the disease process can at times be difficult, since these fungi occasionally are recovered from clinical specimens as contaminants. Documentation of a dematiaceous fungus as the etiologic agent of a mycotic infection necessitates sound evidence that the infection is compatible with a mycosis, that the suspected etiologic agent is seen in clinical specimens, that the morphology of the fungus in the clinical specimens is compatible with the suspected etiologic agent, and that the recovered fungus is properly identified. The repeated recovery of a suspected etiologic agent, especially from more than one type of clinical specimen, is highly significant. The recovery of a fungus from body sites that normally are sterile is important.

Collection and Storage of Specimens

Clinical specimens must be collected aseptically and then promptly transported to the clinical laboratory in a properly labeled sterile container. Specimens collected on swabs or transported to the clinical laboratory in a transport medium are unacceptable for mycological study. An adequate quantity of clinical material is necessary if the information obtained in the laboratory is to be meaningful.

The most frequently submitted specimens for the recovery of dematiaceous fungi include aspirates, biopsy material, scrapings, and tissue specimens. Specimens other than skin scrapings must be protected from dehydration at all times. This protection can be accomplished by ensuring that a few drops of sterile saline or distilled water is added to the specimens at the time of their collection. Biopsy and tissue specimens can be kept moist by placing them between two pieces of sterile gauze moistened with sterile saline or distilled water. Clinical specimens should never be placed on cotton pads, since cotton fibers can be confused with hyphae in the direct microscopic examination. In addition, it is often impossible to recover all of the clinical specimen from among the cotton fibers. Direct microscopic examination of the specimens and subsequent plating must be done promptly.

Direct Examination

Clinical specimens obtained for the recovery of dematiaceous fungi usually do not require extensive processing. If aspirated specimens contain a substantial amount of purulent material, this can be dissolved with N-acetyl-L-cysteine without sodium hydroxide. Tissue specimens and biopsy material should be homogenized in a tissue homogenizer after highly suspicious areas consisting of necrotic, purulent, or caseous material are selectively examined microscopically and inoculated onto isolation media.

Specimens are typically examined microscopically in 10% KOH. The clearing process can be accelerated by gently heating the KOH preparation. The dematiaceous nature of fungal elements in clinical specimens should be determined only by brightfield microscopy. Phase-contrast microscopy is an excellent method for examining specimens, but it does not always permit the demonstration of the dematiaceous nature of these fungi. In some instances, the dark color of these fungi can be seen in tissue sections stained with hematoxylin and eosin. However, if a dematiaceous fungus is suspected, an unstained tissue section should be examined microscopically by bright-field microscopy. A drop of immersion oil can be placed directly onto a paraffin section mounted on a microscope slide, and then the section is examined microscopically.

The etiologic agents of chromoblastomycosis may be filamentous at the surface of the skin. In the deeper subcutaneous tissues, they occur as muriform cells (sclerotic bodies). Muriform cells are typically chestnut brown and variable in size, with thick cross walls arranged in a muriform manner. The cells result from vegetative growth without the elongation seen in hyphae. The presence of muriform cells in clinical specimens is diagnostic of chromoblastomycosis. However, the various etiologic agents of this mycosis cannot be identified solely on the basis of their morphology in tissue.

Phaeohyphomycosis is characterized by the presence in tissue of dematiaceous yeastlike cells, hyphae, or both. The hyphae may be regular and uniform in diameter or irregular in shape with many swollen cells, and they can be either short or very long. The name phaeohyphomycosis is not meant to be restricted to hyphomycetes, but it encompasses all dark hyphae causing disease in tissue, regardless of the taxonomic classification of the etiologic agent. As with chromoblastomycosis, the etiologic agents of phaeohyphomycosis cannot be identified in clinical specimens. These fungi must be grown on laboratory culture medium before they can be identified.

Most mycologists consider it fruitless to directly examine clinical specimens for the yeast form of *Sporothrix schenckii*. The number of yeast cells in the specimen is typically very limited, and their small size and shape are not distinctive. The yeast form of *S. schenckii* is not easily seen in sections of tissue stained with hematoxylin and eosin.

The fungus usually can be seen in tissue sections when the sections first are treated with diastase and then stained by the Gomori or periodic acid—Schiff technique. Fluorescent-antibody-specific conjugates are ideal, although they are available only at certain reference laboratories at the present time. Even though the yeast form may be difficult to see in specimens, there is generally no difficulty in recovering this fungus by the cultivation of clinical materials on media that are routinely used for the isolation of fungi.

Culture and Isolation

Dematiaceous fungi are easily isolated on most routine media. Some of them are sensitive to cycloheximide, and for that reason, Sabouraud dextrose agar (2% glucose) should be used in conjunction with a medium containing cycloheximide. Most dematiaceous fungi grow well at 30°C; some species grow poorly or not at all at 37°C. The majority of the pathogenic dematiaceous fungi usually are visible on isolation media within a week. However, cultures should not be discarded as negative until 6 weeks. Once a dematiaceous fungus is isolated, it must be determined whether or not the isolate is a pure culture. If the culture is not pure, then the fungus must be purified by the isolation of hyphal tips or individual germinating conidia. This purification is extremely important because many of the opportunistic dematiaceous pathogens are polymorphic; that is, they can produce several different kinds of conidia in the same culture. For an accurate identification, it must be known whether the various types of conidia present in a culture were formed by one fungus or by several fungi.

The colony characteristics and microscopic morphology used for the identification of dematiaceous fungi are based upon cultures that are approximately 2 weeks old and have been grown at 25 to 30°C on a medium such as potato dextrose agar or cornmeal agar. These media usually stimulate the formation of conidia. If a suspected pathogen cultivated as described above does not produce conidia, exposure to a naked daylight-type bulb for several days in a 12-hlight, 12-h-dark cycle while the pathogen is growing on a medium such as 2% water agar, sterile wooden sticks, potato dextrose agar, cornmeal agar, or filter paper may stimulate it to form conidia. The culture also can be lyophilized and then regrown, a procedure which often stimulates the development of conidia. Exposure to UV light often stimulates the development of conidia and other kinds of structures.

It has been suggested that pathogenic and nonpathogenic *Cladosporium* isolates can be distinguished from each other by their inability or ability, respectively, to hydrolyze casein, gelatin, or Loeffler serum medium. Although this test may be of some value for designating a *Cladosporium* species as a saprophyte, it cannot be used in place of morphological studies.

When isolates are believed to be *S. schenckii*, they should be subcultured on an enriched medium such as blood agar to determine whether they are dimorphic; that is, whether they grow vegetatively as hyphae at 25°C and as yeast cells at 35°C. The conversion from the mold form to the yeast form is enhanced by incubation of the inoculated medium in a candle jar.

When temperature studies are conducted, it is important to concurrently incubate an additional tube of medium inoculated with the fungus at 25°C to ensure viability of the inoculum. For an isolate to be considered dimorphic, only a few cells of its typical tissue form need to be present;

the entire colony does not have to be converted to its corresponding tissue form.

Identification

The identification of dematiaceous fungi ultimately rests upon their microscopic morphology and, to a lesser extent, upon their gross colonial morphology. The importance of conidium development in defining the numerous genera of dematiaceous fungi makes it essential to determine how a particular fungus forms its conidia. For this reason, slide culture preparations with potato dextrose agar or cornmeal agar are ideal for identification purposes.

Our new understanding of conidium development has resulted in the redefinition of many genera of medically important fungi. Terms such as spore and conidium (plural, conidia) are no longer used interchangeably. Many mycologists consider spores to be propagules that arise either from meiosis (ascospores, basidiospores, oospores, or zygospores) or by mitosis within a sporangium (sporangiospores).

All other asexual, nonmotile propagules are considered conidia. Conidia usually occur on specialized hyphae or hyphal branches called conidiophores. The actual cells that give rise to the conidia are referred to as conidiogenous cells. The distinction between the various kinds of conidiogenous cells is important for the identification of species of dematiaceous fungi. Phialides usually are flask shaped to cylindrical and have an apex that neither increases in length nor changes in diameter as the phialoconidia are formed. A cup-shaped structure called a collarette may be present at the apex of the phialide. In contrast to phialides, the apices of annellides increase in length, become narrower, and have apical rings called annellations. The annellations result when the annelloconidia separate from the apex of the annellide.

Some fungi produce conidia by a blowing-out process. Such conidia are called blastoconidia. They may occur individually or in chains. The term acropetal is used when the conidia at the apex of the chain are the youngest. The term basipetal refers to the condition of a chain of conidia when the youngest conidium is at the base of the chain. A number of the medically important dematiaceous fungi produce conidiophores that are sympodial. In this type of development, a conidium is formed at the apex of the conidiophore. The conidiophore then increases in length by the formation of a new growing point just below and to one side of the conidium. At the apex of this new growth, a second conidium develops. The entire process is repeated, which often results in a conidiophore that has the appearance of a series of bent knees, which is said to be geniculate.

The term anamorph is used to characterize an asexual reproductive structure or form produced by fungi. Occasionally, some fungi seen in the clinical laboratory produce more than one asexual form, or anamorph. An example of this polymorphic nature is *Fonsecaea pedrosoi*, which may form a sympodial anamorph (*Rhinocladiella* form), a phialide anamorph (*Phialophora* form), and an anamorph consisting of branched chains of blastoconidia (*Cladosporium* form). When a single fungus produces more than one anamorph, the term synanamorph can be used to designate any of these concurrently existing forms. The *Scytalidium* form is often associated with the pycnidial fungus *Hendersonula toruloidae*. This form can be referred to as a synanamorph associated with *H. toruloidae*. The *Rhinocladiella*, *Phialophora*, and *Cladosporium* forms are all synanamorphs of *F. pedrosoi*.

A number of medically important fungi have the ability to produce sexual forms. The sexual form of a fungus is referred to as a teleomorph. *Pseudallescheria boydii* is characterized by the formation of cleistothecia; hence, it is a teleomorph. *P. boydii* also may produce two anamorphs, that is, *Scedosporium* and *Graphium* forms. The term holomorph is used to encompass the whole fungus. In this example, the whole fungus consists of the *Pseudallescheria*, *Scedosporium*, and *Graphium* forms. Because fungi are classified by sexual structures, the name used for the teleomorph also is used for the whole fungus. Problems occasionally arise with anamorph-teleomorph connections because a teleomorph may have more than one anamorph and a single anamorph may be produced by several different teleomorphs. For example, *Scedosporium apiospermum* is one anamorph that is produced by more than one *Pseudallescheria* species.

The black yeasts at times are extremely difficult and frustrating to identify. Black yeasts typically represent one growth form or anamorph of polymorphic fungi. The genus *Phaeococcomyces* was established to accommodate isolates that consisted of black, budding yeasts with occasional short elements of pseudohyphae, or toruloid hyphae. The assumption that a black yeast, regardless of whether the colony is initially dematiaceous or not, should be identified as *Aureobasidium pullulans* is incorrect. One of the most frequently isolated black yeasts in the clinical laboratory is the *Phaeococcomyces* synanamorph of *Exophiala jeanselmei*. When fresh isolates of this fungus are transferred from Sabouraud dextrose agar to potato dextrose agar or cornmeal agar, the typical conidiogenous cells and conidia of *E. jeanselmei* rapidly become evident. Dematiaceous, yeastlike co lonies may be formed by such fungi as *A. pullulans*, *E. jeanselmei*, and many other species as well.

Sterile isolates represent a second group of medically important fungi that are especially difficult to identify. They commonly are referred to as members of the form-order Mycelia Sterilia. These fungi have been shown to cause phaeohyphomycosis and mycetoma. When sterile fungi are isolated, they should be exposed to near-UV radiation from a black light (310 to 410 nm) for several days in a 12-h-light, 12-h-dark cycle, incubated at both low and high temperatures, and subcultured onto media such as 2% water agar, hay infusion agar, soil extract agar, cereal agar, potato dextrose agar, cornmeal agar, and sterile, moist, wooden applicator sticks or filter paper. These media may help stimulate the production of conidia or fruiting bodies. The cultures should be kept for several weeks before they are discarded. With time, some of these fungi may develop structures that produce spores or conidia, such as ascocarps, pycnidia, or synnemata, either in the agar or at the colony surface.

Alternaria spp.

Members of the genus *Alternaria* occasionally are implicated as agents of phaeohyphomycosis. These fungi have been associated with infections involving bone, cutaneous tissue, ears, eyes, and the urinary tract. An *Alternaria* sp. and *Alternaria alternata* (synonym, *A. tenuis*) are the only well-documented human pathogens in this genus. The *Alternaria* an amorph of *Pleospora infectoria* has been reported to be a pathogen of humans, but this report has not been convincingly documented.

Alternaria colonies are rapid growing, cottony, and gray to black. The erect conidiophores are dematiaceous, simple or branched, and usually solitary, but they occasionally occur in small groups. The conidia of *Alternaria* spp. develop at the apex of the conidiophore in branching chains, with the youngest conidium at the apex of each chain. The conidia are dematiaceous, muriform, smooth or rough, tapering toward the distal end, and typically with a short cylindrical beak at their apices.

Alternaria isolates are difficult to identify beyond the generic level. If an isolate is recovered that must be identified to species, it should be sent to a specialist.

Aureobasidium spp.

Aureobasidium pullulans has been implicated as an agent of phaeohyphomycosis in humans and other animals. This hyphomycete is capable of causing opportunistic infections and has been reported from skin, nail, subcutaneous, and deeper tissues.

Colonies of *A. pullulans* are smooth, moist, and yellow, white, cream, light pink, or light brown, finally becoming black due to the development of arthroconidia. The conidiogenous cells are undifferentiated from the vegetative hyphae and may be intercalary, terminal, or arising as short lateral branches from the hyphae. The conidia are hyaline, one celled, smooth, ellipsoidal, and variable in shape and size. The conidia develop in a synchronous manner from the conidiogenous cells. Blastoconidia commonly are produced from the conidia that arise from the undifferentiated hyphal cells. A *Scytalidium* anamorph consisting of dematiaceous arthroconidia typically is present.

Hormonema species occasionally are confused with *Aureobasidium* spp. In the genus *Hormonema*, the conidia arise in a basipetal succession from either hyaline or dematiaceous hyphalike conidiogenous cells. In contrast, *A. pullulans* produces its conidia in a synchronous manner. Because of the confusion which has surrounded these two genera, some of the reported cases of infection ascribed to *A. pullulans* may have been caused by misidentified isolates of *Hormonema* spp. This speculation is based upon the fact that several authors have illustrated *Hormonema* spp. under the name *Aureobasidium*.

Cladosporium spp.

Cladosporium bantianum (synonym, *C. trichoides*) and *C. carrionii* are the most important pathogenic members of the genus *Cladosporium*. *C. bantianum* is the most frequently reported etiologic agent of cerebral phaeohyphomycosis, whereas *C. carrionii* occasionally is recovered from patients with chromoblastomycosis. *C. cladosporioides* is of some interest since it was the etiologic agent of a pulmonary fungus ball in one patient. This species has been unconvincingly implicated as a pathogen in eye and nail infections. Occasionally, other *Cladosporium* spp. are reported from cutaneous, eye, and nail infections. Because pathogenesis of *C. bantianum* other than cerebral involvement has not been well defined, it would be appropriate to handle this organism in a safety cabinet.

Cladosporium isolates are rapid growing, velvety or cottony, and usually some shade of olive gray to olive brown or black. From the mycelium, erect, tall, dematiaceous conidiophores arise. At the apex of the branching conidiophore, acropetally branching chains consisting of one- to several-celled, smooth or rough, dematiaceous blastoconidia form; the conidia have a dark hilum (basal scar). The conidia at the bottom of the chains tend to have the appearance of and are commonly referred to as shield cells.

C. bantianum and *C. carrionii* are morphologically similar. *C. carrionii* can be distinguished from *C. bantianum* by its slower growth rate, shorter conidia (2 to 3 by 4 to 5 μm versus 2 to 2.5 by 4 to 7 μm, with some being 3 by 15 to 20 μm), dermotropic nature in contrast to the neurotropic nature of *C. bantianum*, and maximum growth temperature of 35 to 36°C compared with 42 to 43°C for *C. bantianum*. Both of these species may form long chains of blastoconidia. *C. carrionii* can hydrolyze casein and starch, whereas *C. bantianum* apparently does not have this ability.

Curvularia spp.

Curvularia geniculata, *C. lunata*, *C. pallescens*, *C. senegalensis*, and *C. verruculosa* have been implicated in a number of opportunistic infections. Members of this genus have caused endocarditis, eye infections, mycetoma, and pulmonary phaeohyphomycosis.

Curvularia colonies are rapid growing, woolly, and gray to grayish black or brown. The conidiophores are dematiaceous, solitary or in groups, simple or branched, septate, and typically geniculate. The conidia are two to several celled, usually curved, dark with pale ends, solitary, and typically with a dark hilum. The conidia develop from a sympodial conidiophore. Works by Ellis should be consulted if a *Curvularia* isolate must be identified to species.

Drechslera spp.

Several *Drechslera* species have caused opportunistic infections in humans, including meningitis and cutaneous, eye, nasal, and pulmonary infections. The presently recognized pathogenic members of this genus include an unidentified *Drechslera* sp., *Drechslera hawaiiensis*, *D. longirostrata*, *D. rostrata*, and *D. spiciifera*. Alcorn has suggested that some of these species should be classified in the genera *Bipolaris* and *Exserohilum*. Additional study is necessary before this issue can be adequately resolved.

Drechslera species form rapid-growing, woolly, gray-to-black colonies. The conidiophores are dematiaceous, solitary or in groups, simple or branched, septate, and geniculate. The dematiaceous, oblongto-cylindrical conidia are multicelled and develop from a sympodial conidiophore.

Some medical microbiologists have confused *Helminthosporium* spp. with *Drechslera* spp. The conidiophores of *Helminthosporium* spp. are straight, and they stop-lengthening when the terminal conidium is formed. The conidia develop along the conidiophore; hence, the conidiophore is not sympodial. *Helminthosporium* spp. are rarely, if ever, isolated in the clinical laboratory, and members of this genus have not caused phaeohyphomycosis in humans. If it is necessary to identify *Drechslera* isolates, either the works of Ellis or a specialist in this genus should be consulted.

Exophiala spp.

Exophiala jeanselmei, previously known as *Phialophora jeanselmei* or *Phialophora gougerotii* is a relatively common etiologic agent of mycotic subcutaneous abscesses. This dematiaceous hyphomycete may cause either mycetoma or phaeohyphomycosis. *E. moniliae* and *E. spinifera* also have been reported as agents of phaeohyphomycosis, in which they caused subcutaneous cysts. The last member of this genus known to be pathogenic for humans is *E. werneckii* (synonym, *Cladosporium werneckii*), which causes superficial phaeohyphomycosis (synonym, tinea nigra).

The colonial morphology of the members of the genus *Exophiala* is varied. The colonies are slow to rapid growing, often moist and yeastlike at first, becoming woolly with age, and gray to black. Some isolates of *E. werneckii* and the *Phaeococcomyces* synanamorph of *E. jeanselmei* may remain black and yeastlike. Conidiophores are dematiaceous, simple, or hyphalike. The conidiogenous cells are annellides. In *E. jeanselmei*, the annellides are lageniform to cylindrical, tapering to a narrow apex; in *E. moniliae*, they are inflated to elliptical, tapering to a very long, narrow apex; in *E. spini fera*, they are lageniform to cylindrical, tapering to a narrow apex, and they arise from distinct spinelike conidiophores; and in *E. werneckii*, either they are hyphalike or they consist of two-celled yeast cells that are clavate.

The conidia are one-celled in most species and accumulate in balls at the apices of the annellides. With careful study utilizing the oil immersion objective, annellations (rings) usually can be seen at the apices of the annellides.

E. jeanselmei was incorrectly believed by some to belong to the genus *Phialophora*. However, when it was discovered that the conidiogenous cells of *E. jeanselmei* were annellides and not phialides, the fungus was transferred to the genus *Exophiala*. An identical situation occurred when it was discovered that the conidiogenous cells of *E. spini fera* were annellides and not phialides. *E. werneckii* produces one- to two-celled conidia from annellides that exist as either yeast cells or intercalary conidiogenous cells incorporated within the hyphae.

The fungus originally described as *Sporotrichum gougerotii* was considered a morphological variant of *Sporothrix schenckii*. The name *S. gougerotii* is best considered a nomen dubium. Fungi that are currently identified as either *S. gougerotii* or *P. gougerotii* are typically misidentified isolates of *E. jeanselmei*. At one time, these two supposedly different fungi were erroneously distinguished from each other by their tissue morphology. In the sense of some contemporary medical mycologists, *P. gougerotii* is a misapplied name for isolates of *E. jeanselmei* that do not form granules in tissue.

Fonsecaea spp.

The genus *Fonsecaea* contains two species, *Fonsecaea compacta* (incorrectly spelled *compactum* by some) and *F. pedrosoi*. Both of these species are agents of chromoblastomycosis.

Fonsecaea colonies are slow growing, velvety to woolly, and olive to black. *Fonsecaea* isolates are extremely polymorphic. They are characterized by the development of one-celled primary conidia that form on erect, dark, sympodial conidiophores. The primary conidia in turn become conidiogenous cells and form secondary one-celled conidia. This form of development has been incorrectly called *Acrotheca*-like by some. It is actually more similar to the form seen in the genus *Rhinocladiella*. Some of the conidia occur as branching chains of blastoconidia, identical to those found in the genus *Cladosporium*. *Fonsecaea* spp. may produce phialides with the collarettes bearing the balls of one-celled conidia that are typical of the genus *Phialophora*.

F. pedrosoi and *F. compacta* are morphologically distinct. *F. pedrosoi* is differentiated from *F. compacta* by its elongate conidia that occur in loose heads, in contrast to the rounded conidia in compact heads produced by *F. compacta*. As a result of their polymorphic nature, *F. pedrosoi* and *F. compacta* have been placed inappropriately in the genera *Phialophora* and *Rhinocladiella* by some mycologists.

Phaeococcomyces spp.

Phaeococcomyces is a genus that contains black yeasts. Black yeasts are often synanamorphs associated with several of the medically important polymorphic dematiaceous hyphomycetes such as *E. jeanselmei* and *Wangiella dermatitidis*. The genus *Phaeococcomyces*, which was originally named *Phaeococcus*, contains four species that are distinguished from each other primarily on morphological criteria.

Phaeococcomyces exophialae forms slimy, mucoid, slow-growing, smooth colonies that are grayish black. Budding yeast cells which are at first subhyaline are abundant. With age, some of the cells become darker, with thickened cell walls. Some pseudohyphae usually are present. Hyphal development may become dominant in some isolates of this species with subsequent subculture.

Black yeasts occasionally are isolated in the clinical laboratory. They are recognized by their black, mucoid, yeastlike colonies. When grown on cornmeal agar or potato dextrose agar, many black yeasts will rapidly produce the hyphae and conidiogenous cells typical of genera such as *Exophiala* and *Wangiella*. Based upon conidiogenesis, other genera of black yeasts probably will be needed in the future to accommodate this group of fungi.

Phialophora spp.

Members of the genus *Phialophora* are well-recognized etiologic agents of phaeohyphomycosis and chromoblastomycosis. In addition to cutaneous and subcutaneous tissue invasion, some species have caused endocarditis and mycotic keratitis. The pathogenic species of *Phialophora* include *Phialophora bubakii*, *P. parasitica*, *P. repens*, *P. richardsiae*, and *P. verrucosa*.

Phialophora colonies are rapid growing, cottony to woolly, and usually some shade of olive gray. When conidiophores are present, they usually are short. The conidiogenous cells are hyaline to dematiaceous phialides that are cylindrical to flask shaped. At the apices of the phialides, distinct collarettes are present. The conidia are one celled and usually hyaline, and they occur in balls that may occasionally slip down along the phialides in some species. Some species commonly produce intercalary phialides; a yeast form may occur in some isolates.

P. mutabilis and *P. hoffmannii* have been considered species of the genus *Phialophora* for a number of years. Gams and McGinnis recently have reclassified these species in the genus *Lecythophora*. This reclassification was necessary because these fungi produce intercalary phialides with short, lateral, cylindrical necks, which bear the conidia in balls at their apices. Collarettes are present at the tips of the phialides.

Rhinocladiella spp.

Rhinocladiella aquaspersa, previously known as *Acrotheca aquaspersa*, is a rare etiologic agent of chromoblastomycosis. Human cases of chromoblastomycosis caused by *R. aquaspersa* have occurred in Brazil and Mexico.

Colonies of *R. aquaspersa* are rapid growing, velvety, slightly elevated, and olive black. The conidiophores are sympodial, usually darker than the vegetative hyphae, unbranched, and erect. Conidia are one celled, rarely two celled, fusiform, elliptical or obovate, smooth, light brown, and with a dark basal scar. Annellides like those of *Exophiala* spp. and phialides like those of *Wangiella* spp. may be present.

Scedosporium spp.

Scedosporium apiospermum, previously known as *Monosporium apiospermum*, is an anamorph of *Pseudallescheria boydii*, a fungus once classified as *Petriellidiurn boydii* and *Allescheria boydii*. The fungus may cause mycetoma, as well as infections involving the lungs and brain, where the fungus grows in the form of hyphae that look like those produced by *Aspergillus* spp. *S. apiospermum* rapidly produces colonies that are cottony and smoky gray to dark brown. One-celled conidia may occur singly along the hyphae or in clusters at the apices of annellides. The conidia are obovate, truncate, and subhyaline to light black.

S. apiospermum occasionally has a *Graphium* synanamorph present. Several members of the genera *Pseudallescheria* and *Petriella* may produce a *S. apiospermum* anamorph. Therefore, without having the teleomorph present, it is not possible to determine if an isolate of *S. apiospermum* was produced by *Pseu* dallescheria boydii.

Scytalidium spp.

Members of the genus *Scytalidium* have been well documented as opportunistic fungal pathogens of nail, skin, and subcutaneous tissue. The *Scytalidium* synanamorph associated with the pycnidial fungus *H. toruloidea*, as well as *Scytalidium lignicola* and *S. hyalinum*, all have caused disease. *S. hyalinum*, because of its hyaline nature, would best be classified in a genus other than *Scytalidium*.

Scytalidium spp. produce rapid-growing colonies that are at first white, becoming dark gray with age. In *S. lignicola*, arthroconidia of two types are formed. In the first type, the arthroconidia are cylindrical, one celled, and hyaline. In the second type, they are thick walled, yellowish brown, and one or two celled. The *Scytalidium* synanamorph of *H. toruloidea* forms arthroconidia that are brown, cylindrical at first, becoming rounded, barrel shaped or subglobose, and one or two celled.

The monograph on *Malbranchae* spp. by Sigler and Carmichael should be consulted for the identification of *Scytalidium* species and similar hyphomycetes that produce arthroconidia.

Sporothrix spp.

S. schenckii, a dimorphic fungus, is considered the only pathogenic member of the genus *Sporothrix*. Recently, a new species, *Sporothrix cyanescens*, was added to the genus. Some of the isolates upon which the species description was based were isolated from patients with mycosis of human skin. Whether or not *S. cyanescens* is another etiologic agent of sporotrichosis remains to be proven.

Colonies of *S. schenckii* are rapid growing and at first moist, flat, and yeastlike, later developing aerial hyphae. They are initially white, becoming brown to black with age. The conidia are of two kinds in most fresh isolates. Hyaline, one-celled conidia develop solitarily upon denticles along the hyphae; laterally from sympodial, slender, tapering, erect conidiophores; and terminally in clusters at the apices of swollen conidiophores. The second type of conidia are one celled, thick walled, and black. These conidia develop along the hyphae. At 37°C on enriched media, the mold form of *S. schenckii* converts to a yeast form.

The etiologic agent of sporotrichosis originally was described as *S. schenckii*. Later, the fungus erroneously was transferred to the genus *Sporotrichum*. Members of the genus *Sporotrichum* are characterized by the formation of large hyphae with clamp connections and large, one-celled, thick-walled, golden conidia. They are neither dimorphic nor pathogenic for humans and other animals.

Wangiella sp.

W. dermatitidis is an agent of phaeohyphomycosis that typically causes infections involving cutaneous and subcutaneous tissue. The fungus most frequently has been seen in patients living in Japan.

The colonies of *W. dermatitidis* are moist and at first yeastlike, developing some aerial hyphae with age. They are olive to black. Distinct conidiophores are absent. The conidiogenous cells are phialides which do not have collarettes. Some conidiogenous cells appear to possess a group of slightly raised, truncate denticles at their apices that occur as a result of sympodial development. Rare annellides may be produced by isolates of this fungus. The one-celled, light-to-dark, smooth phialoconidia form in balls at the apices of the phialides and then slide down their sides. The phialides develop from conidiophores that are indistinguishable from the hyphae. Most isolates produce an abundant yeast form and large amounts of toruloid hyphae.

The genus *Wangiella* was established to accommodate date the fungus known as either *Hormiscium dermatitidis* or *Phialophora dermatitidis*. The new genus *Wangiella* was necessary because the phialides without collarettes that are typical of *W. dermatitidis* could not be accommodated in any known genus. *W. dermatitidis* can be recognized by its ability to grow at 40°C, whereas similar dematiaceous hyphomycetes do not grow at that temperature.

10

FUNGAL GENETICS

The study of fungal genetics has made a number of significant contributions to our knowledge of genetic processes. The early idea of the connection between genes and enzymes was based on nutritional mutants of *Neurospora*, and rapid progress in bacterial genetics occurred when this approach and selective techniques were extended to *Escherichia coli*. One major advantage which some fungi have over bacteria is the occurrence of meiosis in a closed sac, the *ascus* in the ascomycetes. This allows the genetic effects of a single meiotic event to be studied in detail. There is nothing comparable in bacteria and very often selective techniques are necessary even to detect bacterial recombination. In addition, fusion of nuclei followed by meiosis does not occur in the bacteria. The result of this has been that fungal genetics has made a significant contribution to basic ideas on mechanisms of recombination. In general bacteria are easier to handle for biochemical analysis, and therefore the biochemical basis of recombination is better understood in bacteria than in fungi. This chapter will examine some of the important experiments on recombination in fungi and will include details of the parasexual cycle which occurs in a variety of filamentous fungi.

Neurospora crassa

Neurospora crassa is a haploid organism that grows vegetatively by the formation of a weblike, branching growth called a *mycelium*. For asexual propagation of *Neurospora*, the asexual spore (*conidia*) or pieces of mycelia can be used to inoculate a growth medium to produce a new mycelium. The mycelium consists of cellular compartments separated by a septum with a central hole that permits circulation of

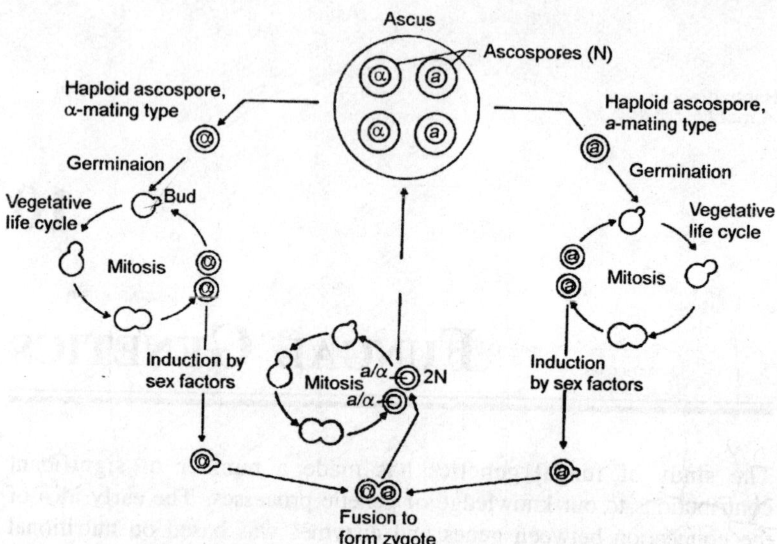

Fig. 10.1. Life cycle of the yeast Saccharomyces cerevisiae.

cell contents, including nuclei, throughout the mycelium. *N. crassa* has two mating types, *A* and *a*, and these are determined by members of an allelic pair. The two mating types look identical and can be distinguished only by the fact that *A* strains mate with *a* strains but not with *A* strains, and *a* strains mate with *A* strains but not with *a* strains. Sexual reproduction occurs by fusion of nuclei of the opposite mating types to form a diploid zygote that has only a transient existence in the life cycle of this organism. The *A/a* zygote undergoes meiosis in an elongating linear tubular structure called an ascus. The two divisions occur in tandem within the developing ascus, producing four haploid nuclei (two *A* and two *a*). Each nucleus divides mitotically, resulting in a linear arrangement of eight haploid nuclei around which spore walls form to produce eight ascospores (four *A* and four *a*). The eight spores of each ascus represent the four meiotic products (each doubled) arranged in an order that reflects exactly the orientation of the four chromatids of each tetrad at metaphase 1 of meiosis: these tetrads are called ordered tetrads. When the haploid ascospores germinate, they give rise to the vegetative mycelium.

The asci themselves are formed within a fruiting body called a *perithecium*. When the asci are "ripe" the ascospores are ejected through the neck of the perithecium and may be collected for random spore analysis. If the asci are removed from the perithecium just

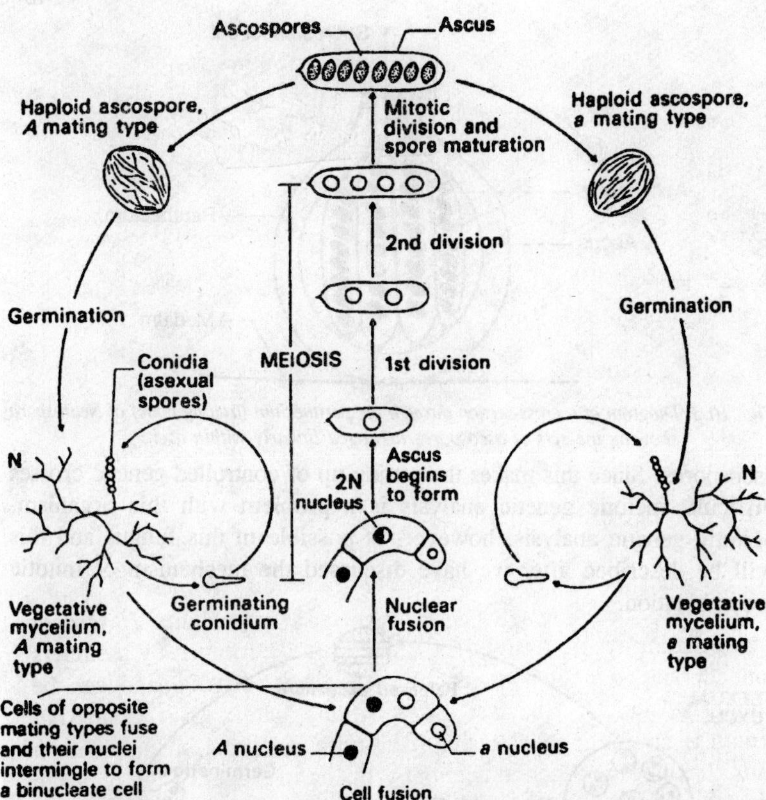

Fig. 10.2. Life cycle of Neurospora crassa.

before they burst, each ascus can be dissected manually for ordered tetrad analysis.

Aspergillus nidulans

This fungus is a haploid organism and has a colourless, multinucleate mycelium. The asexual spores, the conidia, bud off from specialized conidiophores and are uninucleate and dark green in colour in the wild type. When these spores germinate, a mycelium is produced and the vegetative cycle is completed. *Aspergillus* has a sexual cycle in which asci are produced with eight ascospores that, when they germinate, give rise to the vegetative mycelium. Unlike *Neurospora*, which requires fusion of nuclei of two different mating types to instigate the sexual cycle, *Aspergillus* is *homothallic*, meaning that it can and does self-cross. Two nuclei from the same culture can fuse to produce a diploid nucleus, which then undergoes meiosis to produce the

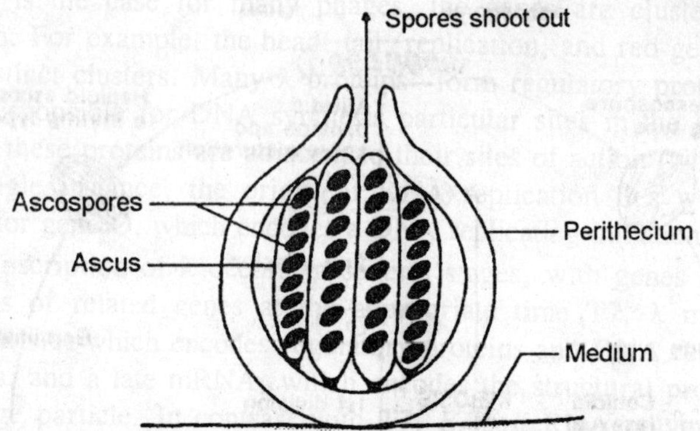

Fig. 10.3. Diagram of a cross-section through the perithecium (fruiting body) of Neurospora showing the sets of ascospores arranged linearly within asci.

ascospores. Since this makes the setting up of controlled genetic crosses difficult, meiotic genetic analysis is a problem with this organism. Mitotic genetic analysis, however, is possible in this fungus and this will be described after we have discussed the mechanism of mitotic recombination.

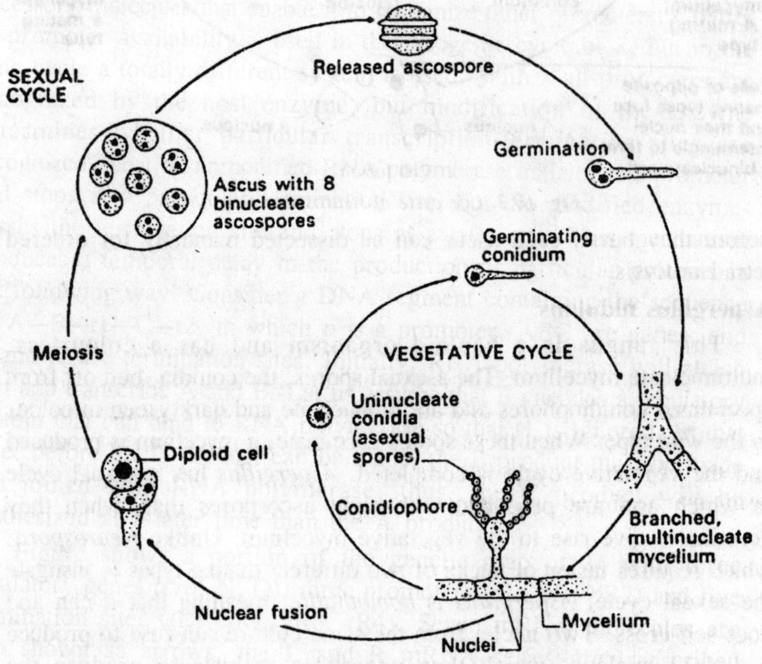

Fig. 10.4. Life cycle of Aspergillus nidulans.

Tetrad Analysis in Sordaria fimicola

Sordaria fimicola is a filamentous fungus which occurs on the dung of plant-eating animals. It produces large linear asci in fruiting bodies known as *perithecia*. Each ascus originates from a cell containing a single diploid nucleus, which undergoes meiosis to produce four nuclei in order. Each nucleus then undergoes mitosis, and the eight nuclei give rise to eight ascospores, normally reflecting the regular distribution of chromosomes which occurs during these divisions. Analysis of the order of ascospores in the asci allows the position of the centromere to be located, and gene order can also be worked out. Abnormal asci have been observed, and these have provided clues to the biochemical mechanism of recombination.

Mapping of the Centromere Distance for the Hyaline Mutation

The easiest mutations to study are those which affect ascospore pigmentation. One of these is the white or *hyaline* mutation *hy* which produces ascospores without the normal black pigment. The order of these ascospores can be studied easily and therefore the segregation of *hy* allele can be followed. *S. fimicola* is homothallic, i.e., there is no mating type, and a single haploid ascospore can grow to produce a mycelium which produces fruiting bodies. These will have ascospores which are all the same colour, either black or hyaline, depending on the parental ascospore colour (*self-fertilized perithecia*). Perithecia can contain ascospores which are all black, all hyaline, or a mixture of black and hyaline. Genetically speaking, we are only interested in the last type, which are known as *hybrid perithecia*. An inspection of these asci shows that there are broadly two types of asci—those which have pairs of ascospores of the same colour. An analysis of the behaviour of the *hy* locus during meiosis reveals that the first type of ascus occurs when no crossing-over has occurred between the locus and its centromere. The second type occurs when crossing-over has occurred between the locus and the centromere. The two types of asci are known as *first-division segregation* and *second-division segregation types* respectively. It follows from this explanation that the greater the distance between the locus and the centromere, the higher will be the frequency of the second-division segregation type. The recombination frequency can be calculated from the formula

$$\% \text{ recombination} = \frac{\frac{1}{2} \times \text{second division segregation asci}}{\text{total asci}} \times 100$$

The reason for the factor of a half is that for every crossover event there will be two recombinant and two parental chromatids. The recombination frequency of 33.3% is the maximum which can be obtained as, when a particular locus is well away from the centromere, each of the six possible ascus types will be equally likely, due to the free recombi-nation of alleles followed by random orientation of the chromosomes in the bivalent at meiosis. Application of the formula gives

$$\frac{1}{2} \times \frac{4}{6} \times \frac{1}{3}.$$

Analysis of this type were performed as early as 1932 on *Neurospora crassa* by Lindegren, who analyzed the mating-type locus following micro-manipulation of asci. This was early evidence that crossing-over occurred at the four-strand stage of meiosis.

Tetrad Analysis in Neurospora crassa

The ability to isolate ordered tetrads (the four products of meiosis) from certain fungi (such as *Neurospora*) allows the distance between a gene and the centromere to be determined. The determination of gene-centromere distance is usually not possible with unordered tetrads or by random progeny analysis. In addition, the isolation of ordered or unordered tetrads provides a different means of mapping the distance between two or more genes.

Gene-centromere distance determination

This requires the isolation of ordered tetrads such as are produced by *N. crassa*. The example considers the mating type alleles *A* and *a*,

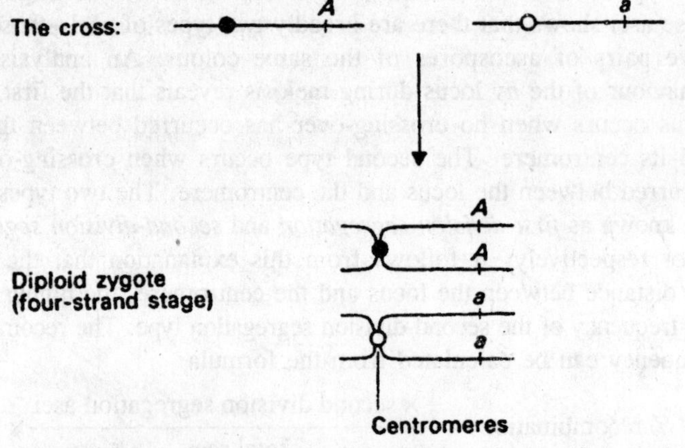

Fig. 10.5. Determination of the gene-centromere distance of the mating-type locus of Neurospora: formation of diploid zygote by crossing a strain of mating-type A with a mating-type a strain.

which are found in linkage group 1. At the zygote stage, each chromosome has doubled and the resulting four-chromatid (tetrad) stage for linkage group 1. Obviously this stage is in a more compact three-dimensional state in the cell but for our purposes the two-dimensional representation is adequate. In this and the other figures that will be discussed in this example, ● will be used to indicate the centromere of the *a* parent. The centromeres are physically alike, of course, but it will be useful to distinguish the two for the sake of discussion. Note that at this diploid zygote stage the chromosome has divided but the centromeres have not.

The meiotic division and the ascus that results when no crossover occurs between the centromere and the mating-type locus. (Note: for simplification the final mitotic division, which merely replicates each of the four meiotic products, is omitted). As can be seen the resulting four ascos-pores directly reflect the arrangement of the four chromatids in the zygote. Since the cen-tromeres do not divide until just before the second meiotic division, there will be a 2:2 segregation (4:4 if we consider the mitotic division that produces eight ascospores) of the centromere from one parent (●) to the cen-tromere of the other parent (○). We say that the centromeres always segregate to different nuclear areas at the first meiotic division; that is, they show *first division segregation*. If no crossover occurs between the gene and its centromere, that gene will also show first division segregation. After the first division, the two chromatids with the *A* alleles had migrated to a different region of the ascus from the two chromatids carrying the *a* alleles. Also, since it is equally probable that the four chromatids in the zygote are inverted, we would expect to find equal numbers of these two types of asci.

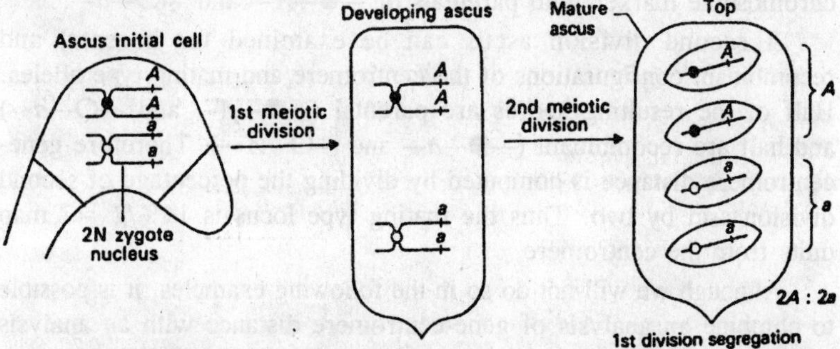

Fig. 10.6. Determination of the gene-centromere distance of the mating-type locus of Neurospora.

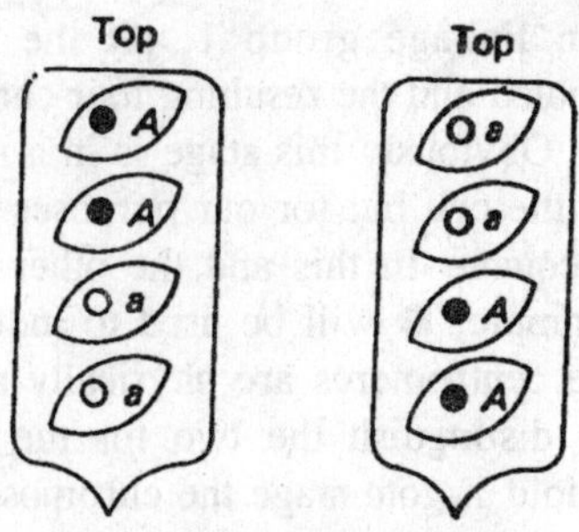

Fig. 10.7. The two possible orientations of mating-type alleles and centromeres (● and ○) in first division segregation asci.

If a single crossover occurs between the mating type locus and its centromere, the centromeres will show first division segregation as before (and thus in a sense we can think of them as being "genes" that always segregate at the first division), and the *A*/*a* alleles will show *second division segregation*. That is, the separation of the allelic genes is delayed until the second meiotic division because of the crossover event. Since the three-dimensional arrangement of the two noncrossover and the two crossover chromatids varies, four second division segregation asci result with equal frequencies. Again the final mitotic division is omitted for the purpose of simplification.

By the analysis of ordered tetrads, the percentage of asci that show second division segregation for a particular allelic pair can be determined. For the mating type locus, this is 14%. As we learned before, the distance between two genes in map units comes directly from the percentage of recom-binant progeny from a particular cross. In the *Neurospora* cross we can consider the centromere to be a chromosome marker with parentals of —●—*A*— and —○—*a*—.

A second division ascus can be examined for parental and recombinant configurations of the centromere and mating type alleles. Half of the resulting spores are parental (—●—*A*— and —○—*a*—) and half are recombinant (—●—*a*— and —○—*A*—). Therefore gene-centromere distance is computed by dividing the percentage of second division asci by two. Thus the mating type locus is 14%/2 =7 map units from the centromere.

Although we will not do so in the following examples, it is possible to combine an analysis of gene-centromere distance with an analysis of map distance between two or more genes if ordered tetrads rather than unordered tetrads are isolated.

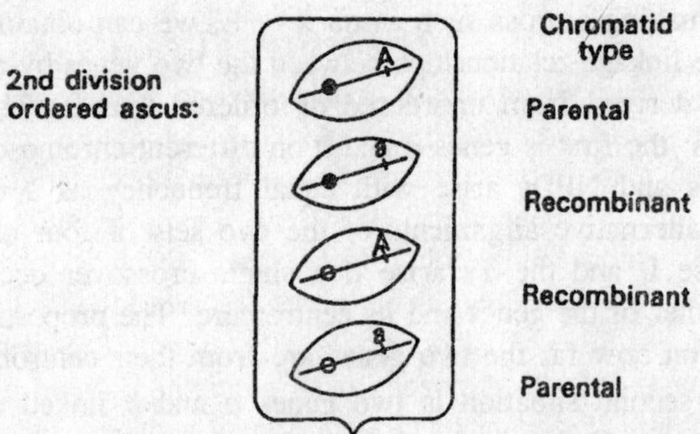

Fig. 10.8. Parental and recombinant configuration of the centromeres and mating-type alleles in a second-division segregation ascus.

Map distance between two genes

With two heterozygous genes in a diploid cell at meiosis there are three possible segregation patterns from a cross. For *ab* x + +, for example, the three types of tetrads are depicted. The parental ditype (PD) tetrad has two types of spores, *a b* and + +, which are both parental. The nonparental ditype (NPD) tetrad has *a* + and + *b* spores, both of which are recombinant (nonparental). The tetratype (T) tetrad has two parental (*a b*, + +) and two recombinant (*a* +, + *b*) spores, hence four types of spores. (Note: the existence of tetratype asci is evidence that crossing-over occurs at the four-strand stage of meiosis).

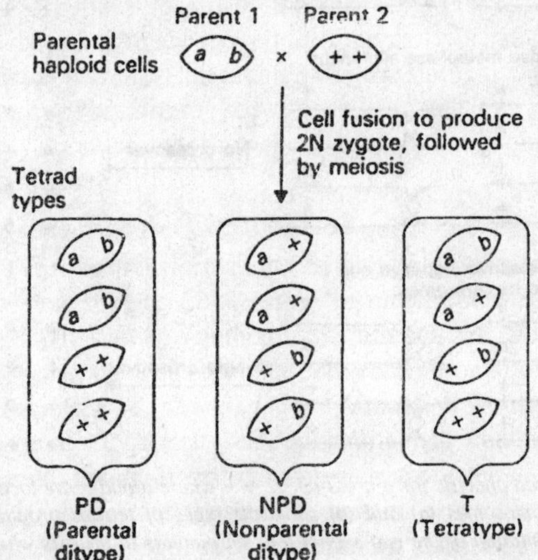

Fig. 10.9. The three types of tetrads that can be produced from a cross ab × + +.

By making a cross such as *ab* x ++, we can obtain information about the linkage relationships between the two genes by analyzing the progeny derived from unordered or ordered tetrads. There are two situations: the first is genes *a* and *b* on different chromosomes. In this case PDs and NPDs arise with equal frequency as a result of the random alternative alignments of the two sets of four chromatids at metaphase I, and the Ts arise if a single crossover occurs between one or other of the genes and its centromere. The proportion of T asci depends on how far the two genes are from their centromeres.

The second situation is two genes *a* and *b* linked on the same chromosome. Ordered tetrads are drawn in the diagrams, but the analysis is the same with unordered tetrads. The possibilities we shall consider are no single, and double crossovers.

If no crossover occurs between the two genes all of the asci are PD asci in which all of the spores are of the parental type. If a single crossover takes place between the two loci all of the asci are T asci in which one-half of the spores are parental and one-half are recombinant. In considering double crossovers since there are four

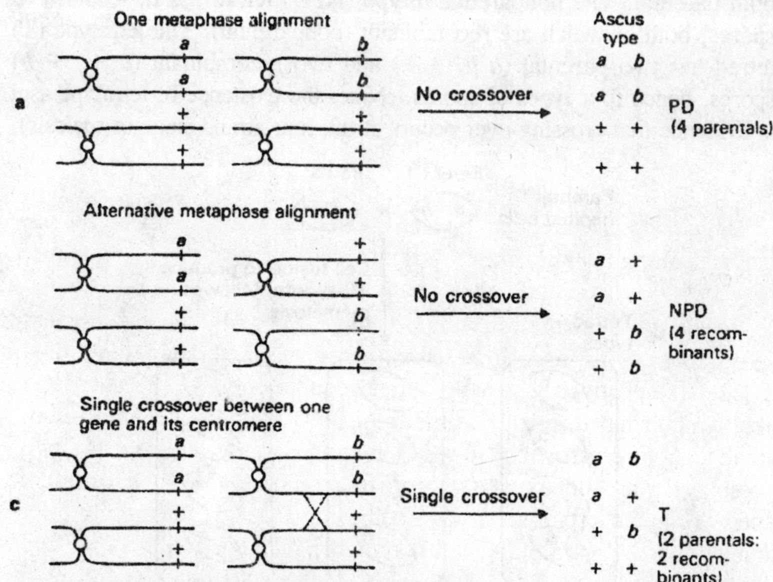

Fig. 10.10. Tetrad analysis for a cross ab × ++ where a and b are located on different chromosomes (a) and (b) show the types of tetrads produced by random orientation of the two sets of four chromatids in meiosis when no crossover occurs; (c) shows the tetrad type produced when there is a crossover between one gene and its centromere.

chromatids in the zygote, it is necessary to consider three types: two-strand, three-strand, and four-strand double crossovers. There are two ways of having three-strand double crossovers, and thus the relative proportion of two-strand : three-strand : four-strand : double crossovers is 1:2:1, respectively. Asci resulting after a two-strand double crossover are all PD asci in which all spores are parental. Asci from three-strand double crossovers and T asci (one-half parental, one-half recombinant, and asci from four-strand double crossovers) are NPD asci.

As has been discussed previously, the distance between any two genes is computed from the formula:

$$\frac{\text{Number of recombinants}}{\text{Total progeny}} \times 100$$

Examining the tetrads, we see that NPD asci contain four recombinant spores and T asci contain two parental and two recombinant spores. Therefore converting the general mapping formula into tetrad terms, the *a-b* distance is equal to:

$$\frac{\frac{1}{2}\text{T} + \text{NPD}}{\text{Total}} \times 100$$

Thus if there were 1000 asci, with 900 PD, 96 T and 4 NPD, the *ab-b* distance would be:

$$\frac{\frac{1}{2}(96) + 4}{100} \times 100 = 52 \text{ map units}$$

However, the formula derived calculates map distances on the basis of recombination frequency rather than on the basis of crossover frequency. A correction can be made for this.

In the theory that was presented, three types of double crossovers were shown: two-strand, three-strand, and four-strand doubles in a ratio of 1:2:1, respectively. The four-strand double crossover frequency is directly determined from the number of NPD asci. In the case of two-strand double crossovers the result is a PD ascus, but here the two crossovers were not counted. From the ratio of the three different types of double crossovers, the number of PD asci that result from double crossovers should equal the number of NPD asci.

There are two types of three-strand double crossovers and each gives rise to a T ascus, as does a single crossover. Thus, for each three-strand double crossover, only one of the crossovers is really being considered in the original formula, and between the two three-

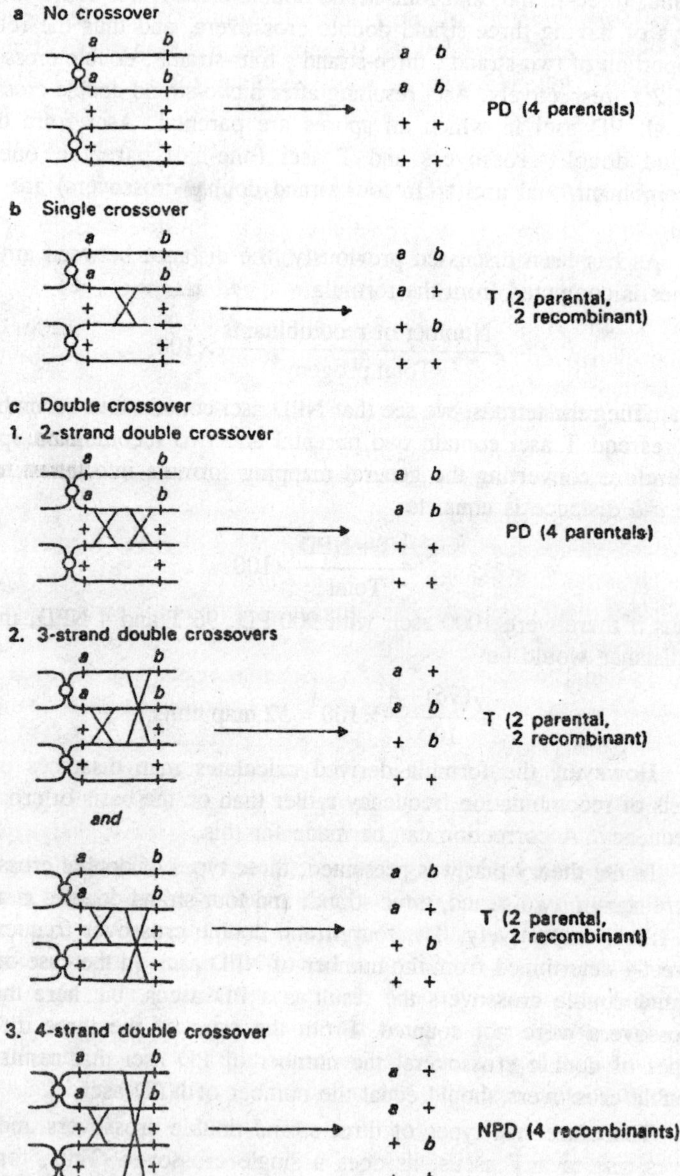

Fig. 10.11. Tetrad analysis for a cross ab × ++ where both genes a and b are located on the same chromosome,

strand double crossover, the equivalent of two crossovers or one double crossover is not being included in the calculations. As we have seen, the NPD frequency is equivalent to a double crossover frequency. The formula for calculating the distance between two genes from tetrad data can then be modified.

Therefore map distances based on crossover frequencies as manifested by the tetrad types, is given by the formula:

$$\frac{\frac{1}{2}T + 3NPD}{Total} \times 100$$

For the numbers given before, this would be:

$$\frac{\frac{1}{2}(96) + 3(4)}{1000} \times 100 = 6 \text{ map units}$$

The result is 6 map units v.s. 5.2 map units for the "old" recombination frequency formula.

The tetrad formula can be used to calculate the map distance between any two genes. In crosses where more than two genes are segregating, the data should be analyzed by considering two genes at a time and sorting the data in PD, NPD, and T ascus types for those two genes.

Tests for Gene Linkage using Tetrad Analysis

Tetrad analysis can be used also to determine whether two genes are linked or unlinked. If two genes are unlinked by being on different chromosomes, the PD frequency will equal the NPD frequency PD asci contain only parental-type spores and NPD asci contain only recombinant type spores. T asci contain half parental and half recombinant spores. Therefore, with PD=NPD, whatever the T value, the recombination frequency (RF) is 50%, which signifies no linkage between the two genes.

The frequency of T asci might provide information, however about whether two unlinked genes are no different chromosomes or far apart on the same chromosome. In the former case, the T asci arise as a result of a single crossover between one or other of the two genes and their respective centromeres, and thus the frequency of T asci will depend on how far away from the centromeres the genes are. In this case the T frequency can vary between 0 (both genes tightly linked to their centromeres) and the limiting frequency of 66.7%. On the other hand, if two genes are very far apart on the same chromosome, there will be large number of crossovers, even-numbered ones occurring with about the same frequency as odd-numbered ones, thus giving rise to equal numbers of PD and NPD asci, respectively. In this situation the T frequency will be 66.7% of all asci, and this will now be explained for a cross *a b* x a^+ b^+ in which the *a* and *b* alleles are initially on chromatids 1 and 2 and the a^+ and b^+ alleles are initially on chromatids 3 and 4. If the two loci are far apart, there will be multiple crossovers between them.

If we consider chromatid 1 and the *a* allele, since multiple crossovers occur between the two loci, there is an equal probability that this chromatid will end up being *b* or b^+. That is the probability it will be *b* is p(*b*)=1/2 and the probability it will be b^+ is p(b^+)=1/2. If we now consider chromatid 2 and compute the probabilities of the genotypes, this will then determine the ascus type. Thus, if chromatid 1 carries *b*, chromatid 2 has a one-third probability to carry *b* and a two-thirds probability to carry b^+. If the former occurs, chromatid 1 will be *a b* and chromatid 2 will be *a b*, and consequently both chromatids 3 and 4 will be a^+ b^+, thereby giving a PD ascus. If the latter is the case, chromatid 1 will be *a b* and chromatid 2 will be *a* b^+, so chromatids 3 and 4 will be a^+ *b* and a^+ b^+ (although which will be which cannot be predicted), thereby giving a T ascus.

In summary, when two genes are unlinked and very far apart on the same chromosome, the result will be a ratio of 1 PD : 1 NPD : 4 T asci. For two genes on different chromosomes, PD=NPD, and the relative number of T will depend on the distance the two genes are from their centromeres. A low T frequency compared with PD and NPD would certainly indicate that the two genes in question are on different chromosomes.

If the two genes are linked, NPD asci can only arise as a result of a four-strand double crossover, which is a very rare event. Therefore in this case the frequency of PD asci will greatly exceed the frequency of NPD asci (i.e. PD >>NPD). The T asci result from single- and three-strand double crossovers, and these will occur with a frequency intermediate between those for PD and NPD.

Abnormal asci

Early ideas on recombination assumed that it was a reciprocal process, and the observation by Lindegren (1953) that 3:1 ratios occurred in the asci of *Saccharomyces cerevisiae*, instead of the expected 2:2 ratios, was not generally accepted until it was confirmed by other workers. Olive (1959) examined 2700 asci of *S. fimicola* and observed 6 asci with 6 black to 2 *hy* spores, and a further 5 asci with 2 black and 6 hyaline spores. This departure from the expected 4:4 segregation was described as *gene conversion*, as apparently one allele was converted into the other during the process. An even more unexpected observation was that for another ascospore mutant (g for grey spores) 5:3 and 3:5 ratios occurred. This implied that the DNA after meiosis was in some way hybrid (or heteroduplex) and that segregation occurred in the final mitosis (*post-meiotic segregation*). Kitani and co-workers analyzed abnormal asci in crosses involving grey ascospores in which loci either side of the grey locus were segregating. They found that 36% of the abnormal asci showed recombination between the outside markers, whereas the distance between these markers was only 4%. There is, therefore, a correlation between crossing over and abnormal segregation, but gene conversion did not always result in crossing-over between outside markers.

Mechanism of Gene Conversion

The above evidence strongly suggested that there was a hybrid DNA intermediate in recombination. Where the two strands of this heteroduplex that mismatching of bases will occur. It is possible that the first double helix has an A-T pair and the mutant double helix has a transversion T-A. The heteroduplex would have an A-A pair which, as adenine is a large molecule, would cause distortion. It is now known from work with bacteria and phages that enzymes exist which can recognize distortion in DNA and can remove a stretch of the DNA molecule. This gap can then be repaired by a DNA polymerase, A hy^+/hy heteroduplex can therefore either be converted into hy^+/hy^+ or hy/hy, or alternatively it may, as in postmeiotic segregation, simply segregate at the next round of replication to give one hy^+/hy^+ and hy/hy duplex.

Model for the Mechanism of Recombination

The data already described led to a variety of models for recombination proposed particularly by Whitehouse and Holliday. The difference between the models depended on the strand which was first nicked to give a single-strand break, and whether or not DNA synthesis occurred during recombination. The model which fits the data and observations best is that of Holliday (1974) as modified by other workers. From this it will be seen that gene conversion can occur in the presence or absence of recombination between outside markers.

Single-strand breaks occur at opposite positions on DNA strands of the same polarity. One of these strands moves towards the other (i) and, following localized DNA degradation, invasion occurs to produce a short stretch of hybrid DNA. Meanwhile DNA synthesis has occurred to fill the gap left by the invading strand (ii). Isomerization of this structure then occurs by the rotation at one end (iii). This is difficult to depict in two dimensions and is best understood by constriction of models using modelling clay or heads. The newly-replicated stretch of DNA joins with the free end of the molecule which was previously invaded. At this point hybrid DNA is opposite non-hybrid DNA and this is known as the *asymmetric phase*. Migration of the cross-configuration can then occur along the molecule, resulting in the symmetrical occurrence of hybrid DNA on both strands. Consequently this is known as the *symmetric phase*. Again isomerization can occur, resulting in different arrangements of the outside markers. Finally the cross-configuration will be resolved by breakage and rejoining, and recombination of outside markers will occur if this happens at stage (iv), or alternatively parental combinations will be obtained if resolution occurs at stage (v). Consequently hybrid DNA and related gene conversion can occur either in the presence or absence of the recombination of outside markers. *Polarity* of gene conversion is thought to result from the initiation of recombination at a particular point which will give a gradient of hybrid DNA. Gene conversion will therefore be more likely nearer to the initiation point and will fall off away from it.

Polarity of Gene Conversion

A detailed examination of recombination within genes has given evidence for higher levels of gene conversion in some regions than in others, i.e. there is *polarity* of gene conversion. This has been found in a number of fungi, including *Ascobolus*, *Saccharomyces*, *Aspergillus* and *Neurospora*. One of the basic techniques in these experiments is to cross two strains containing closely-linked allelic mutations. Rare asci containing wild-type recombinants are then isolated and the ascospores are back-crossed to the original single mutant strain to identify the mutations present in each. The ascospores are expected to contain 4 different genotypes, $m_1m_2^+$, m_1m_2, $m_1^+m_2^+$ and $m_1^+m_2$. However, as will be seen from figure the double mutant is missing and $m_1m_2^+$ is present instead. There is consequently a 2:2 ratio m_1^+/m_1 and a 3:1 ratio for m_2^+/m_2. A series of such crosses in *Ascobolus* showed that particular mutations were more likely to be converted than others, and a ranking order of 188 > 63 > 46 > 137 was drawn up, i.e., in a cross m_{188} x m_{137}, m_{137} would be converted, and so on. The genetic map was also 188 63 46 137 and it was suggested that this polarity was related to the position at which hybrid DNA was most likely to occur. m_{137} would therefore be assumed to be near a terminus, and therefore more likely to participate in hybrid DNA formation, as the crossover could migrate from the initial point.

Genetic Analysis of Aspergillus nidulans

Asci also occur in the ascomycete *Aspergillus nidulans* although in this fungus they are non-linear. It should be noted that there is no mating type and hence one uninucleate spore can germinate to produce fruiting bodies known as *cleistothecia*. In order to cross *Aspergillus* it is necessary to force a *heterokaryon* (a mycelium containing genetically different nuclei) by using nutritional mutants.

Tetrad Analysis of Unordered Asci

As can be seen from figure non-linear asci contain ascospores arranged at random and consequently the location of the centromere cannot be determined as has been described for *Sordaria*. It is still possible, however, to obtain information about the linkage relationships of different loci. The principle can be explained by reference to the segregation of spore colour genes in *A. nidulans*. A cross is made between a yellow-spored strain and a chartreuse-spored strain. Mature asci are isolated, and the ascospores removed with a micro-manipulator. Each ascospore is allowed to germinate and to produce colonies with asexual spores. Three types of asci are obtained as detailed in figure. One type has four yellow colonies and four chartreuse colonies, the *parental di-type* (PDT), and the second type has four green colonies and four pale yellow colonies, the *non-parental di-type* (NPDT). The third ascus type has yellow, chartreuse, green and pale-yellow colonies and is called the *tetratype* (TT). If the two loci are on nonhomologous chromosomes, then there will be free recombination, the PDT and NPDT will be equal in number, and a χ^2 test can be applied to check this. The frequency of tetratypes will depend on the distance between either of the loci and their respective centromeres, as crossing-over must occur in this region to give a TT. When the two loci are both very close to the centromere there will be no TTs. In situations where both loci are 50 or more units from the centromere, the maximum of 67% tetratypes will be obtained.

When the PDTs are significantly greater than the NPDT, this means that the loci are linked. An estimate of linkage can be obtained from the formula

$$\%\ \text{recombination} = \frac{\text{NPDT} + \frac{1}{2}\text{TT}}{\text{total}} \times 100$$

An examination of crossing-over in the bivalents shows the origin of this formula. The NPDT result from crossing-over between all four strands, which is described as a four-strand double crossover. As all strands are recombinant, all of the asci are included in the estimate. Tetratypes, on the other hand, result from a single crossover and only half of the strand are recombinant, hence the number of tetrads must be halved. Tetrad analysis is technically tedious, and data can be obtained more rapidly if random meiotic products can be analyzed. In tetrad analysis 8 ascospores count only as one independent observation, as crossing-over in the ascus automatically fixes the other products. On the other hand, each ascospore counts as a separate observation in random ascospore analysis, so that more accurate estimates of linkage can be made. In organisms such as yeast and the micro alga *Chlamydomonas*, it is difficult to separate tetrads from vegetative cells, so tetrad analysis by micro-manipulation is essential. In *Neurospora* and *Aspergillus*, fruiting bodies allow easy separation of ascospores, so that random ascospore analysis is the preferred method for routine work. However, tetrad analysis has been exceptionally useful to examine individual events at meiosis, and the technique made a significant contribution to the hybrid DNA model for recombination.

Parasexual Cycle

For many years, filamentous ascomycetes such as *Aspergillus nidulans* were considered to have a haploid mycelium with only one diploid nucleus in the life cycle. This occurred in the young ascus which immediately underwent meiosis to produce haploid ascospores. In 1952, however, Roper, using selective techniques, detected a rare diploid mycelium which has proved very useful for genetic work. Since this time, a variety of other fungi have also been found to have a diploid mycelium. The major difference between these diploids and the diploid stages occurring in other life cycles was that there was no regular alternation of diploid and haploid through the agency of meiosis. Diploid strains broken down to haploids by sequential loss of chromosomes in the absence of meiosis. For this reason the phenomenon was called the *parasexual cycle*. Another feature of the cycle was that crossing-over could occur during mitosis of the diploid, *mitotic crossing-over*. This process also occurs in *Drosophila* but it is easier to study in the parasexual cycle.

Evidence for the Occurrence of Diploids

An important technical factor in the isolation of diploids in *A. nidulans* was that the asexual spores, the *conidia*, each have a single nucleus. A second important point was that the spore colour was dependent on the genes carried in the nucleus. Consequently a heterokaryon formed between a yellow and a white strain will produce conidial heads which have yellow or white spores. Occasionally, mixed heads were observed which had both yellow and white spores together. Roper used a double selective technique, forcing a heterokaryon with nutritional mutants and using spore colour mutants to identify the diploid. Heterokaryons were grown and large numbers of conidia were collected to produce a thick suspension (about 10^8 cm^{-3}). These were plated on complete medium at a dilution of 10^{-6} and colonies with either yellow or white conidia were obtained. Plating of the neat suspension and 10^{-1} dilution on minimal medium was not expected to give any colonies, although occasionally the heterokaryon could be re-established. However, it was observed that vigorous prototrophic colonies with green conidia grew on these plates at a frequency of about one in a million spores plated.

It was possible that these green colonies were derived from recombinant ascospores which, of course, would be haploid. These could be eliminated by avoiding areas with fruiting bodies. However, it was essential to prove that the colonies obtained were actually diploid. The first line of evidence was the *phenotype* itself—a green phototrophic strain—formed by complementation at the four heterozygous

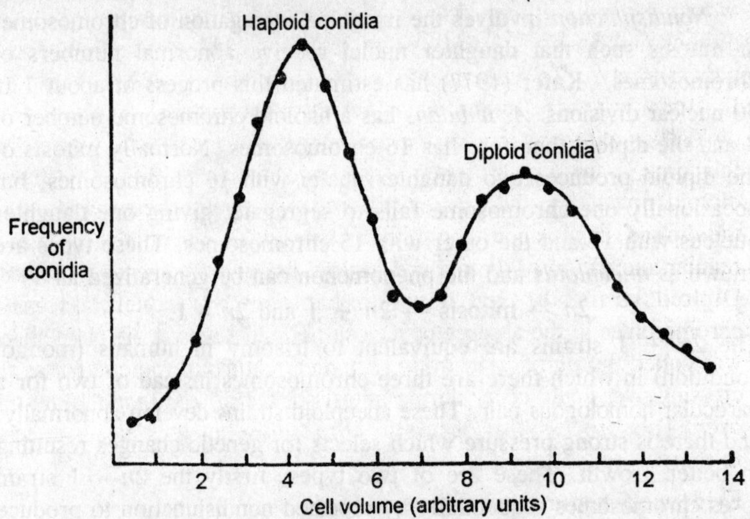

Fig. 10.12. Distribution of the volumes of haploid and diploid conidia of A. nidulans determined by means of a Coulter counter.

loci in the diploid. Simultaneous reverse mutation at two loci or recombination could be dismissed as being unlikely. The second piece of evidence was that the diploid conidia had twice the DNA content of haploid conidia. Both diploid and haploid conidia have a single nucleus, so that these figures relate to the DNA content of the haploid and diploid genomes. Thirdly, measurements of conidial diameter and volume showed that the volume of the diploid was approximately twice that of the haploid. Volumes of conidia obtained from a Coulter counter. Fruiting bodies form on diploid strains but development is abnormal, producing few asci, some of which have 16 ascospores.

The conclusive proof of diploidy came from evidence for heterozygosity and mitotic segregation. Green colonies were observed to produce occasional sectors, patches or spots of white or yellow conidial heads, and nutritional markers were also found to segregate in a similar manner. This gave convincing evidence that genes were masked in the diploid, and that during 'accidents' of growth and mitosis they could be revealed. The next section deals in detail with the processes involved in the production of mitotic segregants.

Basic Principles of the Parasexual Cycle

Variation in the parasexual cycle has been found to depend on three basic processes: nondisjunction, haploidization and mitotic crossing-over.

Nondisjunction involves the irregular segregation of chromosomes at mitosis such that daughter nuclei receive abnormal numbers of chromosomes. Kafer (1977) has estimated this process at about 1 in 50 nuclear divisions. *A. nidulans* has a haploid chromosome number of 8 and the diploid therefore has 16 chromosomes. Normally mitosis of the diploid produces two daughter nuclei with 16 chromosomes, but occasionally one chromosome fails to segregate, giving one daughter nucleus with 17 and the other with 15 chromosomes. These types are known as *aneuploids* and the phenomenon can be generalized as

$$2n \rightarrow \text{mitosis} \rightarrow 2n + 1 \text{ and } 2n - 1.$$

The $2n + 1$ strains are equivalent to trisomy in humans (mongol condition) in which there are three chromosomes instead of two for a particular homologous pair. These aneuploid strains develop abnormally, and there is strong pressure which selects for genetic changes resulting in better growth. These are of two types, firstly the $2n - 1$ strain loses chromosomes sequentially by repeated nondisjunction to produce the haploid chromosome number n. This is known as *haploidization* and it normally occurs in the absence of crossing-over, a point with important genetic consequences. Pontecorvo and Kafer (1958) estimated that about 1 in 200 haploids had undergone one crossover. Secondly the $2n + 1$ strain can lose a single chromosome to give the original diploid number ($2n$). This is known as a *nondisjunctional diploid*. Both of these processes can lead to yellow sectors, although haploidization will give a yellow haploid, and the second process would produce a homozygous yellow diploid.

Mitotic crossing-over can also give rise to homozygous yellow diploids. Compares the processes of mitotic and meiotic crossing-over. It will be seen that meiosis gives yellow haploid strains, while mitosis gives homozygous diploid strains. A further difference is that crossing-over during mitosis is rare. Kafer (1977) estimates the frequency at 0.1% to 0.3% per chromosome arm. For all practical purposes, this means that double and triple crossovers are not observed. This simplifies the analysis of diploids which are heterozygous for a number of loci. Crossing-over results in the association of an allele with a different centromere and, with the correct orientation of chromosomes, two chromatids carrying yellow alleles segregate together into the same nucleus giving a homozygous yellow diploid. Proof that mitotic crossing-over occurs at the four-strands stage of mitosis and that it is reciprocal can be obtained for a diploid heterozygous for fawn and chartreuse spore colour in repulsion. A single crossover gives rise to daughter nuclei which are either homozygous for fawn or for chartreuse spore colour and these can be seen as twin-spots on the mycelium.

The parasexual cycle is therefore an alternation between diploid and haploid stages in the absence of meiosis, but with variation occurring by chromosomal segregation and crossing-over at mitosis.

Application of the Parasexual Cycle

Mapping of genes to linkage groups. The construction of genetic maps of micro-organisms is tedious when there is a large number of chromosomes. Linkage of genes at meiosis will occur only if the intervals are less than 50 map units. To detect this many two- and three-point crosses must be made. Crossing-over does not normally occur at the same time as haploidization, and this has the technical advantage that genes on the same linkage group always segregate together. This applies even to loci which show free recombination at meiosis. A new locus can rapidly be located to a linkage group by isolating a diploid between the new strain and a master strain which has genetic markers an each of the eight linkage groups. Here mutations at a new locus *mor* resulting in a fluffy morphology, always segregated with ribo$^+$ on linkage group VIII while showing free recombination with the other 7 linkage groups. The new locus can therefore be located to linkage group VIII. It should be noted that we talk about linkage groups rather than chromosomes, as no correlation has been made between these groups and the respective chromosomes seen at meiosis. Haploidization is a rare process, and treatment with chemicals is necessary to increase the rate of haploidization: *p*-fluorophenylalanine benlate and chloral hydrate have all been used to interfere with normal mitosis and to increase the rate of nondisjunction and production of haploids.

Mapping of Gene Order and Centromere Location

Genetic loci can be mapped by analyzing the products of mitotic crossing. A good practical exercise is the analysis of a diploid which is heterozygous for acriflavin resistance. The parental haploids show either good growth (*Acr*^{r}) or poor growth (*Acr*^{s}), while the diploid *Acr*^{r}/*Acr*^{s} grows at an intermediate rate. Inoculation of the diploid results in partial growth followed by the appearance of vigorously growing sectors. The intermediate growth of the diploid creates a strong selective pressure for the emergence of strains which are fully resistant to acriflavin. This can occur to give haploids *Acr*^{r} by haploidization or diploids *Acr*^{r}/*Acr*^{s} by either nondisjunction or mitotic crossing over. Diploids derived from mitotic crossing-over provide information about gene order and centromere position. Haploids showed that the *Acr* and *w* loci were in the same linkage group. Mitotic recombinant diploids were either white or green. There are three possible gene orders:

Acr centromere *w*

centromere *Acr w*

centromere *w Acr*.

To obtain homozygous resistant diploids, it is essential to have a crossover between *Acr* and the centromere. As crossovers in this chromosome arm occur only once in each nucleus, we can write down the possible consequences of the three gene orders. Only order (3) can give rise to the observed results and, as there are more white sectors than green, it interval. It should be noted that the distances are relative and not absolute, as we have *selected* for crossing-over in this region.

Three important points can be made about mitotic crossing-over:

1. Loci nearer to the centromere than the crossover point remain heterozygous (proximal loci).

2. Loci further away from the centromere than the crossover point become homozygous (distal loci).
3. Crossovers in one chromosome arm have no effect on loci on the chromosome arm attached to the other side of the centromere. For mitotic crossing-over each chromosome arm is treated as a separate unit.

A more detailed example will illustrate these principles. Let us assume that there are five loci which have been located to a particular linkage group: white spore colour *w* and four loci *a*, *b*, *c*, and *d*. This white locus provides us with a selective system, as we can collect white sectors and then analyze them for ploidy and homozygosity of *a*, *b*, *c*, *d*. The recessive markers must all be in coupling with the *w* allele. The haploids obtained are all either *a*, *b*, *c*, *d*, *w* or a^+, b^+, c^+, d^+, w^+ confirming the location of the markers in the same linkage group. It will be seen that *c* was always homozygous and must therefore be distal to the selected marker *w*. *d* was almost always heterozygous, suggesting that it was very close to the centromere or on the other chromosome are. Another analysis showed that a^+, b^+, c^+, *d*/*d* diploids could be isolated, showing that it was on the left-hand chromosome arm. The frequency of *b* and *a* diploids allows them to be positioned as indicated in the first map: *a* is the least frequent and is therefore closer to the centromere than *b*. The three diploids correspond to the three positions where crossing-over can occur to give rise to a homozygous white-spored diploid.

Analysis of a Translocation

Haploidization is an easy method for detecting translocations, i.e., exchanges between nonhomologous chromosomes. In *A. nidulans* these were very common in the early strains, due to the use of X-rays and ultraviolet radiations to induce mutations. Kafer (1977) has estimated that for survivals of 2-5%, UV induces 15-20% translocations. Diploids can be formed which are heterozygous for a particular translocation. Treatment of these diploids with the usual haploidizing agents gives rise to haploids which are classified for the genetic markers segregating in the cross. It is found that the markers involved in the translocation segregate as one linkage group, reducing the number of linkage groups from eight to seven. Certain classes are missing due to poor growth or inviability, caused by unbalanced chromosome complements. It is possible, using these techniques, to screen strains for translocations and to follow this by meitotic analysis and ascus analysis. Figure shows some of the data obtained for a translocation between linkage group III and VIII. Here a piece of linkage group III has been added on to linkage group VIII. The effects of this are that (i) markers in the two linkage groups segregate together during haploidization, (ii) meiotic linkage of *cha* (VIII) and *s*C(III) can be detected, and (iii) an aneuploid type segregates regularly at meiosis. This type of colony occurs at a frequency of 1 in 3 and has one chromosome arm in duplicate. Reversion of this abnormal colony appears to occur by loss of chromosomal material from either arm, thus restoring the normal haploid complement.

Other Applications of Diploids

The occurrence of a relatively stable diploid in *A. nidulans* has permitted complementation tests to be carried out in detail. One such analysis of the adenylosuccinase (*adA*) locus revealed a phenomenon of *negative complementation*, in which an *adA*/*adA*$^+$ diploid had less than 50% of the wild-type enzyme activity. The partial dominance of the *adA* allele could presumably relate to partial masking of the active site in hybrid enzymes. Another application is a study of mitotic crossing-over within the gene. A diploid which is *pabaA1*/*pabaA2* can give rise to *paba*$^+$ recombinants, which can be detected selectively. This provides a very sensitive system for studying the effects of chemicals, replication, repair systems and radiation on the frequency of intragenic recombination. Chemicals have also been tested for their effects on mutation, nondisjunction and the induction of translocations. Well-marked diploids are ideally suited for studies of this sort. These tests can be used to assess environmental hazards as a supplement to the information obtained from the Ames test.

Occurrence of the Parasexual Cycle

Since the original discovery of diploids in *A. nidulans*, the features of the parasexual cycle have been discovered in a variety of fungi. It has been detected most frequently in the ascomycetes and imperfect fungi, but good evidence for parasexual recombination has been obtained in at least forty species. These include *Phycomyces blakeleeanus* and the slime mould *Dictyostelium discoideum*. There is much variation in the frequency with which the various stages and processes occur. In some cases diploids are transient and cannot be studied in detail. One example of this is *Acremonium* (*Cephalosporium*) *acremonium*, where it is necessary to use protoplast fusion to overcome the technical limitations resulting from the occurrence of uninucleate hyphal cells. In perfect abcomycete fungi it is possible to isolate diploid mycelia from ascospores, although care has to be taken to distinguish these from aneuploids which also occur. In some cases, such as *Saccharomyces cerevisiae*, mitotic crossing-over has been detected but not haploidization. The frequency of mitotic crossing-over in some fungi is much higher than in *A. nidulans*. In spite of extensive searches, diploids have not been isolated in *Neurospora crassa*, and geneticists have been forced to use disomics ($n + 1$) to study dominance relationship. The occurrence of the cycle in a variety of imperfect fungi which are important plant pathogens or which are used industrially is of note. Although proof of the natural occurrence of diploids is difficult to obtain, it is possible that organisms such as *Verticillium* and *Fusarium* show variations due to recombination via the parasexual cycle in natural situation.

FILAMENTOUS FUNGI

The rapid progress in molecular genetics has been extended to the analysis of the filamentous fungi. The reasons for this have been that they are technically easy to work with, have well-characterized genetic systems, can be characterized biochemically and a number of genera produce commercially important compounds such as antibiotics. Progress has been made in isolating temperature-sensitive mutants, in extending the parasexual cycle to industrial fungi, in detecting protoplast fusion and transformation, in analysis of regulation and in the cloning of fungal genes.

Genetic Analysis of DNA Synthesis and the Duplication Cycle in Filamentous Fungi

The basic approach has been to isolate a range of temperature-sensitive mutants, usually by replica plating on to identical media incubated at different temperatures. For *A. nidulans* this has meant incubation at 25^0, 30^0, 37^0, 42^0 and 45^{0}C. The normal optimum temperature is 37^{0}C but mutants can be isolated which fail to grow at, for example, 25^{0}C (cold-sensitive) or 42^{0}C (temperature-sensitive, *ts*). The assumption is that proteins will be affected at random, and that a suitable screening will reveal the type of mutant required. One method of screening is to grow a *ts* mutant at the permissive temperature and then to transfer it to the non-permissive temperature. Analysis can then be made for DNA/RNA ratios, with the expectation of low amounts of DNA in mutants in which DNA synthesis has been affected. Using this approach, a strain with a temperature-sensitive DNA polymerase has been isolated in the maize smut fungus (*Ustilago maydis*), and mutants affected in DNA synthesis have also been isolated in *A. nidulans*. A very profitable approach has been to grow hyphae under various conditions, and then to carry out nuclear staining. Spores, which are uninucleate, are pre-germinated for 11 hours at 32ºC and then transferred to 42ºC for 2 or 4 hours. The hyphae are then stained with aceto-orcein for chromosomal figures or acid fuchsin for spindles. Nuclei of *A. nidulans* undergo semi-synchronous mitosis in each hyphal tip, and therefore a number of nuclei in one hypha are often in mitosis with the chromosomes contracted. Consequently a chromosome mitotic index (CMI) can be calculated from the number of *hyphae* in mitosis rather than the number of nuclei in mitosis. Similar data can be obtained for the spindle mitotic index (SMI). It will be seen that *bim* (blocked in mitosis) mutants have greatly increased CMI and SMI whereas *nim* (never in mitosis have much lower indices. The figure for nuclear division are a measure of the leakiness of the block, and the length of hypha per nucleus (L/N) values indicate the extent to which hyphal growth continues after nuclear division has stopped. Complementation tests and mapping data have indicated that there are 6 *bim* loci, 23 *nim* loci and 5 *nud* mutants. These mutants should enable a detailed analysis to be made of the cytology and biochemistry of mitosis in *A. nidulans*. This organism is particularly favourable as the nuclear membrane remains intact during mitosis, which should allow the isolation of defective spindle fibres and other abnormalities of mitosis.

Detection of the Genetic Loci for Tubulin Synthesis

The spindle fibre of *A. nidulans* consists of approximately 50 microtubules made up of protein subunits called *tubulins*. A very fruitful approach to studying microtubules has been to isolate mutants with altered sensitively to antimicrotubular agents such as colchicine in *Chlamydomonas reinhardi* and benomyl in *A. nidulans*. The analysis of the genetics of tubulin synthesis gives a beautiful example of the potential and usefulness of a combined genetic and biochemical approach using temperature-sensitive mutants. Several mutants of *A. nidulans* have been isolated which are resistant to benomyl, which is known to bind β-tubulins. One particular strain, carrying the mutation *ben A* 11, has been analyzed in some detail. The tubulins of this fungus can be purified by co-polymerization with porcine brain tubulin, and radioactive microbial tubulins can be separated by a two-dimensional gel electrophoresis system. Eight tubulins have been detected by this method, four α-tubulins and four β-tubulins. Strains carrying the *benA* 11 mutation were found to have an abnormal β-tubulin which was separable from the normal tubulins by electrophoresis. The possibility existed that this altered tubulin was a normal tubulin modified by an enzyme coded by the *benA* locus, but this was eliminated by showing that the diploid *benA*11/*benA*11$^+$ possessed both the normal and the altered tubulins. A modifying enzyme would have been expected to alter both tubulins. This finding strongly suggested that *benA*11 was the structural gene for a β-tubulin. The mutation also made the strains carrying it temperature-sensitive for growth, so that revertants could be isolated by plating spores at the, non-permissive temperature. Eighteen revertants were isolated and outcrossed to distinguish between true revertants and suppressor mutations. Four of the strains were true revertants with normal β-tubulin, thus giving further evidence that the *benA* locus coded for β-tubulin. Fourteen strains had suppressor mutations at loci separable from *benA*. An electrophoretic analysis of tubulins from these strains showed that thirteen strains had normal tubulins whereas a strain, later designated *benA*11 *tubA*, was shown to have an altered α-tubulin. Evidently a mutation in an α-tubulin gene could correct for the alteration in β-tubulin caused by the *benA*11 mutation, giving an abnormal but functional microtubule. The production of a *tubA*/*tubA*$^+$ diploid showed the presence of normal and altered α-tubulin, providing similar evidence to that provided for *benA* and β-tubulin. The success of this approach indicates the usefulness of temperature-sensitive mutants and the applications of revertants in the identification of second mutations which can correct the first abnormality. It seems likely that similar approaches can be used for the mitotic mutants and also in other eukaryotic micro-organisms.

Genetic Approaches to the Study of Growth and Wall Synthesis

The genetic analysis of the growth of filamentous fungi provides a good example of the interaction of the genotype with the environment. Mutants can be isolated with altered branching frequency, slower growth and quantitative differences in wall composition. Alteration of the environment can in some cases produce *phenocopies* of these mutants. Thus the presence in the agar medium of the surface-active agent sodium deoxycholate can cause wild-type *A. nidulans* strains to produce small highly-branched colonies which are phenocopies of genetically-controlled compact mutants (*co*). We have already seen examples where the environment is used to modify the expression of temperature-sensitive mutants. This approach has also been used to study the action of mutations affecting branching in *Neurospora* for a septation mutant in *A. nidulans* and for cell-wall mutants of *A. nidulans*. It summarizes some of the systems which have been analyzed in this way, as well as other morphological mutants which are not necessarily temperature-sensitive.

One example will be dealt with in details as it illustrates the procedures used. A temperature-sensitive mutant of *A. nidulans* was isolated which produce distorted hyphae called *balloons* when grown at the non-permissive temperature of 43^0 on minimal medium. This mutant had several distinctive features: (1) the balloons stained intensely with fluorescent brighteners such as photine HV which are known to attach to β-linked carbohydrates; (2) the ballooning phenotype could be eliminated by the addition of mannose to minimal medium, and (3) mannose, in purified wall preparations from ballooning cultures, was reduced. A further interesting feature was that the mutant failed to grow on minimal medium containing 0.9% glucose and 0.1% mannose. Overall this suggested that mannose metabolism was affected in the mutant, designated *mnr man* nose relief, and that this affected the mannose content of wall polymers, ultimately preventing normal polarized growth.

It was further reasoned that the failure of *mnr* strains to grow on glucose/mannose mixtures was due to some competition for phosphorylating enzymes or permeases, resulting in insufficient mannose to replace the wall lesion. An alternative explanation was that the accumulation of toxic concentrations of phosphorylated sugars, due to a genetic block, inhibited growth. What-ever the causes it was theorized that revertants of *mnrA*, in which a second mutation had occurred, could be isolated by plating mutagen-treated spores at high density on to minimal medium containing glucose and mannose. This second mutation should prevent mannose from being used as a carbon source, thus releasing it to relieve the wall lesion.

Alternatively, the second mutation might prevent the accumulation of toxic phosphorylated sugars. Fourteen revertants were isolated, and thirteen were found to be true revertants, whereas one strain was the predicted double mutant. This was able to grow on minimal plus glucose and mannose, but failed to grow on glucose or mannose alone. Genetic analysis has located *mnrA* to linkage group VIII, and new the mutation *manA* to linkage group V. These results have enabled a pathway to be proposed for mannose metabolism in *A. nidulans*. The initial mutation has occurred in phosphomannose mutase, thus preventing glucose from being metabolized to produce mannose for wall growth. Mannose, however, can provide both a carbon and energy source, and a wall precursor. When *mnrA* strains attempted to grow on mixtures of glucose and mannose, then glucose would be metabolized to mannose-6-PO_4 and no further, because of the defective enzyme. Accumulation of this phosphorylated sugar could then cause the inhibition of growth. Alternatively, uptake or phosphorylation of mannose might be inadequate for growth.

The second mutation *manA* has been shown to produce reduced levels of phosphomannose isomerase, which in some way relieves the inhibition of glucose as discussed above. It will be seen that the *manA* mutation prevents glucose from being used to supply mannose as a wall precursor, and similarly prevents mannose from being used as a carbon and energy source. As can be seen *mnr man* and *man* strains have identical phenotypes, i.e. *man* is epistatic to *mnr*. The reason for this is that effectively *man* is earlier in the pathway for mannose synthesis that *mnr*.

Enzyme assays have shown that *mnrA* strains have a thermolabile phosphomannose mutase, strongly suggesting that this is structural gene for this enzyme. The *manA* mutation results in the reduction in level of activity of normal phosphomannose isomerase, which suggests that this mutation affects the regulation of enzyme synthesis. These two mutations allow the specific radioactive labelling of mannose in cultures, as mannose cannot be used as a carbon and energy source, and must be incorporated unchanged into the wall. Autoradiography of [2-^{3}H]-mannose incorporation has shown that the label goes as expected into the tips of growing hyphae, and the analysis of sugar monomers from wall purified from mycelium grown at 43^0 has shown a higher level of specific labelling in the mutant cultures. In the wild-type culture, radioactive mannose can be converted into glucose, glucosamine and galactose, whereas in the double mutant this occurs to only a limited extent.

Industrial Application of the Parasexual Cycle

The detection of the cycle in *Penicillium chrysogenum*, the fungus used to produce penicillin, in 1953 by Pontecorvo and Sermonti led to predictions that the parasexual cycle could be used to increase penicillin yields. This did not materialize for 20 years, as mutation-selection techniques gave the results required by industrialists. Attempts were made to increase penicillin yields but met with failure due to a phenomenon described as *parental genome segregation*. The basic idea was to produce diploids between two different high-yielding strains and to isolate, by haploidization, haploids with an even higher yield. This was not achieved, and strains with the original titres of penicillin were isolated. The most likely explanation for this was that different translocations had been isolated in the two strains, thus restricting free recombination of chromosomes as described earlier.

Ball (1973) succeeded in using the parasexual cycle to increase penicillin yield by a different approach. Instead of taking industrial strains which had been separated for many years, he took one strain which was cloned by isolating colonies from a single spore. Genetic markers were selected which did not influence penicillin yield, and three linkage groups were identified by haploidization. Gene mutations were isolated which increased penicillin yield, and these were mapped to linkage groups by the techniques mentioned earlier. Diploids between different high-yielding mutants were isolated, and recombinant haploids produced. As the strains were closely related, translocations were not present and free recombination occurred, giving haploids with a penicillin yield higher than either of the parental haploids. These techniques can be used to study the pathway of penicillin synthesis, as well as breeding the fungus of characteristics desirable in industrial fermentation. Two types of penicillin mutants have been isolated: those which overproduce penicillin, and those which fail to produce the antibiotic. The aim here is to study the regulation of penicillin synthesis and, if possible, to work out the pathway of synthesis and the enzymes involved.

More recently, protoplasts have been used to hybridize a variety of fungi, and this method has been used to increase the cephalosporin C titres in *Cephalosporium acremonium*. Transformation has also been added to the range of techniques available in the fungi.

Molecular Genetics of Filamentous Fungi: Cytoplasmic Inheritance

The cytoplasmic genetics of filamentous fungi such as *Aspergillus*, *Neurospora* and *Podospora* has been studied for thirty years, but a molecular analysis of cytoplasmic DNA has been possible only since the advent of genetic engineering techniques. Cytoplasmic inheritance can be detected by similar techniques to those used in yeasts except that the filamentous growth habit means that cytoplasmic segregation can often be observed during the growth of heterokaryons, or, in the case of cytoplasmic fusion products, *heteroplasmons*. Thus heterokaryons between strains with genetically different mitochondria often show visual segregation during growth on plates. Analysis of mutants very similar to those in yeasts and mapping of mitochondrial DNA by means of restriction enzymes and other techniques has allowed the construction of maps.

Protoplast and Liposome Fusion

The presence of a thick cell wall in fungi has made a number of biochemical and genetic techniques more difficult. Uptake of antibiotic precursor of ten does not take place and this problem has been solved by adding precursors to artificially-produced liposomes which are then fused with protoplasts in the presence of polyethylene glycol. These protoplasts can be produced from fungi by using a variety of enzyme mixes, particularly glucanases and chitinases which remove the cell wall.

In the presence of osmotic stabilizers, polyethylene glycol and calcium chloride, protoplasts from genetically different strains or even different species can be fused, thus circumventing natural barriers to hyphal fusion and cytoplasmic compatibility. However, for genetic recombination to occur, it is obviously also essential for the nuclei to fuse, so that various parasexual processes can result in recombinant products. This technique has been of particular use in *Acremonium*, which normally has uninucleate hyphal compartments.

Gene Cloning in Filamentous Fungi

The last few years have seen the development of vectors for cloning DNA in filamentous fungi, notably *Aspergillus*, *Neurospora* and *Podospora*. Before this could be achieved it was essential to have a transformation system. This was based on the same principles as that achieved in *Saccharomyces*. It was necessary to have a selective system to detect the transfer of DNA and one such system is illustrated. Protoplasts are produced from an arginine-requiring strain *arg*B, and these are fused as already described with DNA prepared from bacterial strains containing the plasmid PILJ16 which itself contain *Aspergillus* DNA *arg*B$^+$.

Protoplasts are allowed to regenerate on a medium lacking arginine so that *arg*B$^+$ transformants are selected. Examples of three vectors which have been used in similar selective systems. Other vectors are now being developed which have DNA coding for resistance to antibiotics like oligomycin. Transformation frequencies are lower than for yeast systems and vary from one to 5000 transformants per μg of DNA. It appears that most of the vectors produce transformants by integration into the *Aspergillus* chromosome at homologous and/or non-homologous positions, similar to yeast systems. Multiple integration can occur to give a gene dosage effect. So far no autonomously-replicating *Aspergillus* plasmids have been detected, although the integration of pILJ16 appears to be reversible.

Regulation in *Aspergillus nidulans*

The discovery of operons in bacteria leåd to a search for similar systems in the filamentous fungi. *A. nidulans* proved to be a fruitful source of regulatory systems and a selection of these are shown in Table 10.1. The examples quoted are for the utilization rather than the synthesis of metabolites, and the groups of genes with common regulation can be linked or unlinked. A major feature of most of the systems is the presence of a positive regulatory protein. For example, the synthesis of a permease for the uptake of xanthine and uric acid, coded by *uap*A, is inducible in the presence of uric acid and two regulatory proteins coded by *ua*Y and *are*A. The receptor site at which these proteins act is located by the isolation of a *cis*-acting mutation (*uap*100) which results in constitutive synthesis. These sites are usually up-promoter, that is away from the promoter site relative to the initiation codon. Very few systems have yet been characterized at the molecular level, but claims have been published for the isolation of the *ua*Y protein, and the recent cloning of the *prn* operon should result in rapid progress with this system.

Table 10.1. Examples of positively-regulated genetic loci and *cis*-acting regulatory mutations in *Aspergillus nidulans*. The locus regulated by the *cis*-acting mutation is shown in brackets.

Genetic loci	*Linkage*	*Utilization system*	*Locus for positive regulatory protein*	*cis-acting mutations (regulated locus)*
hexA, B uapA, C	Unlinked	Purine	*uaY, areA*	*uap* 100 (*uapA*)
prnD, B, C	Linked	Proline	*prnA*	*prn*[d] (*prnB*)
crnA, niiA	Linked	Nitrate	*nirA, areA*	*nis*-5 (*niiA*)
amdS	—	Acetamide	*intA*	*amdI* (various) (*amdS*)

11

MICROBIAL GENETICS

Genetics is a relatively new field of biology and most associate this science with Johann Gregor Mendel (1822-84), who was the first person to formulate any laws about how characteristics are passed from one generation to the next. This kind of study is often called Mendelian genetics. His work was not generally accepted until 1900, when three men working independently rediscovered some of the ideas that Mendel had formulated thirty years earlier. However, genetics was revolutionized in 1953 when, James Watson and Francis Crick proposed a chemical structure for DNA. Their discovery made it possible to understand more clearly the chemical basis of heredity in both eukaryotic and prokaryotic organisms. Since then the proposed double helix structure for DNA has become the cornerstone for explaining gene function, gene replication, and the nature of mutations. At this time many geneticists began to use microbes as research organisms.

Laboratory mice and fruit flies were abandoned by many geneticists in favour of microbes such as *E. coli*, *Neurospora* (a fungus), and T_4 phage (a bacterial virus). The ease of culturing and short generation times of microbes enabled geneticists to quickly produce large populations and trace inheritable characteristics through hundreds of generations. Although there are a number of advantages to using these organisms, it was soon discovered that microbes demonstrate many genetic characteristics and inheritance patterns not previously found in eukaryotic cells. As these patterns wee explored further, it became clear that transformation, conjugation and transduction would have far-reaching implications in medicine and in the field of eukaryotic genetics.

Fundamentals of Genetics

In both eukaryotic and prokaryotic cells, the molecule that serves as the ultimate agent of chemical control is deoxyribonucleic acid (DNA), while the inheritable material of viruses may be either DNA or ribonucleic acid (RNA). Knowing the Watson-Crick structure for DNA makes possible the definition of a gene both chemically and functionally. A *gene* is a portion of a DNA molecule composed of a specific series of nitrogenous bases that chemically codes for the production of a specific protein or RNA molecule, or serves as an operator in controlling the transcription of RNA within an operon unit. The chemical code for the placement of an amino acid is a specific triplet nucleotide sequence. Since protein molecules contain an average of 300 amino acids, the average gene is made up of about 900 nucleotide pairs. The DNA of *E. coli* is one of the most thoroughly investigated nucleoids and contains about 5×10^6 base pairs. This amounts to approximately 5000 genes, many of which have been identified in their proper sequence.

An organism's DNA constitutes a catalog of genes known as the *genotype* of the organism. The expression of these genes will result in a certain collection of characteristics referred to as the *phenotype*. While the phenotype of an organism consists of its observable characteristics, the genotype is not visible, since it is the DNA chemical code (formula) of an organism. There is not always a total expression of the genotype. Particular genes may not express themselves for a variety of reasons. In some cases, the physical environment will determine if certain genes will have a chance to express themselves. For example, if lactose is not supplied to a bacterial population that can metabolize this sugar, the phenotype will not be seen because the presence of lactose is required to induce the formation of the enzymes needed for its breakdown.

DNA is very stable, thus it is an excellent molecule to serve as the transmitter of chemical codes through generations. The stability of DNA and its resistance to change ensures the continuation of a species even though alterations regularly occur in gene structure. Any permanent change in the nitrogenous base sequence of DNA is called a *mutation*. A single gene may mutate to many different forms. Those forms of a gene that affect the same characteristic but produce different expressions of that characteristic are called *alleles*. For example, in humans there re alleles for eye colours such as blue and brown. In bacteria, there are alleles for enzyme production. In some bacteria the genet that controls the operation of the *lac* operon functions on an inducible basis ("off" and "on"), while others have a different allele of this gene which enables the same operon to function on a constitutive (always "on") basis.

Various mutations may be produced in DNA. A single nitrogenous base may be lost and replaced by a different base. This is known as a *point mutation* and may cause a single amino acid change in a protein. The sugar-phosphate backbone of the DNA may beak, resulting in a change or loss of a sizable portion of the molecule. This kind of event in DNA changes more than just a nucleotide base. If the damaged section is lost and not repaired, the mutation is called a *deletion mutation*. If the damage is required by the insertion of the same piece of DNA in reverse order, it is called an *inversion mutation*. When a totally new base sequence is synthesized to fill the gap, it is called an *insertion mutation*.

Mutations may have any one of a number of effect on the cell in which they occur. In some cases, the change is not harmful. After a single base change (point mutation), the protein synthesized by the gene may still be functional and no phenotypic change may be seen. This may occur because of the nature of the nucleotide code system since, in some instances, the same amino acid can be coded for by more than one triplet codon sequence. Therefore, even if a point mutation takes place, the new triplet codon formed can still call for the positioning of the same amino acid in the protein being synthesized. The protein may also remain active if another very similar amino acid is coded for placement and serves the same function in the completed molecule. Other more extensive mutations may result in the severe reduction of enzyme activity. This reduction may be due to a distortion of the active site on the enzyme surface. Poorly constructed enzymes may also be more susceptible to environmental changes. Minor fluctuations in temperature, ion concentration, or oxidizing agents my alter the enzyme's three-dimensional shape and cause inactivation. In some cases, the mutation may result in the complete loss of gene activity and death of the cell if an essential gene is inactivated.

Mutations may occur either spontaneously or be caused by agents such as radiation and chemicals. Anything that causes a permanent changes in the DNA of a cell is called a *mutagenic agent*. Mutagenic agents include x-rays, mustard gas, and nitrous acid. Naturally-occurring spontaneous mutations are at a relative low rate in microbes. About one in one mission (1×10^{-6}) bases may undergo a natural reorganization of chemical bonds within the DNA. For example, thymine may quickly and temporarily shift its internal bonds into a different arrangement. If this rearrangement takes place when thymine is base-paired with adenine (A-T), an error in base pairing will occur at the moment of DNA duplication. Because of this spontaneous bond rearrangement, normal base pairing (A-T) cannot take place, and the guanine-containing nucleotide is hydrogen bonded into the sequence by mistake (G-T).

Points mutations such as this may also be caused by mutagenic agents. Ultraviolet radiation induces mutations by causing the formation of bonds between thymine nitrogenous bases located next to one another on the same strand of the DNA double helix. These lined bases are

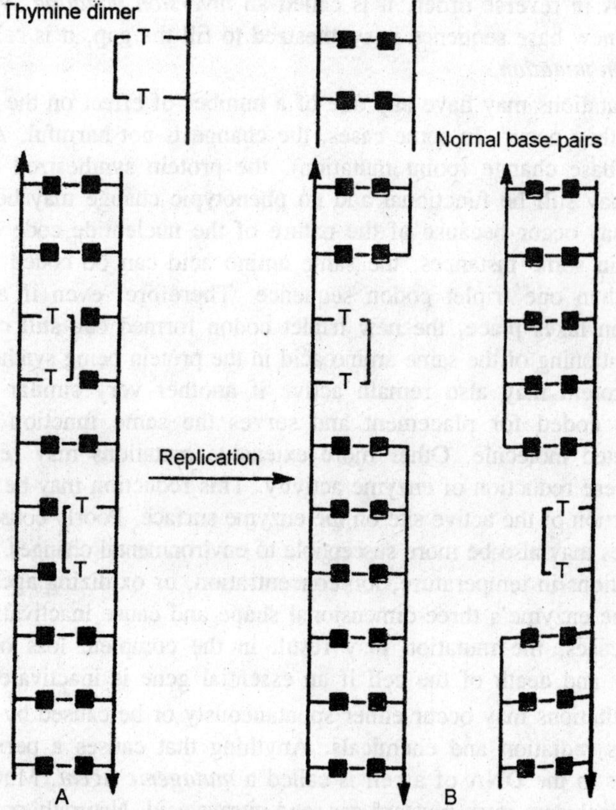

Fig. 11.1. The DNA strand shows several thymine dimers formed by exposure to ultraviolet radiation. After DNA duplication, several "gap" occur in the duplicated strands, since normal base pairs are not able to be formed.

known as thymine *dimers*. This change in bonding results in a *frameshift* mutation since the dimer will be skipped during replication and the transcription portion of protein synthesis. Reading-frameshift mutations might be compared to "skipping" the two latters *p* and *e* in the word "independence." It would then read "independence" and make no sense. This type of mutation and others may also be caused by chemicals such as nitrous acid, acridine dyes, and alkylating agents. They induce changes that result in mispairing, reading-frameshifts, and deletions. However, many bacteria have the ability to repair some of these changes. The repair process known as *photoreactivation* requires the presence of visible light immediately after dimer formation. The visible light energy is able to break the dimer and reestablish the original base sequence before a permanent change is fashioned into the DNA. Another, more complex, repair system, called *dark repair*, may also correct ultraviolet damage. In this series of reactions, visible light is not required. The damaged section of DNA is "clipped out" of the strand enzymatically, the "hole" is enlarged to ensure a better "patch job," and a new segment of DNA is synthesized. Neither of these repair mechanisms is foolproof and many types of mutagenic agents may produce a variety of alterations in DNA structure and function that are easily recognized as phenotypic changes.

Mutagenesis and Carcinogenesis

Mutations that show themselves as phenotypic changes are easily observed in microbial populations. The rapid growth, rate and short generation time of microbes makes possible the identification of the few cells that have experienced a change in their DNA as a result of exposure to a mutagenic agent. By exposing a population of microbes to a test chemical and counting the number of mutants that appear after culturing, it is possible to determine whether the chemical is a mutagen and at what concentration permanent changes are produced in the DNA. Because many mutagenic agents are also *carcinogenic* (cancer causing) to the eukaryotic cells of humans and animals, the carcinogenicity and mutagenicity of a chemical can be tested in much the same way. Such a carcinogenicity test has been developed by Bruce Ames and is known as the *Ames test*. The evidence that many cancers are caused by exposure to man-made and natural chemicals in our environment is mounting. Between 60 and 90 percent of cancer cases are believed to be environmentally induced by carcinogenic agents such as radiation and chemicals. Cosmetics, cleaning agents, food additive, industrial materials, adhesives, and many other synthetic materials all contain new chemicals that are constantly being introduced into the environment at levels never before experienced. Each year approximately 70,000 different chemicals become commercially available and ultimately may be ingested, inhaled, or absorbed into the body. The Ames test is a short-term test (2 days) of the potential carcinogenicity of many of these chemical compounds. Ames and his coworkers have produced a mutation in *Salmonella* that prevents its growth under normal conditions. However, if another mutation is caused by a test chemical, the microbe will be able to grow no agar plates. If the chemical does not cause this mutation, there will be no growth.

The test is run by exposing the bacteria to a number of dilutions of the suspect chemical, the microbes are mixed with a liver extract to simulate animal conditions, and the mixture is plated out on agar plates. The colonies growing on the plate are counted and compared to a control of a known carcinogen. A positive Ames test (appearance of colonies) indicates that the chemical can cause mutations and may be carcinogenic, however, a negative test may not mean that the chemical is safe to use because some carcinogenic agents do not cause mutations, and no carcinogenic test is perfect. Bacterial and human DNA are not the same and may not be affected by a chemical in the same way. To substantiate the result of the Ames test, long-term animal tests, which may take a year or longer, should be run on the suspect chemical. A number of chemical companies, including Dow and Dupont, have adopted the Ames test as part of their testing procedures. They have taken the position that new chemicals should be "guilty" of being carcinogens until proven "innocent". Tests on chemicals already in use are also being run; however, there is a resistance to ban them from use without running both the short-term Ames test and long-term animal test. For example, Tris, the flame retardant used in children's pajamas and other clothing, was identified by Ames and his associates in 1975 as a carcinogenic chemical that could be absorbed into the body. However, until animal tests confirmed the short-term Ames test, the chemical continued to be used. These animal tests required almost two years of research and evaluation. The same experience has also occurred with the chemicals used in hair dyes. Pretesting new chemicals before they are mass produced and marketed may significantly reduce the incidence of cancer.

Effects of Selection

When a mutation occurs in a population of microbes, the total number of cells containing the change may be relatively small (for example, on the order of 1×10^{-10}). However, natural and unnatural (artificial) selection may quickly increase the percentage of the population that contains the new gene. Natural selection occurs as a result of differential reproduction. If a microbe containing the mutation is better able to survive and reproduce than those lacking this change, the mutant will increase naturally in frequency through time. Selection might also affect the frequency of mutants in a mixed culture of penicillin-resistant and susceptible microbes. By adding penicillin to the culture, the drug will act as a selective agent and rapidly destroy the susceptible cells. In only a few hours, the entire population might be composed of penicillin-resistant microbes.

The combined action of natural and artificial selection on resistant cells has resulted in an increase in drug-resistant pathogens throughout the world. The selection of resistant mutants, however, does not totally explain the emergence of resistant organisms. Other genetic mechanisms involving the acquisition of new genes play an important role in this phenomenon. Selective agents do not act directly on the genotype of microbes. Selection is based on how the agent interacts with the phenotype of the cell. If the genes are inactive for any reason, the microbe will be adversely affected by the selective agent. Thus, a mutation that is beneficial to the microbe, but not phenotypically expressed, can be lost from the population. Since the phenotype of a cell is the result of the expressions of its functioning gene combinations of genes that are generated in a cell population. Any process that results in the integration of new combinations of genes together in a single cell is called *genetic recombination*. When fertilization (conjugation) occurs in eukaryotic organisms, genes donated by each of the parents are recombined into a single cell—the zygote. Since unique genes (due to mutations) and gene package (due to crossing-over and independent assortment) are brought together in a single cell (genetic recombination), this new individual will have gene combinations (a genotype) not found in either parent.

In prokaryotes there is no true fusion of cells. However, a portion of a DNA molecule from a donor may be transferred to a recipient cell to form a partial zygote called a *merozygote*. The incoming segment of DNA, the *exogenote* or *extrachromosomal*, DNA, may join into the recipient DNA, the *endogenote*, by breakage and reunion. This process results in a recombinant strand of DNA since the original genes of the recipient have been replaced by the new exogenote genes. As a result, the genetic makeup of the cell has new genes and new

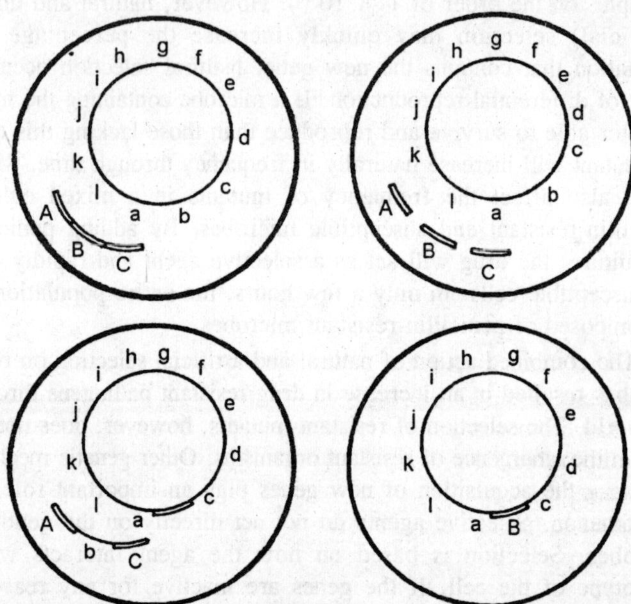

Fig. 11.2. The process of genetic recombination involves the replacement of old genes with new genes from some outside source.

gene combinations that may enable it to survive better in a changed environment. We now know that the exogenote DNA does not always join with the endogenote by genetic recombination, but may remain free in the cytoplasm of the recipient. This separate loop of DNA may function in the host without replicating. As a result, only one cell in the population will contain the exogenote at any particular time. As the population increases, the genetic uniqueness of this single microbe may be overshadowed by the other cells. However, self-replicating exogenotes may remain in the population indefinitely. These separate loops of genetic material replicate on their own and are distributed to the newly forming cells during binary fission in which the same manner as the host chromosomal genes. Cells which contain functional self-replicating exogenotes demonstrate a great amount of genetic variation not seen in cells that lack these extra genes. The possession of these genes may be a selective advantage because of increased genetic variety. It is also an advantage to be able to transfer exogenote and endogenote genes to other cells to maintain this variety. The three methods of transferring exogenote and endogenote DNA that have been identified in bacteria are transformation, conjugation and transduction.

TRANSFORMATION

The *transformation* process occurs when the recipient cell takes in a segment of "naked" DNA from the environment. This extrachromosomal DNA may have been released from a donor cell while it was alive or after it died. The first evidence of this process was seen in 1928 when Fred Griffith noted that nonencapsulated *Streptococcus pneumoniae* ("R" strain) could be changed to the capsule-producing form ("S" strain). This was done by culturing live "R" strain cells in media containing dead capsule-producing streptococci. Today a number of other genera have been identified that may also undergo transformation. *E. coli*, *Haemophilus*, *Bacillus*, and *Pseudomonas* are all able to be transformed. However, cells do not simply "take in" extra-chromosomal DNA. Those microbes with the genetic ability to take up naked DNA and be transformed are called *competent cells*. Competent cells may operate naturally or they may be artificially stimulated to take in the DNA by altering the environment in which the culture is being grown. For example, *E. coli* can only become competent by being cultured in media with a high concentration of calcium ions.

Transformation occurs in three stages. A large segment of DNA is first bound to a special receptor site on the surface of the competent cell. The segment is then cut into smaller, more manageable pieces by a DNAase enzyme released by the recipient. Finally, the attached segment of DNA is actively moved into the cell where it is prepared for recombination with the endogenote. The transformation process plays an important role in forming new gene combinations and creating genetic variety in microbes; however, conjugation probably plays a more important role.

CONJUGATION

As in the case of transformation, the ability to undergo conjugation is genetically determined. *Conjugation* in prokaryotes is the transfer of genes from one cell to another by direct contact. The genes that control the process are located on an extrachromosomal piece of DNA called a plasmid. A *plasmid* is a small, circular piece of extrachromosomal DNA that self-replicating and contains a limited number of genes. Plasmids that control such characteristics as fertility are called F factors, and those that contain genes for transferable drug resistance are known as R factors. To date no Gram-positive cells have been shown to conjugate; however, many F factor-containing Gram-negative bacteria may undergo plasmid-controlled conjugation.

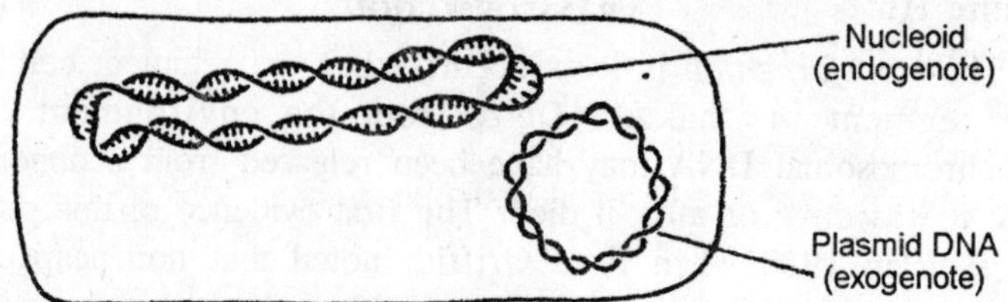

Fig. 11.3. Plasmids are "extra" loops of DNA that are not essential to the life of the cell.

Microbes which contain the F factor are designated F$^+$, or males, since they serve as donors. Those that lack the F factor are designated as F$^-$, or females, since they serve as recipients. When the F factor is donated to an F$^-$ cell, the female becomes a male, or F$^+$. The first stage in plasmid-controlled conjugation involves the attachment of the two cells. The sex pilus, a filamentous structure extending from the cell wall of the F$^+$ cell, attaches to the cell wall of an F$^-$ bacterium. A conjugation tube is formed between the cells and as this tube is formed, the plasmid is replicated inside the donor cell. This process takes place in the same way that the host nucleoid is duplicated. One of the F factor loops remains attached to the inner surface of the F$^+$ cell while the other plasmid is free to move through the conjugation tube into the recipient cell. After the transfer has been completed and the cells separate, both the recipient and the donor contain plasmids. The percentage of F factor-containing bacteria in a population increases if the microbes are crowded into close contact. Plasmid-controlled conjugation occurs more easily and successfully within the Gram-negative enteric species found as normal flora in the intestinal tract. These bacteria show a great amount of genetic variety. Bacterial populations that lack this close contact usually have a lower rate of conjugation, fewer F factors, and less genetic variety.

In some cases, the F factor plasmid may become integrated into the endogenote DNA. These cells have been carefully studied, and the frequency with which they genetically recombine with F$^-$ cells is 1000 times greater in comparison of F$^+$ cells. Therefore, these cells are called *Hfr* bacteria, or *high-frequency recombinants*. During conjugation between an Hfr and an F$^-$ cell, all of the genetic material in the Hfr undergoes DNA duplication. One of the loops is then broken within the Hfr gene sequence and being to move through the conjugation tube into the F$^-$ recipient. The entire length of DNA rarely moves into the recipient, since the two cells are being held together by such a fragile connection. Therefore, it is highly unlikely that the recipient will receive the entire Hfr sequence and be converted to an Hfr or F⁻ cell. More importantly, the F⁻ cell will receive new genes from the Hfr cell that can be genetically recombined with the endogenote. When this recombination occurs, the recipient will have new genes and gene combinations not previously found in the population. After reproduction by binary fission, a complete line of offspring will be produced that have these new characteristics. Any population of microbes produced by asexual reproduction from a single parent cell is called a *clone*.

Genetic variation may also be produced in conjugating populations if the Hfr cell converts to another form containing a plasmid known as F' ("F prime"). An F' plasmid is formed when the Hfr segment detaches from the host DNA loop and mistakenly carries with it some of the adjacent host genes. When this happens, the newly formed F⁻ plasmid contains endogenote genes from the host cell that may be transferred to other F⁻ cells during regular conjugation. These genes may then be recombined with the genes of the recipient cell. Plasmid-

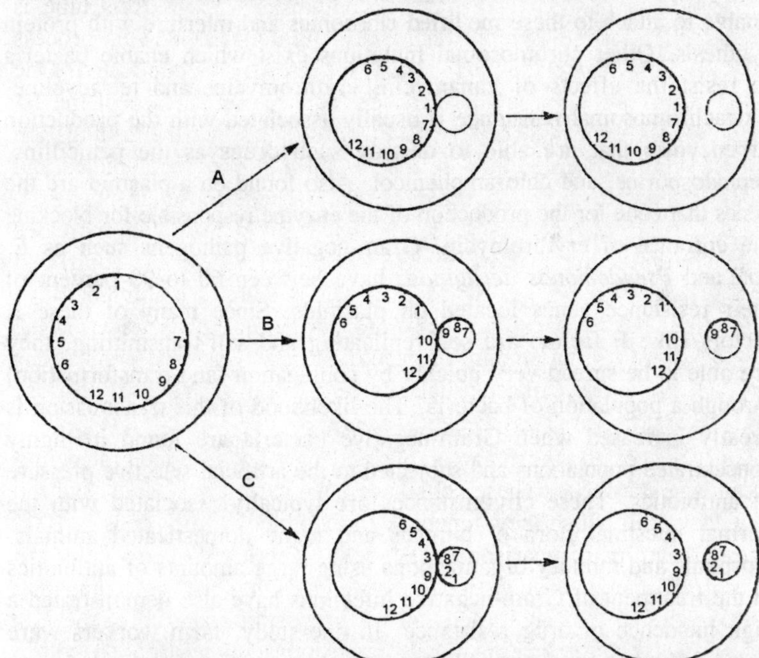

Fig. 11.4. (A) The first set of diagrams illustrates the elimination of an integrated F' factor strand in an Hfr cell. The next two series (B, C) show how it is possible to have endogenote genes incorporate into the F factor. F factors that contain genes from the host DNA are known as F' plasmids.

controlled conjugation generates genetic variation in a number of ways. The most important outcome of this form of gene transfer is the formation of new gene combinations. If these combinations provide the cell with a selective advantage, such as resistance of drugs, the recombinant microbe may produce a clone that could have a great medical significance.

Gene Transfer, Plasmids, and Drug Resistance

The ultimate source of all alleles is the process of mutation. These changes may occur in either the endogenote or exogenote of a cell. Genes transmitted from one cell to another may increase in frequency in a microbial population. Genes for drug resistance that are found on the endogenote provide the cell with *chromosomal resistance*. Those found on plasmids provide *extrachromosomal resistance*. For example, chromosomal resistance to streptomycin occurs with a frequency of about 10^{-10} and primarily centers on the cell's ability to modify ribosomal structure. Drugs such s streptomycin are unable to attach to these mc lified ribosomes and interfere with protein synthesis. Other chromosomal mutations exist which enable bacteria to resist the effects of kanamycin, erythromycin, and tetracycline. Extrachromosomal resistance is usually associated with the production of enzymes that are able to destroy such drugs as the penicillins, cephalosporins, and chloramphenicol. Also found on a plasmid are the genes that code for the production of the enzyme responsible for blocking the entrance of erythromycin. Gram-negative pathogens such as *E. coli* and *Pseudomonas aeruginosa* have between 60 to 90 percent of their resistance genes located on plasmids. Since many of these R factors (like F factor) are self-replicating and self-transmitting, they are able to be spread very quickly by conjugation (and transformation) through a population of bacteria. The likelihood of this transmission is greatly increased when Gram-negative bacteria are found in highly concentrated populations and subjected to the artificial selective pressure of antibiotics. These circumstances are typically associated with the normal intestinal flora of humans and many domesticated animals. Hospitals and military organizations using large amounts of antibiotics in the treatment of Gram-negative infections have also demonstrated a high incidence of drug resistance. In one study, farm workers were found to harbor an abnormally large percentage of antibiotic-resistant, Gram-negative bacteria in their fecal material. These farmers regularly supplemented their pig feed with antibiotics. This same heavy selective pressure is placed on populations of Gram-negative pathogens in hospital environments. As more patients are treated with larger dose of broad spectrum antibiotics to control already resistant microbes, the frequency of resistance in pathogens such as *Proteus*, *Serratia*, *Pseudomonas*, and *Enterobacter* has drastically increased. The fact that patients and staff alike are confined to the hospital environment only serves to increase the possibility of exchanging drug-resistant bacteria and increases the severity of the problem. Furthermore, increased resistance may also increase the virulence of pathogens and make the treatment of Gram-negative infections more complicated. Gram-positive pathogens such as *Staphylococcus aureus* have also experienced a dramatic increase in drug resistance. However, conjugation has not been observed, and a third method of plasmid transfer is needed to explain the high incidence of drug resistance in this group. The *transduction* process relies on viruses to carry the genes for drug resistance from one bacterium to other of the same species.

TRANSDUCTION

Many bacterial viruses (bacteriophages) have a *lytic life cycle*. In this cycle, the virus adsorbs to the surface of the host cell and injects its nucleic acid core through the outer covering. Once inside, the nucleic acid takes command of the host's metabolism to synthesize more virus particles. After the synthesis is complete, the host cell is ruptured to free new virions. The lytic cycle takes place very quickly (about 40 minutes) and there is no delay from the time of initial penetration to lysis of the host. During the lytic cycle, the DNA of the host is broken into small segments that are about the same size as virus nucleic acid. Occasionally, in the case of certain types of bacteriophages, during the assembly of the complete virion a small segment of host DNA is incorporated into the virus protein coat in place of the phage genome. If the DNA found in this virus coat, for example, contains a gene for drug resistance, it may, on infection of another cell, be transferred to that cell and integrated into the endogenote, making the recipient cell drug resistant. The integration resembles the events that take place during transformation. Because any segment of host DNA may be transferred in this way, the process is called *generalized transduction*.

In the *lysogenic*, or *latent*, *life cycle*, lysis of the cell does not take place after the nucleic acid has penetrated the host. The virus integrates into the host chromosome, where it is called a *prophage* (previrus), and it replicates as part of the host chromosome. In this condition, the host cell is called a *lysogenic bacterium*, because it is capable of being destroyed at a later time if the virus prophage becomes active and reenters the lytic cycle. The delay of host destruction that is characteristic of the lysogenic cycle results in the formation of many generations of lysogenic bacterial cells. During the period between virus infection and lysis, these virus-infected bacteria may function as active pathogens. The potential for *specialized transduction* of bacteria begins when the prophage shifts into the lytic cycle and accidentally incorporates an adjacent section of the host DNA into the virus core. When the modified prophage emerges from the host chromosome, it takes command of cell metabolism and synthesize more protein coat and nucleic acid core. Since there is a different gene on the plasmid, it too is synthesized and incorporated into all the new virions. When the lytic cycle is completed, a hundred or so of the virions may be released from the host cell. Each of these contains the new gene. The virions may infect other bacteria of the same susceptible species and integrate this gene into the endogenote, thus transduction them to recombinant forms. If the gene being transferred by transduction is for drug resistance, it is easy to see how portions of a bacterial population may develop this genetic trait.

Relatedness and Genetic Engineering

The genetic material of a prokaryotic cell is not a simple, single, static piece of DNA. Rather, this material is a complex of separate units, many of which may be dynamically interchanged among one another and exchanged between different cells. At one time or another, plasmids prophages, and nucleoids have many common properties. They are circular double helices of DNA capable to self-replication, and they can produce proteins that affect the phenotype of the cell in which they are located. All are capable of interacting with one another.

Many have been found to induce their own transfer by conjugation or the transfer of other genetic units. As more is learned about these various elements, the question of a possible evolutionary relationship arises. It is possible that they all arose through a process of interconvertibility. If plasmids were to combine, resulting in an element essential for the functioning of a cell, such a cointegrated element could have become a nucleoid. If a prophage lost its ability to integrate into the host chromosome and become a protein-coated virus, it could have become a plasmid and possibly a plasmid might have gained the genes needed to form a protein coat and become a virus. The exact sequence of these proposed events may never be known; however, there is no doubt that a very close relationship exists among these genetic elements. The knowledge gained from research into these structural and functional relationships has in recent years led to the development of tools and techniques for the controlled laboratory recombination of genes from a variety of different organism. These processes are more commonly referred to as *gene manipulation*, or *genetic engineering*.

Plasmid-controlled conjugation is one of the most widely used gene transfer methods producing genetic recombinants in the laboratory. Detailed research into the structure and function of DNA has enabled microbiologists to link together specified gene sequences and existing plasmids. Plasmids produced in a test tube from the separate genetic elements of unrelated organisms are called *chimeras* (from Greek mythology, a creature with the head of a lion, body of a goat, and tail of a serpent). If the gene segments linked together are from related organisms, they are called *hybrid* plasmids. The term *composite* plasmid is used to tube. One of the first hybrid plasmids was produced in 1973 by Stanley Cohen and Annie Chang at Stanford University. This plasmid contained genes from two different *E. coli* cells. Chimeras have been produced by combining genes from *Staphylococcus aureus* and *E. coli*

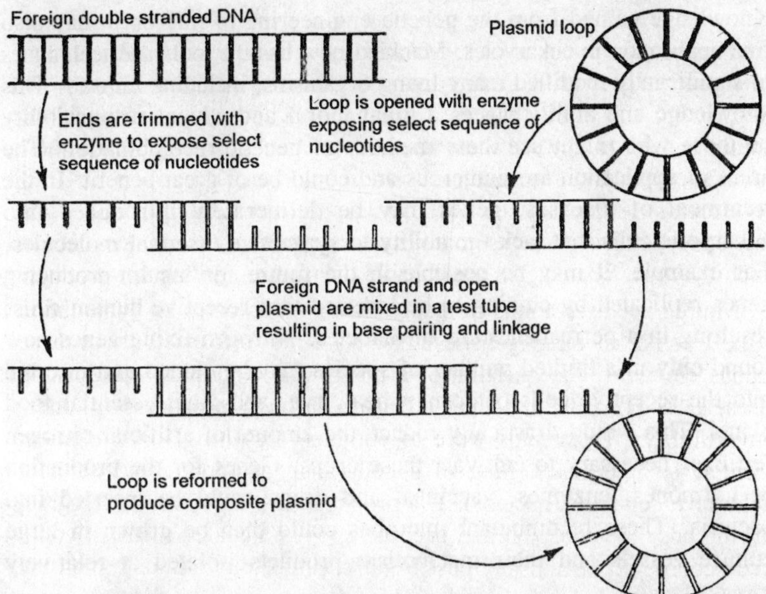

Fig. 11.5. The series of events shown here are all able to be carried out in a test tube. Composite plasmids produced by this method may be inserted into host cells by transformation.

and *Xenopus laevis*, a toad. After a composite plasmid is produced in a test tube, the methods of transformation are used to insert the plasmid into a host cell. This is accomplished by chemically altering the environment of the competent cell and stimulating it to take in the plasmid DNA. Once inside the synthesizing proteins and replicating. In many instances, a plasmid located inside a host cell overreproduces itself. This results in the formation of a cell that contains numerous replicas of the plasmid. Scientists have taken advantage of this reproduction and cloned large amounts of genes specifically introduced into the cell on composite plasmids. These cells can be disrupted and the desired genes isolated in pure form. These may then serve as a further source of DNA for genetic manipulations in other cells, or they may be studied separately. Some of the genes that have already been cloned include histone genes from sea urchins, the gene for mouse mitochondrial DNA, the alpha and beta globin genes of rabbits, and genes for ribosomal RNA from a number of different eukaryotic cells.

The variety of genes that may be artificially introduced into bacteria appears to be endless, as long as essential procedures are followed. This means that the potential for permanently altering the genetic makeup of an organism is within the realm of possibility. Knowledge gained from the genetic engineering of microbes may also find application in eukaryotes. Mankind now has the tools and techniques to significantly modified many living organisms, including himself. This knowledge and ability places a great moral and ethical responsibility on those who might use these methods of genetic recombination. The areas of application are numerous and could be of great benefit. In the treatment of diseases, genes may be deliberately introduced into eukaryotic cells that lack the ability to synthesize essential molecules. For example, it may be possible in the future for insulin-producing genes replicated by cloning to be inserted into receptive human cells, resulting in a permanent cure of diabetes. Nitrogen-fixing genes now found only in a limited number of species may be cloned and inserted into the receptive cells of corn, wheat, rye, and other essential food plants. This would drastically reduce the amount of artificial nitrogen fertilizer necessary to cultivate these crops. Genes for the production of hormones, enzymes, vaccines, and drugs could be inserted into bacteria. These recombinant microbes could then be grown in large culture vessels and their metabolism products isolated at relatively low cost.

There was a concern within the scientific community about the possible misuse or indiscriminate application of recombinant DNA work.

Without adequate guidance and control measures, scientists considered the possibility that new strains of disease-causing viruses could be produced that might be difficult, if not impossible, to control. Multiple-drug resistant plasmids could be produced in combination with toxigenic or other virulence factors that might be hazardous to the public. In order to safeguard scientists and the public from possible contamination with recombinant forms, however, special containment procedures have been established. In addition, scientists have developed special strains of hot bacteria that are extremely difficult to grow except in very selective environments. R Curtiss has developed an *E. coli* strain that cannot synthesize the murein (peptidoglycan) layer of the cell wall except under special laboratory conditions; is sensitive to bile salts, detergents, and ultraviolet light; is unable to serve as a recipient in matings that have any of the five most common plasmids found in enteric bacteria; and is resistant to transduction by certain *E. coli* phase. With all these built-in growth limitations, this strain of *E. coli* could not likely survive outside the laboratory.

The successful development of genetic recombination techniques has enabled many scientists to manipulate the gene of a variety of cell types for the benefit of mankind. This has opened the door for the development of a new field known as *biotechnology* and spawned great interest on the part of the chemical-pharmaceutical industry. The range and variety of anticipated genetic processes and products may be limited only by the scope of the imagination. While most of the hoped-for products are 5 to 20 years in the future, a number of chemical-pharmaceutical companies have actively entered the area of biotechnology by employing or establishing research scientists skilled in the manipulation of genes. Their goal is to combine technological and marketing skills to produce gene products that are readily available to the public.

Gene products currently begins produced include human insulin and growth hormones, foot-and-mouth vaccine for cattle, enzymes for the production of chemicals, and new bacterial strains useful in oil-recovery operations. It has been estimated that such biotechnological products will grow to a value of over $1.4 billion by the years 1985. A few companies currently involved in biotechnological research include Eli Lilly Pharmaceutical, Benentech, Shell Oil, Biogen, DuPont, Standard Oil, Abbott Laboratories, Dow Chemical, and Bristol-Myers.

12

YEAST AS THE GENETIC MATERIAL

It has been seen that the chromosomes of bacterial and eukaryotic organisms are too large to be isolated easily, and cannot be analyzed by the techniques which have been used so successfully with plasmids and virus chromosomes. However the mitochondrial DNA from yeast is the exception. It has a molecular weight 50 million, it can be isolated and analyzed in essentially a very similar fashion. Consequently the mitochondrial DNA of yeast is probably the most completely mapped and characterized piece of eukaryotic DNA. More recently, a second extrachromosomal molecule has been detected in yeast, the so-called 2 μm circle, molecular weight 3.9 million. This has been used to transmit yeast genes into *E. coli* and back into yeast by a high-frequency transformation system. Here we shall discuss the mapping of the mitochondrial genome and the discovery of intervening sequences in genes.

Non-chromosomal Inheritance in Yeast

The character differences have because the essential feature for the genetic analysis of any chromosome. The study of the genetics of east mitochondria has relied heavily on the fact that respiratory-deficient mutants could be isolated which failed to grow on glycerol or lactate. These substrates, unlike glucose, have to be respired, as they cannot be fermented. Consequently, mutants with defective mitochondria failed to grow on media containing glycerol, but grew on glucose media. The classical evidence for extrachromosomal inheritance can be demonstrated by reference to the petite mutants of yeast. It has been observed that segregation of chromosomally-controlled petites, but for mitochondrial petites 4:0 or 0:4 ratios of normal to petite colonies are obtained, depending on the type of petite mutation. Neutral petites often result from prolonged treatment with ethidium bromide, which is known to inhibit mt DNA synthesis, and they have been shown in some cases to have a complete lack of mitochondrial DNA. Suppressive petites have lost sequences of mitochondrial DNA of varying lengths. The resistance of colonies to antibiotics such as erythromycin and chloramphenicol has proved to be a very useful market in crosses which have attempted to map these genes within the mitochondrial genome. Wild-type strains are only sensitive to these antibiotics when they are growing on glycerol or other non-fermentable substrates.

Table 12.1. Gene loci mapped in the mitochondrial genome. Previous names for gene loci are shown in brackets.

Gene symbol	*Phenotype*	*Component affected (where known)*
*ana*1 (*mik* 1)	Antimycin resistance	mt membrane?
cap 1, 2	Chloramphenicol resistance	mt ribosome
cya 1, 2, 3, 17 (*oxi*)	No growth on glycerol	Cytochrome c oxidase subunits
cyab 1 (*cob*)	No growth on glycerol	Cytochrome b subunit
ery 1, 2	erythromycin or	mt ribosome
spi 1	spiramycin resistance	
oli 1, 2, 3	Oligomycin resistance	mt ATPase subunits
par 1	Paromomycin resistance	
var 1		Protein of unknown function
omega (ω)		Polarity of transmission of markers

The third class of mutants was isolated by selection of strains which failed to grow on glycerol media, but which did not produce such small colonies as petites. These mutants had point mutations affecting specific components of the respiratory chain, e.g., cytochrome coxidase or cytochrome *b*. Complementation tests are difficult for mitochondrial mutants, as recombination can occur relatively freely. However, allelism is normally assumed if recombinants occur at less than 0.5% in ascospores. Table summarizes some of the mutants isolated and their defects, where these are known. Terminology is rather confused, so alternative gene symbols are given in brackets so that data can be compared from different sources.

Cytoduction

Cytoduction is a useful technique for studying cytoplasmic inheritance. It deals with the fusion of cells without the fusion of nuclei (*karyogamy*). A mutation at the *kar* locus eliminates the nuclear fusion which normally occurs to produce a diploid cell. Interchange of cytoplasmic organelles can occur without nuclear exchange and then the cells segregate. This allows the geneticist to study the interaction between chromosomal genes and cytoplasmic organelles as well as being a diagnostic test for cytoplasmic genes. The technique has been used to study the two-micron circular plasmid as well as mitochondrial DNA and the dsRNA killer factor.

Mapping of the Mitochondrial Genome

Various methods have been applied to construct maps of the yeast mitochondrial genome. The methods of classical genetic mapping can be used, and data from a four-point cross in which selection was made for *cya* are shown. In similar crosses, linkage was shown between *cap* and *ery*, but in general crosses of this type are difficult, as recombination is relatively high and complicated by the segregation of a polarity locus ω. Deletion mapping has been used to order genes, although this method has problems, because not all deletions are simple and may be multiple deletions spaced along the chromosome. Spontaneous deletions, however, are thought to be reliable, and an analysis of the functions eliminated in a series of petite strains has been made. Each of the petite genotypes can be explained by the deletion of a single segment from a circular map.

Mitochondria are the semi autonomic organelles as they have DNA in them. They are capable of independent protein synthesis, which means that the mitochondrial genome codes for 15S and 21S ribosomal RNA, as well as for a complete set of mitochondrial tRNAs. tRNA genes have been mapped by DNA-RNA hybridization using defined deletion mutants. Radioactively-labelled tRNAs will only bind to fragments retaining the gene which codes for them. By this method most of the tRNAs have been mapped to between *cap*1 and *par*1. A second approach is to look for the synthesis of rRNA in different deletion mutants *in vitro*. The synthesis of 21S RNA was correlated with the retention of *ery* and *cap*, while the 15S RNA synthesis ability correlated with *par*.

Restriction endonuclease maps have been constructed using *Pst*1 (one fragment), *BamH*I (3 fragments), *Hind*II-III (9 fragments) and *EcoR*I (9 fragments). Eventually it should prove possible to correlate the genetic map with this restriction map. Already an analysis under the electron microscope of a restriction fragment and 21S RNA has shown that the gene coding for this RNA molecule has a 1000 base pair insertion. This is known as an intervening sequence (or *intron*), and similar effects have been observed for other mitochondrial genes, and indeed for a variety of eukaryotic genes. The coding sequence is known as an exon. It has been found that the complete DNA sequence is transcribed into RNA, and then the intron is 'spliced out' before the mRNA is translated. The discovery of this quite unexpected gene organization has caused considerable speculation, and various ideas have been put forward to explain its occurrence. It has been suggested that introns are regulatory elements, or that they enhance variation and speed of evolution. Another possible function is that they divide up functional areas of proteins, which permits reassortment and recombination of exons in such a way as to produce new functional proteins. Much of this is speculation and further work is needed.

Mapping of the Yeast Mitochondrial DNA

Several genes have been reported coding for various proteins. There are about 25 transfer RNA genes, 16 of which are grouped between the 21S ribosomal RNA and cytochrome c oxidase III. A major feature of yeast mitochondrial DNA is variability both in length and in internal structure. In different strains the DNA varies from 75 to 80 kb, and there is also variation in the presence or absence of introns. Some introns contain open reading frames which codes for 'maturases', proteins involved in the splicing of mRNA precursors, for example the intron 2 in cytochrome b. Other interesting features of yeast mitochondrial DNA are that there are regions which are AT rich, and that the genetic code used by the mitochondrion is different from the normal dictionary.

Transformation in Yeast

It has always been a source of frustration to fungal geneticists that there was apparently no fungal equivalent of the F, λ and T4 systems found in *E. coli*. Mycoviruses do occur, but these are difficult to work with, and in any case contain double-stranded RNA. The availability of cloned yeast DNA, particularly a derivative of *ColE*1 and *leu*2^+ genes, meant that *leu*2^- protoplasts of yeast could be saturated with DNA and a search made for *leu*2^+ transformants. Hinnen et. al. (1978) did in fact find transformants at a frequency of 1 in 10^7 regenerated protoplasts, but the plasmid and the *leu*$^+$ genes had integrated into the yeast chromosome. Hybridization and restriction endonuclease mapping techniques were used to demonstrate that *ColE*1 DNA had also integrated in the yeast chromosome.

This was very interesting, but the plasmid apparently did not replicate autonomously, so the system could not be used as an easy cloning system for DNA. As long ago as 1967, however, a plasmid had been detected in yeast. This was the two-micron circle already mentioned, which had a contour length of 1.95 μm and had 5.9 kilobases. There were between 50 and 100 plasmid copies per cell, and the molecule contained an inverted repeat sequence 600 base pairs long. The consequence of this was that a recombinant event within the repeat sequences could convert one molecule into an alternative form. The molecule could therefore exist as type A or type B, with a slight difference in sequence and a different pattern of restriction endonuclease fragments. Type A gave rise to fragments of 2.3 kb and 3.6 kb with *Eco*RI digestion, whereas type B gives fragments of 3.8 kb and 2.1 kb. Evidence for the inverted repeats came from denaturation/renaturation experiments in which single-strand molecules were found to reanneal in the repeated region to give two single-strand loops separated by a double-stranded region. Stable propagation of the plasmid requires an

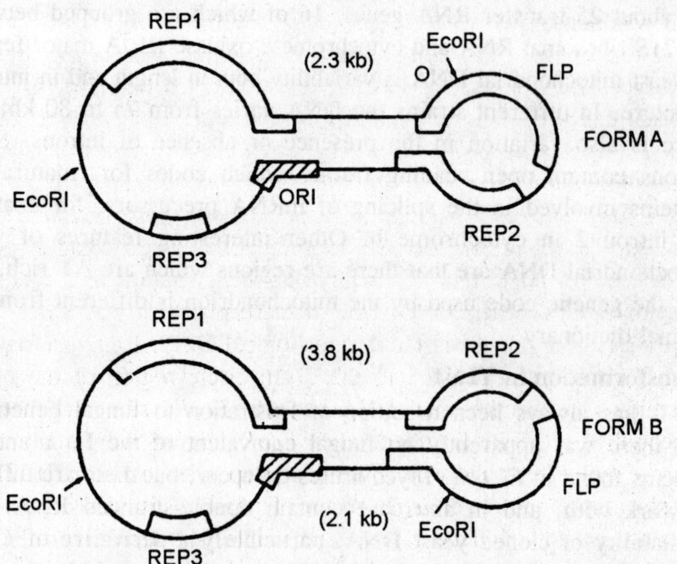

Fig. 12.1. Genetic and molecular map of the two-micron circular plasmid of S. cerevisiae to show the components involved in replication and stability. REP1 and REP2 are trans-active loci whereas REP3 is cis-active, ori is the replication origin and FLP is the 'flip' locus involved in recombination between forms A and B.

origin of replication and three loci REP1, REP2, and REP3. REP1 and REP2 can act in the *trans*-configuration and are considered to code for proteins responsible for stable transfer, whereas *ori* and REP3 only act in *cis*, suggesting that they are the sites at which the protein interacts with the DNA. A further locus FLP codes for a protein involved in conversion of form A to form B.

Thus we see that the plasmid is a replicon which persists, replicates and recombines in the yeast cell. The scene was set to exploit the plasmid for transformation. Beggs (1978), first produced chimaeric (hybrid) plasmids between the two-micron circle and a R bacterial plasmid which coded for tetracycline resistance, pMB9. The presence of the *Tc'* marker facilitated the selection of transformed clones, and the presence of a bacterial plasmid ensured that the hybrid plasmid would replicate in *E. coli*, thus allowing cloning of the hybrid plasmid. Two-micron circle DNA from yeast and the pMB9 plasmid were partially treated with *Eco*RI so that most molecules had only one cut. The DNA types were mixed and ligated, and used to transform *E. coli*. Selection was made for *Tc'* and the clones screened for the presence of the larger hybrid plasmids. A variety was identified and, as it was not known whether cloning would inactivate the activities of the 2-µm circle, four different hybrid plasmids were used. Sheared wild-type DNA from yeast (6 to 15 kb) was tailed with poly(dA). The hybrid plasmids were converted into linear molecules by treatment with *Pst*1 and then tailed with poly (dT). The two types of molecule were allowed to anneal and the DNA was used to transform a *recA*1, *leuB*6 $hsdR^-$ $hsdM^+$ strain of *E. coli*. Clones were selected which were leu^+ Tc', and two plasmids were identified which could transform the *E. coli* strain to leu^+ *Tc'* with equal efficiency at high level (10^5 transformants per µg DNA). One of these plasmids was a derivative of pJDB41 and was given the designation pJDB219. It had an insertion of 1.2 kb of yeast DNA into the 2-µm circle region of the hybrid plasmid. The chromosomal DNA had one *EcoR*1 site, and therefore appeared as a 0.8 kb fragment when *EcoR*1 digests were analyzed by agarose gel electrophoresis. The next step was to use pJDB219 to transform protoplasts of $leu2^-$ strain of yeast. This was done in the presence of polyethylene glycol, and viable leu^+ transformants were obtained at levels of between 5 x 10^{-4} to 3 x 10^{-3} transformants per viable cell.

The extraction of plasmid DNA was made from these transformed cells, *EcoR*1 digested and run on agarose gels. *c*RNA was prepared from the original *EcoR*1 digested pJDB219 plasmid and used as a labelled probe to test for the presence of pJDB219 in the *leu*$^+$ yeast clones. All of the expected bands were found, but two further low-molecular-weight bands were also detected. This suggested that perhaps further recombination or deletions had occurred. As a final proof, plasmid DNA extracted from yeast cells was used to transform the original *E. coli* strain, and it was found that transformation occurred at the same efficiency for *leu*$^+$ and *Tc*^{r}.

It has been shown that hybrid plasmid can readily transform yeast cells, and that the organism has considerable potential as a molecular cloning vehicle. The techniques usually reserved for *E. coli* can now be exploited in a eukaryotic microorganism.

Development of Yeast Plasmids and Artificial Chromosomes

The development of a series of yeast plasmids was due to the success of this cloning technique. These vary to their copy number, stability and transformation efficiency. The first plasmids in the transformation experiments described above used integrative plasmids called YIp which contain essentially only bacterial sequences and the yeast gene *leu*. Addition of the 2-micron circle to this by Beggs produced an episomal plasmid YEp with a higher stability, copy number and transformation frequency. Stability could be increased further by the addition of an autonomously replicating sequence (*ars*) from chromosomal DNA which produced a YRp plasmid. Now-a-days it has been possible to add a centromeric sequence to produce YCp which has a low copy number like a nuclear chromosome. A major advance has been the constriction of artificial chromosomes by the addition of telomeres, a centromere, an *ars* sequence and selectable genes HIS and LEU. *Telomeres* are a device to protect the linear chromosomes from nuclease attack and to prevent rejoining of reactive DNA strands. In yeast they are tandemly repeating units of a simple irregular sequence 5'(C_{1-2}A)3' replicated approximately 30 times. The origin of these man-made chromosomes is complex, and originally involved using telomeres from *Tetrahymena*. Short chromosomes of less than 20 kb showed mitotic instability, whereas chromosomes longer than 55 kb showed mitotic and meiotic stability.

Cloning of the *DEX* Gene in Yeast

The cloning of yeasts has become possible due to the availability of a range of yeast vectors. One example is the cloning of the *DEX* gene coding for amyloglucosidase, an enzyme which allows the yeast to utilize dextrins normally left in beer after a fermentation. It was first necessary to produce a gene bank or genome library, as described. The yeast DNA from a *DEX*$^+$ strain was digested with *Sau3A* and the plasmid pJDB207 was digested with *BamH*1. *Sau3A* has a tetranucleotide target GATC and controlled partial digestion produced random fragments of about 5 kb which were representative of the complete genome. *BamH*1 recognizes the target G↓GATCC and cuts the plasmid at one site within the tetracycline-resistant gene. The choice of these two restriction enzymes meant that there was a four-base-pair homology between the single strands of the yeast and plasmid DNA. Ligation of the two types of DNA fragments resulted in a mixture of regenerated circular plasmids, some of which would contain the DEX gene. The plasmid was treated with a phosphatase to prevent regeneration of the original plasmid. The mixture was then used to transform *E. coli* cells and selection was made for *amp*^{R} *tet*^{S} cells. 50 000 clones were selected, which on a probability basis would be likely to contain DNA representative of the complete yeast genome. These cells then formed the DNA library, which was stored at 20^0 in 15% glycerol. DNA from these cells was used to transform yeast cells and selection was made for Leu$^+$. Colonies produced were screened for the production of *amyloglucosidase* which was detected by the presence of a halo around colonies on starch-containing media.

Epidermal Growth Factor (EGF)

The review of literative has developed a considerable interest in yeast as a host for cloning and expression of human DNA. One difficulty is that proteins produced internally are difficult to purify. However, it is possible to exploit natural export systems in yeast such as the *a* and α mating type pheromones. The basic strategy is to insert the EGF DNA into the α-factor gene which has previously been cloned on a shuttle vector very similar to those described earlier. One such recombinant vector is pYαEFG-21 in which the EGF-gene is fused in frame with the gene leader sequence coding for lys-agr-(glu-ala)$_3$. This hydrophobic sequence is necessary for export of the peptide through the membrane. Active EGF is in fact exported, but the majority of the peptides had (glu-ala)$_3$ at the amino-terminal end, suggesting that the normal levels of proteolytic enzymes could not cope with processing the large amounts of fusion protein produced.

Control of Yeast Mating Type

Mating in *S. cerevisiae* is controlled by the MAT locus and there are two alleles *a* and α. Some strains were observed to change mating type very frequently, and cloning of the relevant regions revealed a

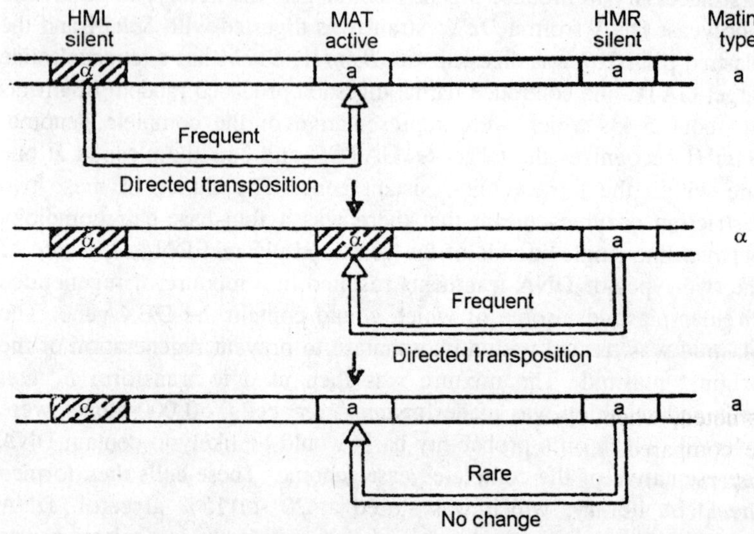

Fig. 12.2. Molecular structure of the MAT locus and the silent cassettes HMR and HML.

phenomenon of frequent switching controlled by 'silent loci' HML and HMR which contain copies of either *a* or α, but which are not expressed. When the HML locus has the α allele and the MAT locus has a, there can be directed transposition which results in a substitution and a change of mating type. This has been termed the *cassette* model, as the silent cassette can replace the active allele. The reverse of this process can occur from HMRa. It is however, rare for an *a* cassette to replace an *a* allele at MAT. The MAT, HML and HMR loci have all been cloned and sequenced. A number of sequences are identical at each locus, notably W, X Z1 and Z2, HMRa lacks the W region, and there is 105-base pair difference between Y*a* and Yα. Space does not permit a detailed description of the events involved in the switching mechanism. Although the process of switching is sometimes compared with transposition, it should be noted that there are some differences. At MAT locus initiates the switch, and the inverted repeats and sequences related to the transposition even are not observed.

13

EUKARYOTIC MICROORGANISMS

As noted, when the ribosomal RNA (rRNA) sequences of organisms are compared, the living world can be divided into three divisions: *Bacteria*, *Archaea*, and *Eucarya*. In this chapter we will consider the *Eucarya*. These organisms, which include the algae, fungi, protozoa, multicellular parasites, and insect vectors, have one feature in common: they are all eukaryotic organisms. The basic cell structure of the *Eucarya* is distinctly different from that of the *Bacteria* and *Archaea*. They are included in a textbook of microbiology because many members of these groups are microscopic and are studied with techniques that are similar to those used to study bacteria and archaea. In addition, many of these organisms cause disease in humans as well as plants and animals.

Classification using gross anatomical characteristics of algae, fungi, protozoa, and even some multicellular parasites has always been problematic. Now, however, with modern techniques that examine these organisms at the molecular and ultrastructural levels, it has been discovered that some of the organisms that were traditionally grouped together were more dissimilar than similar. Instead, they arose at various times along a continuum of evolution. Therefore, in classification schemes that describe an accurate evolutionary history of organisms and are based on molecular and ultra structure examination, the words *algae*, *fungi*, and *protozoa* are no longer really accurate. For the purposes of this book, however, we will use term algae to describe the photosynthetic members and fungi and protozoa to describe the non-photosynthetic members that are discussed in this chapter. In addition, a discussion of arthropods and helminths is included because these eukaryotes are also implicated in humans disease.

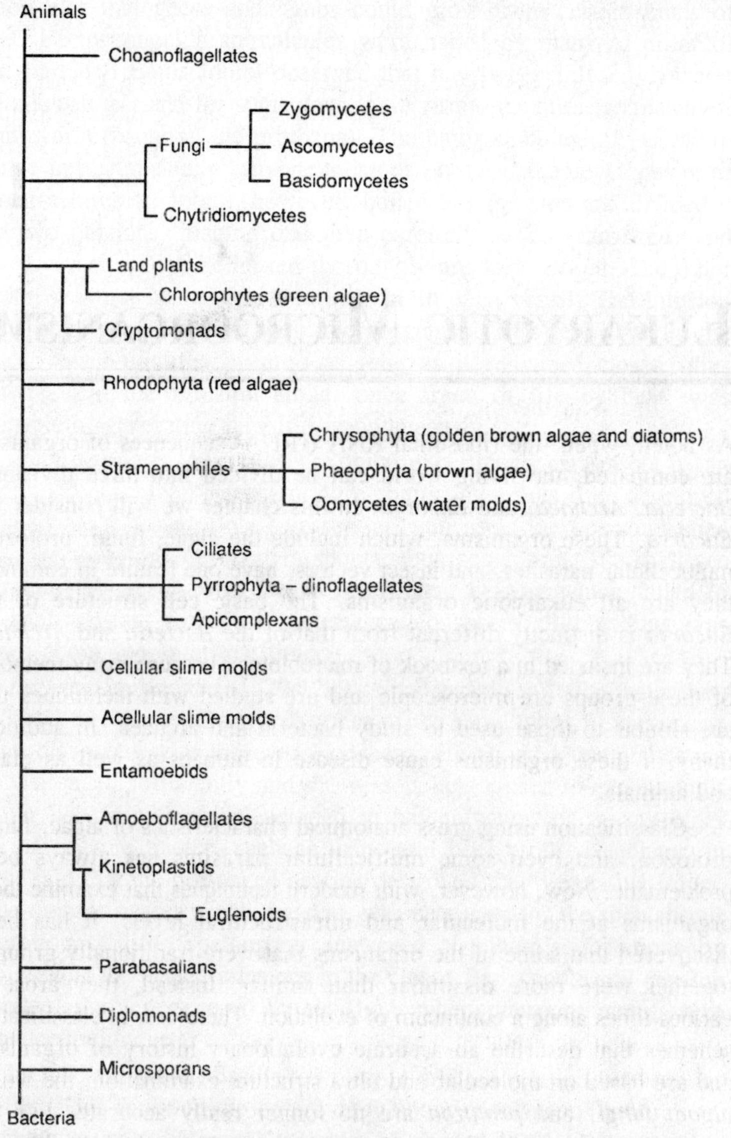

Fig. 13.1. A phylogeny of the Eukaryotes based on ribosomal RNA sequences comparison.

Algae

The algae are a diverse group of eukaryotic organisms that share some fundamental characteristics but are not related on the phylogenetic tree. These organisms are studied by *algologists* in a field known as *algology*. Algae are organisms that use light energy to convert CO_2 and H_2O to carbohydrates and other cellular products with the release of oxygen. Algae contain chlorophyll *a*, which is necessary for photosynthesis. In addition, many algae contain other pigments that extend the range of light waves that can be used by these organisms for photosynthesis. Algae include both microscopic unicellular members and macroscopic multi cellular organisms. Algae, however, differ form other eukaryotic photosynthetic organisms such as land plants in their lack of an organized vascular system and their relatively simple reproductive structures.

Algae do not directly infect humans, but some produce toxins that cause paralytic shellfish poisoning. Some of these toxins do not cause illness in the shellfish that feed on the algae but accumulate in their tissues and when eaten by humans cause nerve damage.

As one of the primary producers of carbohydrates and other cellular products, the algae are essential in the food chains of the world. In addition, they produce a large proportion of the oxygen in the atmosphere.

Classification of Algae

As just noted, *algae* is not a strict classification term; nevertheless, organisms considered under the general heading of algae are grouped for identification by a number of properties. These include the principal photosynthetic pigments of each group, cell wall structure, type of storage products, mechanism of motility, and mode of reproduction. The names of the different algal groups are derived form the major colour displayed by most of the algae in that group. Note that the different algal groups lie in different places along the evolutionary tree.

Algal Habitats

Algae are found in both fresh and salt water, as well as in soil. Since the oceans cover more than 70% of the earth's surface, aquatic algae are major producers of oxygen as well as important users of carbon dioxide. Unicellular algae make up a significant part of the *phytoplankton* (*phyto* means "plant" and *plankton* means "drifting"), the free-floating, photosynthetic organisms that are found in marine environments. More oxygen is produced by the phytoplankton than by all forests combined. Phytoplankton is a major food source for many animals, both large and small. Microscopic animals in the *zooplankton* (*zoo* means "animals") graze in this phytoplankton, and then both the zooplankton and phytoplankton become food for the benthic whales,

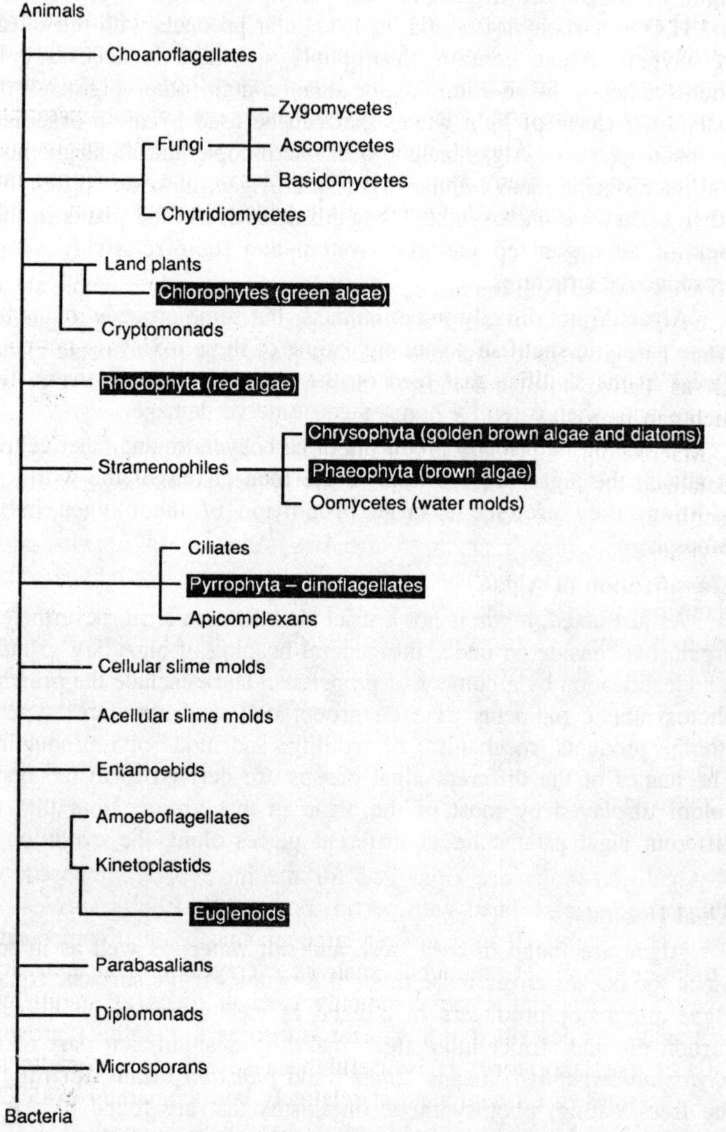

Fig. 13.2. Phylogeny of algae. The algal groups are highlighted.

some of the world's largest mammals. The unicellular algae of the phytoplankton are well adapted to this aquatic environment. As single cells, they have large, adsorptive surfaces relative to their volume and can move freely about, thus effectively using the dilute nutrients available.

Because one or more algal species can grow in almost any environment algae often grow where other forms of life cannot thrive. Frequently, algae are among the first organisms to become established in barren environments, where they synthesize the organic materials necessary for the subsequent invasion and survival of other plants and animals. They are often found on rocks, preparing the surface for the growth of more complex members of the biological community.

Structure of Algae

Algae can be both microscopic and macroscopic. Microscopic algae can be single-celled organisms floating free of propelled by flagella, or they can grow in long chains of filaments. Some microscopic algae such as *Volvox* form colonies of 500 to 60,000 biflagellated cells, which can be visiblé to the naked eye.

Macroscopic algae are multicellular organism with a variety of specialized structures that serve specific function. Some possess a structure called a *holdfast*, which looks like a root system but primarily serves to anchor the organism to a rock or some other firm substrate. Unlike a root, it is not used to obtain water and nutrients for the organisms. Nutrients and water surround the organisms and do not need to be drawn up form the soil. The stalk of an alga, known as the *stipe*, usually has leaflike structures of *blades* attached to it. The blades are the principal photosynthetic portion of a alga, and some also bear the reproductive structures. Many large algae have gas-containing *bladders* or floats that help them maintain their blades in a position suitable for obtaining maximum sunlight.

Cell walls

Algal cell walls are rigid and for the most part composed of cellulose, often associated with pectin. Some multicellular species of algae such as some red algae contain large amounts of other compounds in their cell walls. Compounds such as carrageenan and agar are harvested commercially and commonly used in foods as stabilizing compounds. As described earlier, agar is also used to solidify growth media in the laboratory. It is useful because although it melts at 100°C, it stays in a liquid state at relatively low temperature (45°C-50°C) so that nutrients can be added, and yet is solid at room temperature to act as a growing surface in a Petri dish.

Diatoms are algae that have silicon dioxide incorporated into their cell walls. When these organisms die, their shells sink to the bottom of the ocean, and the silicon-containing material does not decompose. Deposits of diatoms that formed millions of years ago are mined for a substance known as diatomaceous earth, used for filtering systems, abrasives in polishes, insulation, and many other purposes.

Eukaryotic cell structures

As is true in all eukaryotes, algae have a membrane-bound *nucleus*. The genetic information is contained in a number of chromosomes that are tight packages of DNA with their associated basic protein. These chromosomes are enclosed in a nuclear membrane.

In addition, algae have other organelles in their cytoplasm such as *chloroplasts* and *mitochondria*. Chloroplasts contain chlorophyll as well as other light-trapping pigments such as carotenoids and phycocyanin. Photosynthesis occurs in the chloroplast. Respiration and oxidative phosphorylation occur in the mitochondria.

Algal Reproduction

Most single-celled algae reproduce asexually by binary fission, as do most bacteria. The major difference between prokaryotic and eukaryotic fission involves events that take place with the genetic material within the cell. It has been discussed already that in prokaryotic fission, the circular DNA replicates and each daughter cell receives half the original double strand of DNA plus a newly replicated strand. In eukaryotic organisms with multiple chromosomes, after the DNA is replicated, the chromosomes go through a nuclear division process called *mitosis*. This process ensures that the daughter

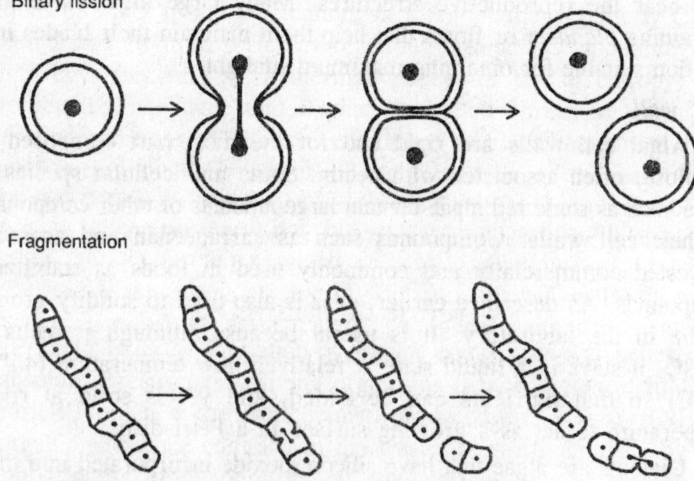

Fig. 13.3. Binary fission is an asexual reproduction process in which a single cell divides into two independent daughter cell.

cells receive the same number of chromosomes as the original parent. Some algae, especially multicellular filamentous species, reproduce asexually by fragmentation. In this type of reproduction, portions of the parent organisms break off to form new organisms, and the parent organism survives.

Sexual reproduction also regularly occurs in most algae. During the process known as *meiosis*, *haploid* cells with half the chromosome content are formed. These cells are called *gametes* and when they fuse together they form a *diploid* cell with a full complement of chromosomes known as a *zygote*. Gametes are often flagellated and highly motile. Many algae alternate between a haploid generation and a diploid generation. Sometimes, as is the case with *Ulva* (sea lettuce), the generations look physically similar and can only be told apart by microscopic examination. In other cases, the two forms look quite different.

Paralytic Shellfish Poisoning

Although algae do not directly cause disease in humans, they do so indirectly. A number of algae produce toxins that are poisonous to humans and other animals. Several dinoflagellates of the group Pyrrophyta cause red tides, or algal blooms in the ocean. Red tides were reported in the Bible along the Nile and today seem to be spreading worldwide. In the warm waters of Florida and Mexico, red tides result form an abundance of *Gymnodinium breve*. This dinoflagellate discolours the water about 17 to 75 km from shore and produces *brevetoxin*, which kills the fish that feed on the phytoplankton. It is unclear why algae suddenly grow in such large numbers, but it is though that sudden changes in conditions of the water are responsible. The runoff of fertilizers along the waterways and coastlines may also be the cause of red tides. In addition, an upwelling of the water often brings more nutrients as well as the cyst that are resting, resistant stages of the *G. breve* from ocean bottom to the surface. When these cysts encounter warmer waters and additional nutrients, they are released from their resting state and begin to multiply rapidly. Persons eating fish that have ingested *G. breve* and thus contain brevetoxin may suffer a tingling sensation in their mouths and fingers, a reversal of hot and cold sensations, reduced pulse rate, and diarrhea. The symptoms may be unpleasant but are rarely deadly, and people recover in 2 to 3 days.

Red tides caused by the dinoflagellate of the genus *Gonyaulax* are much more serious. *Gonyaulax* species produce *neurotoxins* such as *saxitoxin* and *gonyautoxins*, some of the most potent no-protein poisons known. Shellfish such as clams, mussels, scallops, and oysters feed on these dinoflagellates without apparent harm and, in the process, accumulate the neurotoxin in their tissues. Then when humans eat the shellfish, they suffer symptoms of paralytic shellfish poisoning including general numbness, dizziness, general muscle weakness, and impaired respiration. Death can result from respiratory failure. *Gonyaulax* species are found in both the North Atlantic and the North Pacific. They have seriously affected the shellfish industry on both coasts over the years. At least 200 manatees died along the coast of Florida in the spring of 1996 as a result of red tide poisoning.

Another dinoflagellate, *Pfiesteria piscida*, usually is found as a non-toxic cyst or ameba in marine sediments. This organism changes when a school of fish approaches. The fish secrete a chemical cue that alerts the *Pfiesteria* to transform into a flagellated zoospore. It then release two toxins. One stuns the fish and the other causes its skin to slough away. The *Pfiesteria* then enjoys a meal on the fish's red blood cells and sexually reproduces. The toxins are so potent that researchers working with them in a laboratory were seriously affected. This organisms now must be studied using the same precautions that are used to study the virus that causes AIDS.

Another algal toxin found in some species of diatoms has more recently been recognized to cause paralytic shellfish poisoning. This position is domoic acid and is most often associated with mussels and other shellfish and crabs that accumulate the poison. Persons eating the shellfish suffer nausea, vomiting, diarrhea, and abdominal cramps as well as some neurological symptoms such as loss of memory. State agencies constantly monitor for algal toxins, and it is wise to check with the local health department before harvesting shellfish for human consumption. Algal toxins may be present even when the water is not obviously discoloured. Cooking the shellfish does not destroy these toxins.

Protozoa

Along with the algae, the protozoa constitute another group of eukaryotic organisms that traditionally have been considered part of the microbial world. Protozoa are microscopic, unicellular organisms that lack photosynthetic capability, usually are motile at least at some stage in their life cycle, and reproduce most often by asexual fission.

Classification of Protozoa

As with algae, classification of protozoa according to rRNA and ultrastructure shows that hey are not a unified group, but appear along

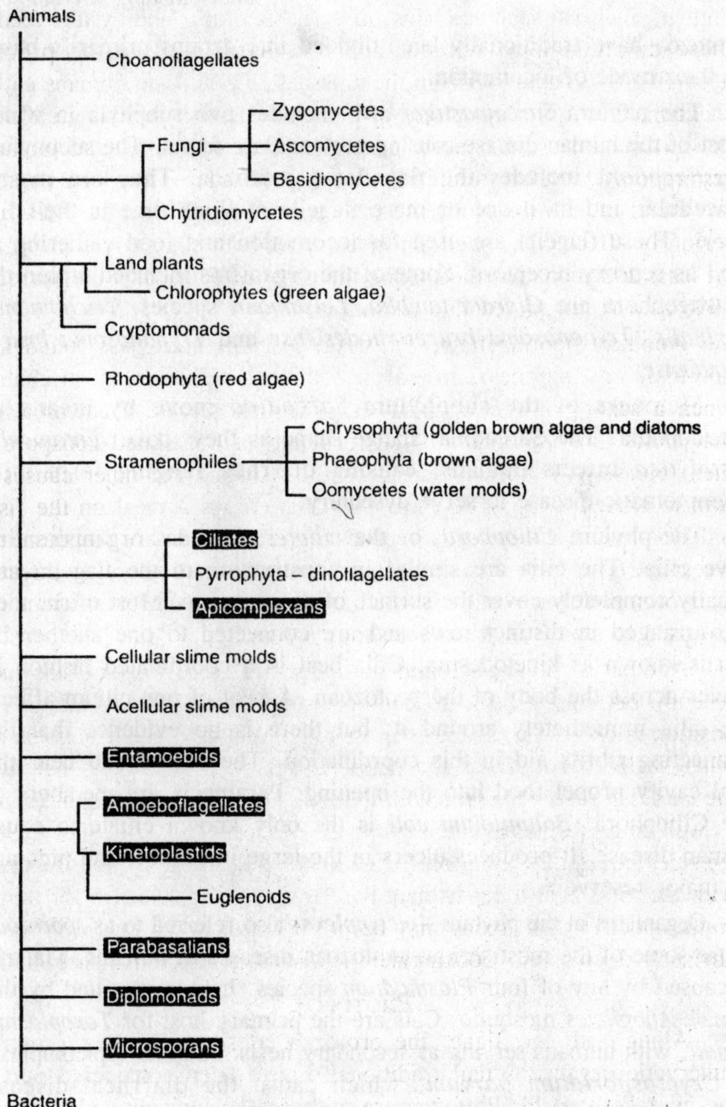

Fig. 13.4. A phylogenetic scheme of eukaryotic based on rRNA sequence comparison. The protozoan groups are highlighted.

the evolutionary continuum. The primary reason that they are lumped together in a field known as *protozoology* is because they are all single-celled eukaryotic organisms that lack chlorophyll. Some members of this group cause disease. We will concentrate on these organisms.

Protozoa have traditionally been divided into groups primarily based on their mode of locomotion.

The phylum *Sarcomastigophora* includes two subphyla in which most of the human disease-causing protozoa are found. The subphylum *Mastigophora* includes the flagellated protozoa. They are mostly unicellular and have one or more flagella at some time in their life cycle. These flagella are used for locomotion and food gathering as well as sensory receptors. Some of the organisms included within the Mastigophora are *Giardia lamblia*, *Leishmanii* species, *Trichomonas vaginalis*, *Trypanosoma brucei rhodesiense* and *Trypanosoma brucei gambiense*.

Members of the subphylum *Sarcodina* move by means of pseudopodia. The Sarcodina change shape as they move. *Entamoeba histolytica* infects humans, causing diarrhea ranging form mild asymptomatic disease to serve dysentery.

The phylum *Ciliophora*, or the *ciliates*, includes organisms that have cilia. The cilia are similar in construction to the flagella and usually completely cover the surface of an organism. Most often, they are arranged in distinct rows and are connected to one another by fibrils known as kinetodesma. Cilia beat in a coordinated fashion in waves across the body of the protozoan. A beat of one cilium affects the cilia immediately around it, but there is no evidence that the connecting fibrils aid in this coordination. The cilia found near the oral cavity propel food into the opening. Paramecia are members of the Ciliophora. *Balantidium coli* is the only known ciliate to cause human disease. It produces ulcers in the large intestines, and pigs are its major reservoir.

Organisms in the phylum *Apicomplexa*, also referred to as *sporozoa*, cause some of the most serious protozoan diseases of humans. Malaria is caused by any of four *Plasmodium* species. It is transmitted by the female *Anopheles* mosquito. Cats are the primary host for *Toxoplasma gondii*, with humans serving as secondary hosts. Another Apicomplexa is *Cryptosporidium parvum*, which cause the diarrheal disease cryptosporidiosis.

The phylum *Microspora* includes the intracellular protozoa that infect immunocompromised humans, especially persons with AIDS. There are other protozona phyla such as Labyrinthomorpha, Ascetospora, and Myxozoa, but they are not implicated in human disease and so we will not consider them here. They are most often found in marine habitats and are parasitic on fish and other sea life.

Protozoan Habitats

A majority of protozoa are free-living and found in marine, freshwater, or terrestrial environments. They are essential as decomposers in many ecosystems. Some species, however, are parasitic, living on or in other host organisms. The hosts for protozoan parasites ranges from simple organisms, such as algae, to complex vertebrates, including humans. All protozoa require large amount of moisture, no matter what their habitat.

In marine environments, protozoa make up part of the zooplankton, where they feed on the algae of the phytoplankton and are an important part of the aquatic food chains. On land, protozoa are abundant in soil as well as in or on plants and animals. Special protozoan habitats include the guts of termites, roaches, and ruminants such as cattle.

Protozoa are an important part of the food chain. They eat bacteria and algae and, in turn, serve as food for larger species. The protozoa help maintain an ecological balance in the soil by devouring vast numbers of bacteria and algae. For example, a single paramecium can ingest as many as 5 million bacteria in one day. Protozoa are important in sewage disposal because most of the nutrients they consume are metabolized to carbon dioxide and water, resulting in a large decrease in total sewage solids.

Structure of Protozoa

Cell wall

Protozoa lack the rigid cellulose cell wall found in algae. Most protozoa do, however, have a specific shape determined by the rigidity or flexibility of the material lying just beneath the plasma membrane. *Foraminifera* have distinct hard shells composed of silicon or calcium compounds. The foraminiferans, which secrete a calcium shell, have through the course of millions of year formed limestone deposits such as the white cliffs of Dover on England's southern coast.

Eukaryotic cell structure

Protozoa are eukaryotic organisms and as such have a membrane-bound nucleus as well as other membrane-bound organelles such as mitochondria. Protozoa are not photosynthetic and thus lack chloroplasts.

Protozoa have specialized structures for movement such as *cilia*, *flagella*, or *pseudopodia*. Eukaryotic flagella and cilia are distinctly different in construction from prokaryotic flagella. Protozoa are grouped by their mode of locomotion. For example, the Mastigophora have flagella, and Ciliophora have cilia during the least some part of their life cycle, and Sarcodina use pseudopodia for movement.

Feeding in protozoa

Since protozoa live in an aquatic environment, water, oxygen, and other small molecules readily diffuse through the cell membrane. In addition, protozoa use either pinocytosis or phagocytosis to obtain food and water.

Protozoan Reproduction

The life cycle of protozoa are sometimes complex, involving more than one habitat or host. Morphologically distinct forms of a single protozoan species can be found at different stages of the life cycle. Such organisms are said to be *polymorphic*. This polymorphism is comparable in some respects to the differentiation of various cell types that form plant and animal tissues. The ability to exist in either a *trophozoite* (vegetative or feeding form) or *cyst* (resting form) is characteristic of many protozoa. Certain environmental conditions, such as the lack of nutrients, moisture, oxygen, low temperature, or the presence of toxic chemicals may trigger the development of a protective cyst wall within which the cytoplasm becomes dormant. Cysts provide a means for the disposal and survival of protozoa under adverse conditions and can be compared to the bacterial endospore. Protozoan cysts, however, are not as resistant to heat and other adverse conditions as are bacterial endospores. When the cyst encounters a favourable environment, the trophozoite emerges. Thus, a number of parasitic protozoa are disseminated to new hosts during their cyst stage.

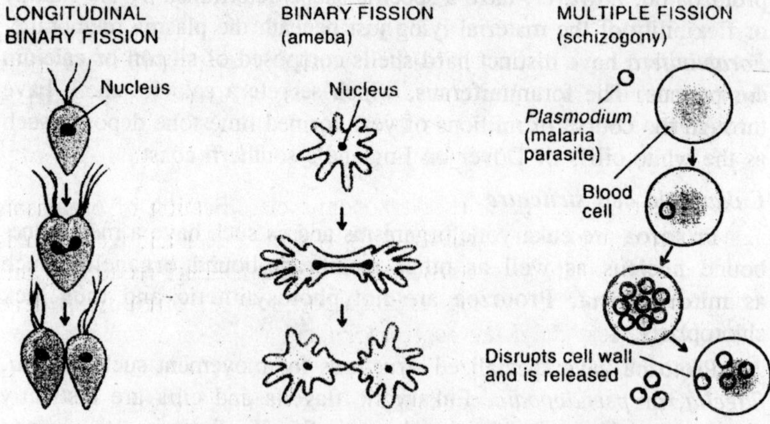

Fig. 13.5. Various forms of asexual reproduction in protozoa.

Both asexual and sexual reproduction are common in protozoa and may alternate during the complicated life cycle of some organisms. Binary fission takes place in many groups of protozoa. In the flagellates, is usually occurs longitudinally, and in the ciliates, it occurs transversely. Since some protozoa possess both cilia and flagella, their method of asexual reproduction determines in which group they are classified.

Some protozoa divide by multiple fissions, or *schizogony*, in which the nucleus divide a number of times and then the cell produces many small single-celled organisms, each one capable of infection. Multiple fission of the asexual forms in the human host results in large numbers of parasites released into the host's circulation at regular intervals, producing the characteristic cyclic symptoms of malaria.

Protozoa and Human Disease

The major threat posed by protozoa results from their ability to parasitize and often kill a wide variety of animal hosts. Human infections with the protozoans *Toxoplasma gondii*, which causes toxoplasmosis, *Plasmodium* species, which are responsible for malaria, *Trypanosoma* species, which cause sleeping sickness, and *Trichomonas vaginalis*, which causes vaginitis, are common in many parts of the world. Malaria has been one of the greatest killers of humans through the ages. AT least 300 million people in the world contract malaria each year, and 3 million die of it. Protozoan infections of animals are so common that it is difficult to estimate their extent or overestimate their economic importance. Some parts of tropical Africa are uninhabitable, due to large part to the presence of the teste fly, the carrier of the trypanosomes that cause African sleeping sickness. Humans have no natural defense mechanisms against this infection. Some animals, including cows and horses, are reservoirs for these organisms.

Fungi

The term fungi descries a taxonomic classification of organisms but no longer includes organisms such as slime molds and water molds that had traditionally been considered to be fungi. Fungi require organic compounds for energy and as a carbon source, often from dead organisms. Most fungi are aerobic or facultatively anaerobic. Only a few fungi are anaerobic.

A large number of fungi cause disease in plants. Fortunately, only a few species cause disease in animals and humans. As modern medicine has advanced to treat once-fatal diseases, however, it has left many

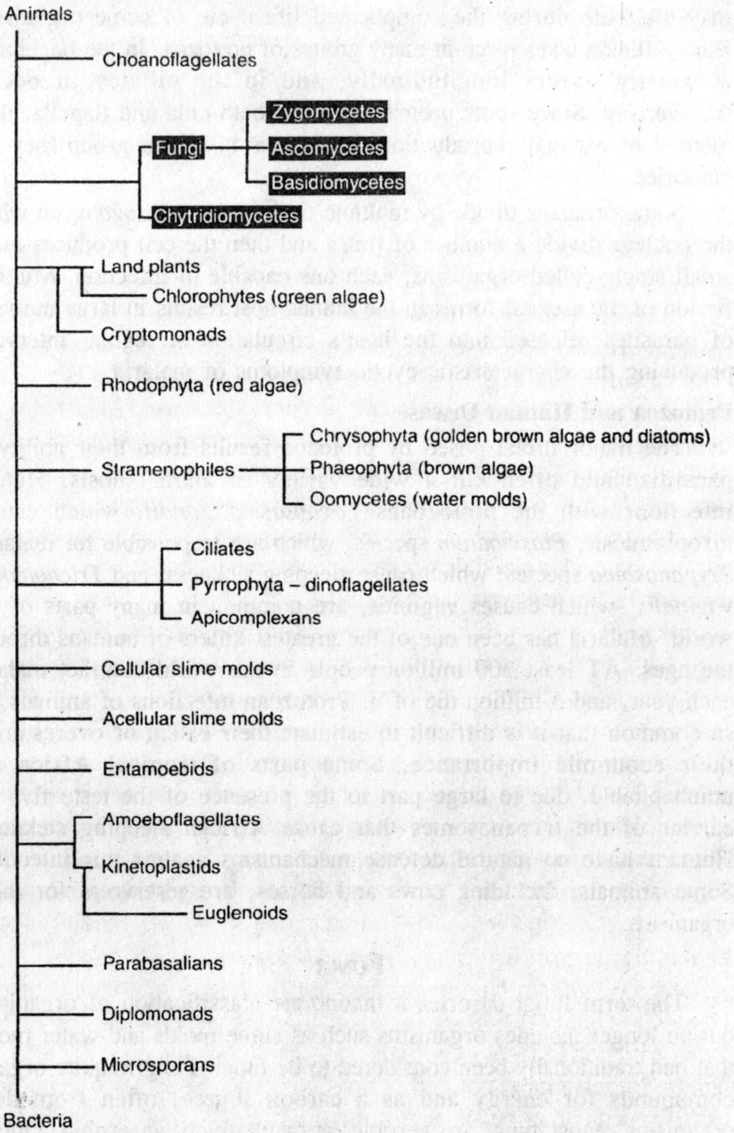

Fig. 13.6. A phylogenetic scheme of eukaryotic based on rRNA sequence comparison. The fungal groups are highlighted.

individuals with impaired immune systems. It is these immunocompromised individual who are most vulnerable to the fungal diseases. The study of fungi is known as *mycology*, and a person who studies fungi is known as a *mycologist*. Along with bacterial, fungi are the principal decomposers of carbon compounds on earth. This decomposition release carbon dioxide into the atmosphere and nitrogen compounds into the soil, which are then taken up by plants and converted into organic compounds. Without this breakdown of organic material, the world would quickly be overrun with organic waste.

Classification of Fungi

Many fungi are microscopic and can be examined using basic microbiological techniques; others are macroscopic. All fungi have chitin in their cell walls and no flagellated cells at any time during their life cycle. There are four groups of true fungi, the *Zygomycetes*, *Basidiomycetes*, *Ascomycetes*, and *Deuteromycetes* or *Fungi Imperfecti*. The classification of the first three groups is based on their method of sexual reproduction. For the fourth group, the Deuteromycetes, sexual reproduction has not been observed, and so these fungi have traditionally been lumped together. With additional rRNA analysis, however, most Deuteromycetes can now be placed in one of the other three fungal groups. Most are either Ascomycetes or Basidiomycetes who have lost the sexual part of their life cycle.

The Zygomycetes include the common bread mold (*Rhizopus*) and other food spoilage organisms. The Ascomycetes include fungi that cause Dutch elm disease and rye smut. Smuts got this common name because the black spores that are produced give the appearance of soot. The Basidiomycetes include the common mushrooms and puffballs. Many medically and economically important species of fungus, including the one that produces penicillin, belong to the Deuteromycetes. In addition to these four groups of fungi, the *Chytridiomycetes* are a close relative on the evolutionary scale. They have flagellated sexual spores and more variable life cycles, however, than the fungi. Most of these organisms live in water or soil, and a few are parasitic. Black wart disease of potatoes is caused by a chytridiomycete.

Common grouping of fungal forms

When people talk about fungi, they frequently use terms such as yeast, mold, and mushroom. These terms have nothing to do with the classification of fungi but instead indicate their morphological forms.

Yeasts are single-celled fungi. Yeasts can be spherical, oval, or cylindrical and are usually 3 to 5 μm in diameter. Some years reproduce by binary fission, whereas others reproduce by budding, in which a small outgrowth on the cell produces a new cell.

Molds are filamentous fungi. A single filament is known as a *hypha* (plural, *hphae*) and a collection of hyphae growing in one place is known as *mycelium*. Hyphae develop from fungal spores. A fungal reproduction spore is typically a single cell about 3 to 30 μm in diameter, depending on the species. When a fungal spore lands on a suitable substrate, it germinates and sends out a projection called a *germ tube*. This tube grows at the tip and develops into a hypha. The cells divide and form new cells. In some fungi, the cell wall does not completely close off one cell form another. In that case the cells become multinucleated. The white mass seen inside the potato or on moldy bread is an example of a mycelium. Only a small portion of the mycelium is actually visible on the surface of the bread; the rest is buried deep within. Some mycelia appear above the surface of the substrate as a mushroom, or puffball. These macroscopic structures produce reproductive spores. Some large mushrooms are edible.

Hyphae are well adapted to absorb food. They are narrow and threadlike. With their high surface-to-volume ratio, they can absorb large amount of nutrients. Hyphae release enzymes that break down the material into readily absorbed smaller organic compounds. In addition, these enzymes act to repel the growth of other hyphae near it. As a result, hyphae spread throughout the food source, ensuring that each hypha will have access to adequate nutrients.

Parasitic fungi have specialized hyphae called *haustoria*, which can penetrate animal or plant cell walls to gain nutrients. Saprophytic fungi sometimes have specialized hyphae called *rhizoids*, which anchor them to the substrate.

Dimorphic fungi

Dimorphic fungi are capable of growing either as yeastlike cells or as mycelia, depending on the environmental conditions. Many of the fungi that causes disease in humans are dimorphic. Certain fungi such as *Coccidioides immitis* grow in soil as molds. When their spores, which are readily carried in the air, are inhaled into the warm, moist environment of the lungs, they develop into the yeast form of the organism and cause disease.

Fungal Habitats

Fungi are found in virtually every habitat on the earth where organic materials exist. Whereas algae and protozoa grow primarily in aquatic environments, the fungi are mainly terrestrial organisms. Some species occur only on a particular strain of one genus of plants, whereas others are extremely versatile in what they can attack and use as a source of carbon and energy. Materials, such as leather, cork, hair, wax, ink, jet fuel, and even some synthetic plastic like the polyvinyls can be attacked by fungi. Some species can grow in concentrations of salts, sugars, or acids strong enough to kill most bacteria. Thus, fungi are often responsible for spoiling pickles, fruit preserves, and other foods. Some fungi are resistant to pasteurization and others can grow at temperatures below the freezing point of water, rotting bulbs and destroying grass in frozen ground. Fungi are found in the thermal pools at Yellowstone National Park, in volcanic craters, and in lakes with very high salt content, such as the Great Salt Lake and the Dead Sea.

Fungal reproductive cells, or spores, are found throughout the earth. They also occur in tremendous numbers in the air near the earth's surface as well as at altitudes of more than 7 miles. Although not as resistant as bacterial endospores, fungal spores are generally resistant to the ultraviolet rays of sunlight. Sunlight will, however, sometimes kill fungal vegetative cells. Fungal spores are a major cause of asthma.

Growth requirements of fungi

Most fungi prefer a slightly moist environment with a relative humidity of 70% or more, and various species can grow at temperatures ranging from –6°C to 50°C. The optimal temperature for the majority of fungi is in the range of 20°C to 35°C.

The pH at which different fungi can grow varies widely, ranging from as low as 2.2 to as high as 9.6, but fungi usually grow well at an acid pH of 5.0 or lower. This explains why fungi grow well on fruits and many vegetables that tend to be acidic.

As heterotrophs, fungi secrete a wide variety of enzymes that degrade organic materials, especially complex carbohydrates, into small molecules that can be readily absorbed. Most fungi are aerobic, but some of the yeasts are facultative anaerobes and carry out alcoholic fermentation. Facultatively anaerobic fungi live in the intestine of certain species of fish and help degrade algae. Some fungi living in the rumen of cows and sheep are known to be obligately anaerobic. They are important in the digestion of the plant material that these animals ingest.

Fungal Disease in Humans

Fungi cause disease in humans in one of four ways. First, a person may develop an allergic reaction to fungal spores or vegetative cells. Second, a person may react to the toxins produced by fungi. Third, the fungi may actually grow on or in the human body, causing disease or *mycoses*. Fourth fungi can destroy the human food supply, causing starvation and death.

Allergic reaction in humans

Medical mycologists study fungi that affect humans, including fungi that cause allergic reactions. Allergic diseases such as hay fever and asthma can result form inhaling fungi or their spores if exposed humans have become sensitized. Sometimes, severe, long-term allergic lung disease results from these allergic reactions.

Effect of fungal toxins

For their hallucinogenic properties, certain mushrooms have long been used as part of religious ceremonies in some cultures. The lethal effects of many mushrooms have also been known for centuries. The poisonous effects of a rye smut called *ergot* were known during the Middle Ages, but only recently has the active chemical been purified form this fungus to yield the drug ergotamine, which is now used to control uterine bleeding, relieve migraine headaches, and assist in childbirth.

Some fungi produce toxins that are carcinogenic. The most thoroughly studied of these carcinogenic toxins, produced by species of *Aspergillus*, are called *aflatoxins*. Ingestion of aflatoxins in moldy foods, such as grains and peanuts, has been implicated in the development of liver cancer (hepatoma) and levels of aflatoxins in foods such as peanuts, and if a certain level is exceeded, the food cannot be sold.

Mycoses

Fungal diseases are called *mycoses*. The names of the individual diseases often begin with the names of the causative fungi. Thus, *histoplasmosis*, a disease seen worldwide, is a mycosis caused by the fungus *Histoplasma capsulatum*. Similarly, *coccidioidomycosis* is a mycosis caused by *Coccidioides immitis*, a fungus unique to certain arid regions of the Western Hemisphere. Diseases caused by the yeast *Candida albicans* are called *candidiasis* and are among the most common mycoses. Infections can also be referred to by the parts of the body that they affect. *Superficial mycoses* affect only the hair, skin, or nails. *Intermediate mycoses* are limited to the respiratory tract or the skin and subcutaneous tissues. *Systemic mycoses* affect tissues deep within the body.

Symbiotic Relationships Between Fungi and other Organisms

Fungi form several types of symbiotic relationships with other organisms. For example, lichens result from the association of a fungus with a photosynthetic organism such as an alga or a cyanobacterium. These associations are extremely close and, in some cases, the fungal hyphae actually penetrate the cell wall of the photosynthetic partner. The fungus provides the protection and growing platform for the pair. In addition, the fungus absorbs water and minerals for the association. The photosynthetic member supplies the fungus with organic nutrients. It is possible to grow each partner of the lichen association separately. Usually the algal partner is able to grow well when separated, but the fungal partner does not. Because of this association, lichens can grow in extreme ecosystems where neither could survive on its own. Lichens are often a good indicator of air quality since they are very sensitive to sulfur dioxide, ozone, and toxic metals. You will not find very many lichens in cities with air pollution.

Mycorrhizae, fungal symbioses of particular importance, are formed by the intimate association between fungi and roots of certain plants such as the Douglas fir. It is estimated that 80% of the vascular plants have some type of mycorrhizal association with their roots. There are more than 5,000 species of fungi that form mycorrhizal associations. By increasing the absorptive power of the root, they often allow their plant partners to grow in soils where these plants could

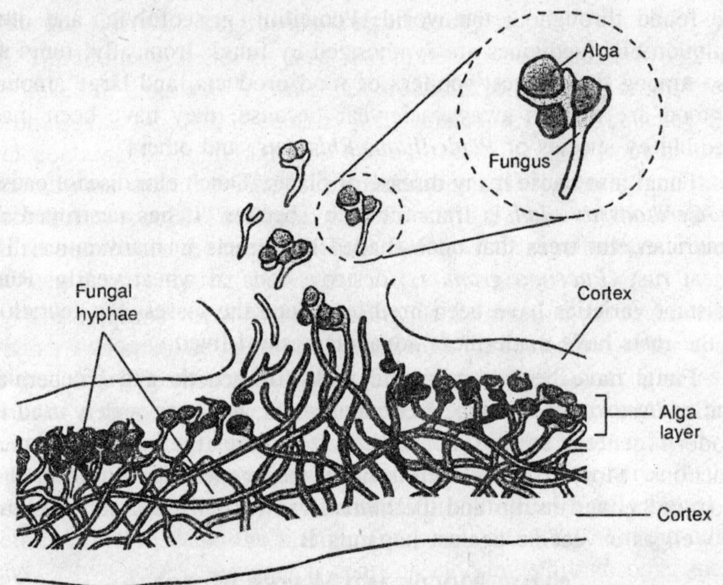

Fig. 13.7. Diagram of a lichen, consisting of cells of a phototroph, either an alga or a cyanobacterium, entwined within the hypae of the fungal partner.

not otherwise survive. With the world facing major shortages of food and forest products, a better understanding of these symbiotic relationships is urgently needed. Many trees, including the conifers, are able to live in sandy soil because of their extensive mycelial mycorrhizas. In Puerto Rico, for example, the pine tree industry most perished before the proper fungi were introduced to form mycorrhizal relationships with the trees. Now the industry is flourishing. Similarly, orchids cannot grow without mycorrhizal association of a fungus that helps provide nutrients to the young plant.

Certain insects also depend on symbiotic relationships with fungi. For example, leaf cutter ants are estimated to bring about 15% of the tropical vegetation into their nests to use for food. The leaves are used as food by a fungus that the ants cultivate in their nests. The fungus removes the poison form the leaves of the plant, and then the ants use the fungus as their food source.

Economic Importance of Fungi

Many fungi are important commercially. The yeast *Saccharomyces* has long been used in the production of wine, beer, and bread. Other fungal species are useful in making the large variety of cheese that are found throughout the world. Penicillin, griseofulvin, and other antimicrobial medicines are synthesized by fungi. Ironically, fungi are also among the greatest spoilers of food products, and large amounts of food are thrown away each year because, they have been made inedible by species of *Penicillium*, *Rhizopus*, and others.

Fungi also cause many disease of plants. Dutch elm disease caused by *Ceratocystis ulmi* is transmitted by beetles. It has destroyed the American elm trees that once shaded the streets in many cities. The wheat rust (*Puccinia graminis*) destroys tons of wheat yearly. Rust-resistant varieties have been bred to reduce the losses, but mutations in the rusts have made these advantages short-lived.

Fungi have been very useful tools for genetic and biochemical studies. *Neurospora crassa*, a common mold, has been widely used for modern genetic studies as well as for investigating biochemical reactions. More recently, yeasts have been genetically engineered to produce human insulin and the human growth hormone somatostatin, as well as a vaccine against hepatitis B.

Slime Molds and Water Molds

The slime molds and water molds used to be considered types of fungus. They are, however, completely unrelated to the true fungi and are good examples of *convergent evolution*. Convergent evolution occurs when two organisms develop similar characteristics because of adaptations to similar environments and yet are not related on a molecular level.

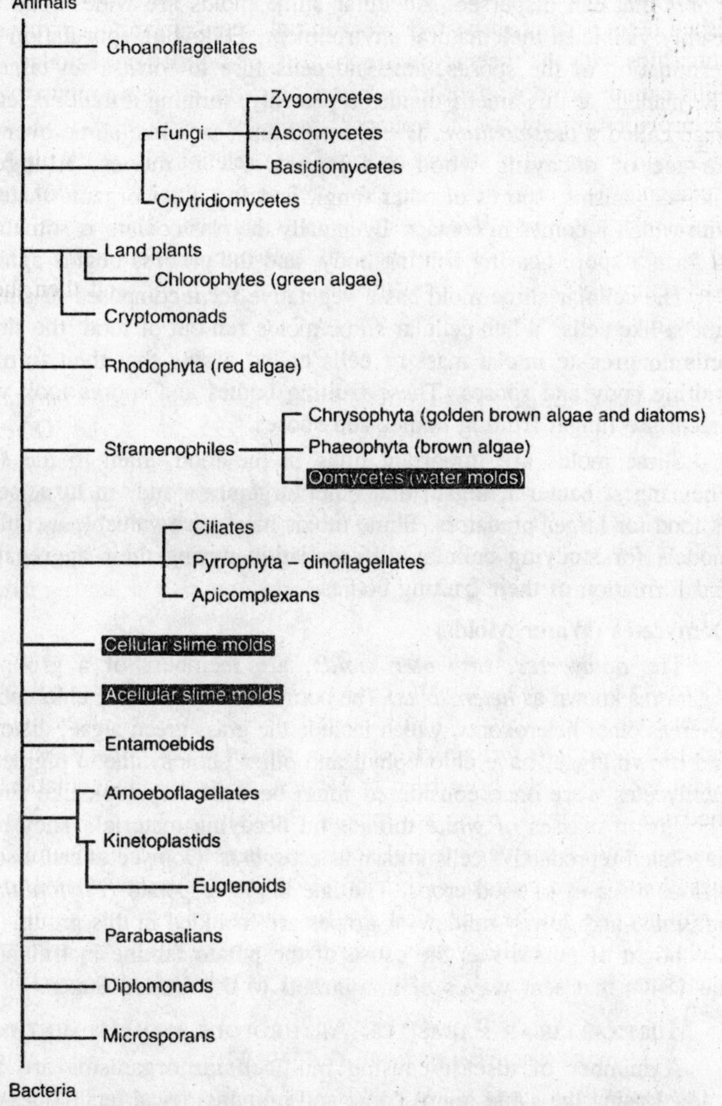

Fig. 13.8. A phylogenetic scheme of eukaryotes based on rRNA sequence comparisons. The slime molds and water molds are highlighted.

Acellular and Cellular Slime Mods

There are two groups of slime molds, the *acellular slime molds* and the *cellular slime molds*. The slime molds are terrestrial organisms living on soil, leaf little, or the surfaces of decaying leaves or wood. They are non-motile, and reproduction depends on the formation of spores that can dispersed. Acellular slime molds are widespread and readily visible in their natural environment. Following sporulation and germination of the spores, ameboid cells fuse to form a myxameba. The nucleus of this ameba divides repeatedly, forming a multinucleated stage called a *plasmodium*. The plasmodium oozes like slime over the surface of decaying wood and leaves. As it moves, it ingests microorganisms, spores of other fungi, and any other organic material with which it comes in contact. Eventually the plasmodium is stimulated to form a spore-bearing fruiting body, and the process begins again.

The cellular slime mold has a vegetative form composed of single, ameba-like cells. When cellular slime molds run out of food, the single cells congregate into a mass of cells called a *slug* that then forms a fruiting body and spores. These fruiting bodies and spores look very much like fungal fruiting bodies and spores.

Slime molds are important links in the food chain in the soil. They ingest bacteria, algae, and other organisms and, in turn, serve as food for larger predators. Slime molds have been valuable as unique models for studying cellular differentiation during their aggregation and formation of their fruiting bodies.

Oomycetes (Water Molds)

The *oomycetes*, or *water molds*, are members of a group of organisms known as *heterokonts*. The oomycetes do not have chlorophyll, whereas other heterokonts, which include the grass green algae, diatoms, and brown algae, have chlorophyll and other photosynthetic pigments. Oomycetes were once considered fungi because they look like fungi. They form masses of white threads on decaying material. They have flagellated reproductive cells known as *zoospores*. Oomycetes cause some serious diseases of food crops. The late blight of potato (*Phytophthora infestans*) and downy mildew of grapes are included in this group. The late blight of potato was the cause of the potato famine in Ireland in the 1840s that sent waves of immigrants to the United States.

Multicellular Parasites: Arthropods and Helminths

A number of disease-causing multicellular organisms are also studied using the same microscopic and immunological techniques that are used to study microorganisms and viruses. As a result, they are included here. Most of the medically important multicellular parasites fall into one of two groups: *arthropods* and *helminths*. The arthropods are more highly advanced on the evolutionary scale and include the insects, ticks, lice and mites. Their main medial importance is that they serve as *vectors* that may transmit microorganisms and viruses to humans. The helminths, which include the *nematodes* (roundworms), *cestodes* (tapeworms), and the *trematodes* (flukes), are more primitive animals. In only a few instances do they transmit microbial infections to their host animal. Instead, they cause disease by invading the host's tissues or robbing if of nutrients.

Most multicellular parasites have been well controlled in the industrialized nations, but they still cause death and misery to many millions in the economically underdeveloped areas of the world. Our need to know about these problems has come about because more people are traveling farther, more people are moving from one place to another, and more goods are being exchanged worldwide. A clear example of this occurred in New York City in the summer of 1999 when West Nile fever was contracted by a number of people. At least 61 persons suffered serious disease and seven people died. A significant number of crows died at the same time and were found to be carrying the diseases. In addition to birds and people, horses, cats, and dogs were also found to carry the virus. It is not clear how the virus arrived in New York city, but it appears could have been carried by a traveler from Africa, West Asia, or the Middle East, where it is commonly found. It could possibly have been brought by an imported bird form the same areas. Worldwide travel makes us more vulnerable to disease from other parts of the world.

In addition, worldwide climatic conditions are changing and bringing increases in certain insect populations to areas that were previously free of them. As a result, more cases of multicellular parasitic infections are being seen by physicians in the United States than previously.

Arthropods

The anthropods include insects, ticks, fleas, and mites. Arthropods act as vectors for transmitting diseases. In some instances, an arthropod such as fly simply picks up a pathogen on its feet from some contaminated material such as feces and then lands on food that is then eaten by humans, thus transmitting the pathogen. In this case, the fly acts as a *mechanical vector*. In other cases, such as with *Plasmodium* sp., the cause of malaria, the vector, a mosquito, is a host for the organism before it transfers that organisms to a human through a bit on the skin. In this case, the vector acts as an essential part of the life cycle of the organism and is known as *biological vector*. The pathogen actually multiples in number within the vector.

Mosquitoes

The female mosquito needs the blood of a warm-blooded animals for the proper development of her eggs. To get this, she needs to bite such an animal. The mosquito can take in as much as twice its body weight in blood, thus giving it a relatively good chance of picking up infectious agents such as malarial parasites circulating within the host's capillaries. The anatomy of a mosquito is particularly adapted to transmit disease. The mouthparts of the female mosquito consist of sharp stylets that are forced through the host's skin to the subcutaneous capillaries. One of these needlike stylets is hollow, and the mosquito's saliva is pumped through it. The saliva increases blood flow and prevents clotting as the victim's blood is sucked into a tube formed by the other mouthparts of the insect. The saliva can also cause allergic reactions (the itch of a mosquito bite). After the mosquito has taken more than one blood meal, she can transmit disease from one animal to the next. Viruses found in the blood of the first animal are then transmitted to the next, and so on.

Mosquitoes in an area of arthropod-borne disease can be trapped and identified microscopically, and the blood they have ingested can be tested to see on which kinds of animals the different species are feeding. Precise identification of species and subspecies of these genera is important because different species of mosquitoes differ greatly in their breeding areas, time of feeding, and choice of host. Identification depends largely on microscopic examination of antennae, wings, claws, mating apparatus, and other features. Correct identification is often essential in designing specific control measures.

Fleas

Fleas are wingless insects that depend on powerful hind legs to jump from place to place. Points of importance in identifying fleas include the spines (combs) about the head and thorax, the muscular pharynx, the long esophagus, and the spiny valve composed of rows of teethlike cells. Fleas are generally more of a nuisance than a health hazard, but they can transmit the bacterium *Yersinia pestis*, which causes plague, and a rickettsial disease, murine typhus, to humans. Larval fleas have a chewing type of mouth for feeding on organic matter. They ingest eggs of the common dog and cat tapeworm, *Dipylidium caninum*, serving as its intermediate host. Children acquire this tapeworm when they accidently swallow fleas. Fleas can live in vacant building in a dormant stage for many months. When the building becomes inhabited, the fleas quickly mature and hungrily greet the new hosts.

Lice

Like fleas, lice are small, wingless insects that prey on warm-blooded animals by piercing their skin and sucking blood. The legs and claws of lice, however, are adapted for holding onto body surfaces and clothing rather than for jumping. Human lice generally survive only a few days away from their hosts.

Pediculus humanus, the most notorious of the lice, is 1 to 4 mm long, with a characteristically small head and thorax, and a large abdomen. This louse has a membrane-like lip with tiny teeth that anchor it firmly to the skin of the host. Within the floor of the mouth is a piercing apparatus somewhat similar to that of fleas and mosquitoes. *Pediculus humanus* has only one host—humans—but easily spreads from one person to another by direct contact or by contact with personal items, especially in areas of crowding and poor sanitation.

There are two subspecies, popularly termed head lice and body lice. Body lice can transmit trench fever, which is caused by the bacterium, *Bartonella quintana*; epidemic typhus, which is caused by the bacterium. *Rickettsia prowazekii*; and relapsing fever, caused by the bacterium *Borrelia recurrentis*. Trench fever occurs episodically among severe alcoholics and the homeless of large American and European cities.

The crab louse, *Phthirus pubis*, is commonly transmitted among young adults during sexual intercourse. It is not a vector of infectious disease, but it can cause an unpleasant itch.

Ticks

Ticks are arachnids. Arachnids differ from insects in their lack of wings and antennae, and their thorax and abdomen are fused together. Although like insects the immature ticks have three pairs of legs, the adults have four pairs. *Dermacentor andersoni*, the wood tick, is the vector for Rocky Mountain spotted fever caused by the bacterium *Rickettsia rickettsii*. Another tick, *Ixodes scapularis*, transmits with its saliva *Borrelia burgdorferi*, the spirochete that causes Lyme disease. In addition, the saliva of several genera of ticks can produce a profound paralysis, especially in children on whom the tick feeds for several days. Paralyzed humans and animals usually recover rapidly following removal of the tick.

Mites

Mites, like ticks, are arachnids. They are generally tiny, fast moving, and live on other surfaces of animals and plants. *Demodex folliculorum* and *D. brevis* are elongated microscopic mites that live in the hair follicles or oil-producing glands usually of the face, typically without producing symptoms. Other species of mites cause human disease.

The disease scabies, caused by a mite, *Sarcoptes scabiei*, is characterized by an itchy rash most prominent between the fingers, under the breasts, and in the genital area. Scabies is easily transmitted by personal contact, and the disease is commonly acquired during sexual intercourse. The female mites burrow into the outer layers of epidermis, feeding and laying eggs over a lifetime of about 1 month. Allergy to the mites is largely responsible for the itchy rash. The diagnosis can only be made by demonstrating the mites, since scabies mimics other skin disease. Treatment of scabies is easily accomplished with medication applied to the skin. *Sarcoptes scabiei* is not known to transmit infectious agents.

Mites of domestic animals and birds can cause an itchy rash in humans, as can mites sometimes present in hay, grain, cheese, or dried fruits. The dust mites that often live in large numbers in bedrooms can sometimes cause asthma when the mites and their excreta are inhaled. The mites of rodents can transmit rickettsial disease to humans. Rickettsial pox, caused by *Rickettsia akari* transmitted by mouse mites, is a mild disease characterized by fever and rash. Epidemics occur periodically in cities of the eastern United States. Serious rickettsial disease of other parts of the world such as scrub typhus are transmitted by rodent mites.

Helminths

In addition to the arthropods that can lead to disease in humans, the other group of multicellular animals that causes human disease are the *helminths*. In humans, the helminths that cause disease generally belong to one of three classes: the *nematodes*, or roundworms; the *cestodes*, or the tapeworms; and the *trematodes*, or the flukes.

These multicellular parasites have been controlled in the developed nations, but they continue to kill many millions in underdeveloped parts of the world. Helminths enter the body in a number of ways. They may be eaten in contaminated food, be passed through insect bites, or directly penetrate the skin. They cause disease by invading the host tissues or robing the host of nutrients. Some helminths have complex life cycles, involving one or more intermediate hosts where the sexually mature forms occur.

Nematodes or Roundworms

The nematodes or roundworms have a cylindrical tapered body with a tubular digestive tract that extends from the mouth to the anus. There are both male and female nematodes. Nematodes include a large number of species. Many nematodes are free-living in soil and water. Others are parasites of human and other animals and plants and produce serious disease.

The nematodes that cause disease can be divided into two groups—the ones that inhabit the gastrointestinal tract of the host, and the ones that are found in the blood and other tissues of the host. Generally, diagnosis of worm infestation depends on microscopic identification of the worms or their ova (eggs), or on blood tests for antibody to the worms.

Cestodes or tapeworms

Cestodes or tapeworms have flat, ribbon-shaped bodies that are segmented. The head (scolex) of the tapeworm has suckers for attachment and sometimes has hooks. Directly behind the head is a region that produces the reproductive segments (proglottids). Each segment has both male and female sex organs. The tape worm does not have a digestive system but rather absorbs nutrients directly. Tapeworms are often associated with beef, lamb, pork, and fish. Transmission of these organisms to humans often occurs when the flesh of these animals is eaten either uncooked or undercooked. Some tapeworms are transmitted to humans from ingesting fleas infected with dog or cat tapeworms.

Trematodes or flukes

Trematodes or flukes are bilaterally symmetrical, flat, and leaf-shaped. They have suckers that hold the organism in place as well as suck fluids form the host. Most species are hermaphroditic (have both sex organ in the same worm). Most trematodes have a complicated life cycle, which may include one or more intermediary hosts. Usually, the worms begin with a larval form developing within the egg. These larvae escape into the environment, where they are taken up by one

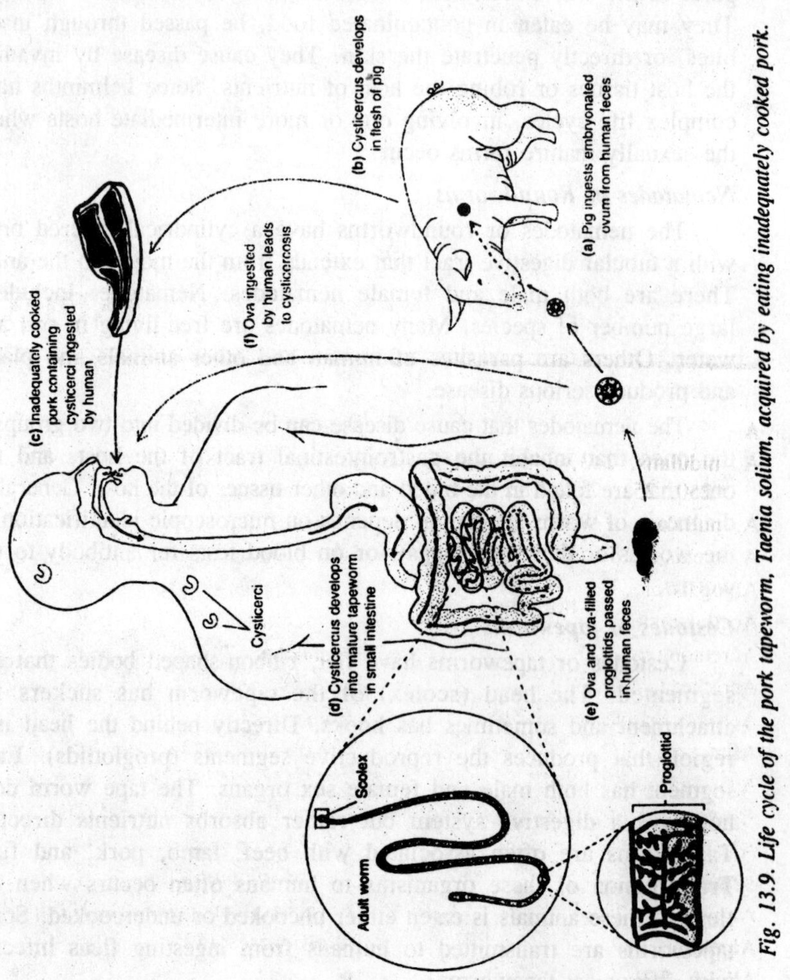

Fig. 13.9. Life cycle of the pork tapeworm, Taenia solium, acquired by eating inadequately cooked pork.

or more intermediate hosts such as a snail. Eventually, the last stage is a tail-bearing larva known as a *cercaria*, which is released from the snail and is ready to attach to the susceptible host. For example, if a human is wading in water and the cercaria of *Schistosoma manosoni* have been discharged into that water, the cercaria can penetrate the skin and work its way through the circulation to the liver and intestine, where it matures and lays eggs.

INDEX